AF675356

Energy Processes, Systems and Equipment

Energy Processes, Systems and Equipment

Editors

Sławomir Pietrowicz
Izabela Sówka
Piotr Kolasiński

MDPI • Basel • Beijing • Wuhan • Barcelona • Belgrade • Manchester • Tokyo • Cluj • Tianjin

Editors
Sławomir Pietrowicz
Wrocław University of Science and Technology
Poland

Izabela Sówka
Department of Environment Protection Engineering, Faculty of Environmental Engineering, Wroclaw University of Science and Technology
Poland

Piotr Kolasiński
Department of Thermodynamics and Renewable Energy Sources, Faculty of Mechanical and Power Engineering, Wrocław University of Science and Technology
Poland

Editorial Office
MDPI
St. Alban-Anlage 66
4052 Basel, Switzerland

This is a reprint of articles from the Special Issue published online in the open access journal *Energies* (ISSN 1996-1073) (available at: https://www.mdpi.com/journal/energies/special_issues/processes_systems_equipment).

For citation purposes, cite each article independently as indicated on the article page online and as indicated below:

LastName, A.A.; LastName, B.B.; LastName, C.C. Article Title. *Journal Name* **Year**, *Volume Number*, Page Range.

ISBN 978-3-0365-1883-1 (Hbk)
ISBN 978-3-0365-1884-8 (PDF)

© 2021 by the authors. Articles in this book are Open Access and distributed under the Creative Commons Attribution (CC BY) license, which allows users to download, copy and build upon published articles, as long as the author and publisher are properly credited, which ensures maximum dissemination and a wider impact of our publications.
The book as a whole is distributed by MDPI under the terms and conditions of the Creative Commons license CC BY-NC-ND.

Contents

About the Editors

Sławomir Pietrowicz started his research and teaching work at Wrocław University of Science and Technology, Poland in 1997. In the years 2000-2002, he conducted research at Shizuoka Univerisity, Japan as the beneficiary of the Monbusho Scholarship. In 2004 obtained his Ph.D. Over the years he collaborated with various scientific centers (CERN Switzerland, CEA/Saclay France, ESS Sweden, KEK Japan), between 2009 and 2011 worked/stayed at CEA, Saclay, France as a researcher in the cryogenics field. After habilitation in 2014, he was appointed Professor at Wroclaw University of Science and Technology. He was the Vice-Dean for Science and International Cooperation at Faculty of Mechanical and Power Engineering, Wroclaw University of Science and Technology. At present he is the Head of the Department of Thermodynamics and Renewable Energy Sources Department. His research topics cover various fields i.e. thermodynamics, cryogenics, RES, CFD, two-phase flow in capillary tubes.

Izabela Sówka has been active at the Wrocław University of Science and Technology, where she is currently employed as a professor in the Department of Environment Protection Engineering. She is the initiator and president of the Center for Sustainable Development and Climate Protection. She is also a member of many workgroups and committees at the academical, regional and national level. Among her other achievements one ought to mention the membership of the Environmental Engineering Committee of the Polish Academy of Sciences and vice-chairmanship of the State Council for Environmental Protection and Scientific Council at the Institute of Environmental Engineering of the Polish Academy of Sciences in Zabrze. Her research topics are related to identifying air pollution and its sources, air quality assessment, analysis of the health impacts of air pollution, carbon footprint, global atmospheric fenomena assessment, RES.

Piotr Kolasiński is connected with the Wrocław University of Science and Technology. He obtained his professional experience initially as an assistant at the Department of Thermodynamics of Institute of Power Engineering and Fluid Mechanics, where he was involved in research on the potential application of multi-vane expanders in ORC (Organic Rankine Cycle) systems. In 2010 after obtaining his Ph.D., he held an internship at the Technical University of Munich (TUM), Germany, in which he took part in a training program in the field of nuclear energy. In 2017 he obtained habilitation and in 2019 he was appointed Professor at Wroclaw University of Science and Technology. He is the Head of Laboratory of Thermodynamics and Thermal Properties of Materials. His research topics are related to volumetric machines, organic Rankine cycle (ORC) systems, waste energy recovery, renewable energy, thermodynamics and heat transfer.

energies

Editorial

Energy Processes, Systems and Equipment

Izabela Sówka [1,*], Sławomir Pietrowicz [2,*] and Piotr Kolasiński [2,*]

[1] Department of Environment Protection Engineering, Faculty of Environmental Engineering, Wrocław University of Science and Technology, 27 Wybrzeże Wyspiańskiego Street, 50-370 Wrocław, Poland
[2] Department of Thermodynamics and Renewable Energy Sources, Faculty of Mechanical and Power Engineering, Wrocław University of Science and Technology, 27 Wybrzeże Wyspiańskiego Street, 50-370 Wrocław, Poland
* Correspondence: izabela.sowka@pwr.edu.pl (I.S.); slawomir.pietrowicz@pwr.edu.pl (S.P.); piotr.kolasinski@pwr.edu.pl (P.K.); Tel.: +48-71-320-25-60 (I.S.); +48-71-320-36-17 (S.P.); +48-71-320-23-39 (P.K.)

Abstract: The scientific and technical issues related to energy harvesting and conversion are inseparably bound to the issues of environmental protection. Energy conversion systems and devices that are applied for converting the chemical energy contained in different fuels into heat, electricity, and cold in industry and housing are sources of different gases and solid particle emissions. Thus, the development of different technologies for energy conversion and environmental protection that can be jointly applied to cover growing energy needs has become a crucial challenge for scientists and engineers around the world. Progress in the precise description, modeling, and optimization of physical and chemical phenomena related to these energy conversion systems is a key research and development field for the economy. Legal and social issues that are affecting key aspects and problems related to the energy conversion and power sector are also significant and worth investigating. The aim of Energy Processes, Systems and Equipment Special Issue is to publish selected high-quality papers from the XV Scientific Conference POL-EMIS 2020: Current Trends in Air and Climate Protection—Control Monitoring, Forecasting, and Reduction of Emissions (29–31 March 2021, Wrocław) and other papers related to the field of energy conversion.

Keywords: energy conversion; fuels; energy storage; ecological and legal aspects

check for updates

Citation: Sówka, I.; Pietrowicz, S.; Kolasiński, P. Energy Processes, Systems and Equipment. *Energies* **2021**, *14*, 1701. https://doi.org/10.3390/en14061701

Received: 24 January 2021
Accepted: 16 March 2021
Published: 18 March 2021

Publisher's Note: MDPI stays neutral with regard to jurisdictional claims in published maps and institutional affiliations.

Copyright: © 2021 by the authors. Licensee MDPI, Basel, Switzerland. This article is an open access article distributed under the terms and conditions of the Creative Commons Attribution (CC BY) license (https://creativecommons.org/licenses/by/4.0/).

Since the beginning of humankind, energy has been the most important need for each human and living being. Consequently, the most important human energy need is the primary chemical energy contained in food. However, modern humans also have a very high demand for high-quality energy products such as electricity, high- and low-temperature heat, and mechanical power. For many years, it has been observed that this demand is constantly growing, which is undoubtedly due to the increasing number of energy receivers and the ease of access to various energy products [1,2]. The increase in energy demand applies to all areas of the economy, i.e., industry, transport, and households [3,4]. For this reason, in many countries, investments related to the implementation of new large power plants and smaller dispersed energy systems along with the development of innovative technological solutions that minimize environmental impact are necessary [5–11]. The development of different means of energy conversion that can be applied to cover growing energy needs and to increase the power and efficiency of generating sources has become a crucial challenge for scientists and engineers around the world, making the power industry, in which operation is based on subsequent energy conversion processes, one of the most important fields of the local, national, and global economy today.

The constantly growing energy generation is reflected in the increasing consumption of different energy carriers, such as fossil fuels, and over the last few decades, increasing attention has been paid to existing fuel resources and the possibility of their depletion in the future. Attention was paid to the development of technologies enabling energy harvesting from alternative sources (renewables and waste) [12]. Increased activity in this area led to the development of many innovative and efficient energy conversion technologies over the last few decades, the use of which minimizes the negative impact on the environment. International legal regulations related to the necessity of alternative energy usage have been introduced, and social campaigns aimed at educating consumers about the need for energy consumption rationalization and measures of energy consumption minimizing have also been launched.

Despite the growing usage of alternative energy sources, many large power plants worldwide are still based on fossil fuels [13]. Consequently, problems related to environmental protection, i.e., emissions of greenhouse gases and solid pollutants released during the combustion of different fuels, are of great importance. The need to reduce the emission of harmful substances into the atmosphere has been conditioned by international legal obligations (EU directives, EURO standards, and others) [14,15]. Works on highly efficient technologies that reduce the emission of these substances are setting trends in modern research in the field of energy conversion. Among others, effective electrostatic precipitators [16], heavy metal capture [17], NOx reduction [18], and oxy-fuel technologies [19] can be mentioned. However, in the case of industry, meeting stringent emission standards is becoming increasingly technically difficult, as well as more cost- and energy-intensive. One of the possible solutions is to limit the share of manufacturing processes that require fossil fuel combustion. In many systems, however, this task is very difficult or impossible to perform for technical (i.e., in selected technological processes, the energy obtained from fossil fuel combustion is the only energy source that can be applied) and economic reasons because it is connected with expensive investments. For many developing countries that do not have an infrastructure based on the use of renewable energy or nuclear power, the possibility of minimizing the share of fossil fuel combustion processes is very limited.

In addition to the increased use of renewable and waste energies, other prospective means of reducing fossil fuel consumption are activities related to the diversification of energy systems [20], supporting prosumer activities [21], better energy management [22], and increasing the efficiency of energy conversion processes and energy consumption [23]. It should be remembered that the use of alternative energy resources is not always easy. In some cases (e.g., use of hydrogen), the harvesting technology is still imperfect [24]. Another problem is social resentment towards some technologies, such as municipal waste incineration plants [25]. Establishing proper cooperation between machines and devices using alternative energy sources and the energy grid of the country is also an important challenge. In particular, this concerns matching the energy supply and demand characteristics, as the place of energy generation is not always near to where the demand occurs. It is then necessary to transfer energy to the consumer through an energy grid which has specified requirements related to the quality and quantity of transferred energy. Especially for systems harvesting energy from alternative sources, these requirements are not always met [26,27]. However, the efficient and sustainable use of alternative energy sources is necessary. In addition to many of the positive aspects related to environmental protection, there are other important reasons. Energy safety is one of them. By using alternative energy and local fossil fuel sources, there is the possibility of virtual power plant system implementation [28]. Such a system could cover the local energy demand and supply the energy in the case of a fuel crisis, natural disaster, or the failure of the national energy system. It is expected that energy consumption will grow in the future, which may have a negative impact on the natural environment. Therefore, special focus should be paid to the development of emerging clean energy conversion technologies that can potentially be applied in the near future for energy generation. These technologies include large scale systems, such as, for example, combined cycle gas turbines (CCGT) or thermonuclear reactors like the ITER (international thermonuclear experimental reactor), as well as smaller power systems that can be applied in energy clusters (e.g., hydrogen systems, solar panels, batteries, and other energy storage devices).

All in all, progress in the precise description, modeling, and optimization of physical and chemical phenomena related to energy conversion processes bound to large and dispersed power systems is a key research and development field for the economy. Detailed analysis related to individual input parameters and components that are directly or indirectly influencing the energy conversion chain should be conducted to achieve the final effect of a highly efficient and environmentally friendly power industry. Legal and social issues that are affecting the key aspects and problems related to the energy conversion and power sector are also significant and worth investigating. The aim of Energy Processes, Systems and Equipment Special Issue is to publish selected high-quality papers from the XV Scientific Conference POL-EMIS 2020: Current Trends in Air and Climate Protection—Control Monitoring, Forecasting, and Reduction of Emissions (29–31 March 2021, Wrocław) and other papers related to the field of energy conversion.

Author Contributions: Conceptualization, I.S., S.P., and P.K.; writing—original draft preparation, I.S., S.P., and P.K.; writing—review and editing, I.S., S.P., and P.K. All authors have read and agreed to the published version of the manuscript.

Funding: This research received no external funding.

Conflicts of Interest: The authors declare no conflict of interest.

References

1. Bercu, A.-M.; Paraschiv, G.; Lupu, D. Investigating the Energy–Economic Growth–Governance Nexus: Evidence from Central and Eastern European Countries. *Sustainability* **2019**, *11*, 3355. [CrossRef]
2. Mardani, A.; Streimikiene, D.; Nilashi, M.; Arias Aranda, D.; Loganathan, N.; Jusoh, A. Energy Consumption, Economic Growth, and CO_2 Emissions in G20 Countries: Application of Adaptive Neuro-Fuzzy Inference System. *Energies* **2018**, *11*, 2771. [CrossRef]
3. Lu, W.-C. Electricity Consumption and Economic Growth: Evidence from 17 Taiwanese Industries. *Sustainability* **2017**, *9*, 50. [CrossRef]
4. Hu, Y.; Guo, D.; Wang, M.; Zhang, X.; Wang, S. The Relationship between Energy Consumption and Economic Growth: Evidence from China's Industrial Sectors. *Energies* **2015**, *8*, 9392–9406. [CrossRef]
5. Wang, S.; Li, W.; Dincer, H.; Yuksel, S. Recognitive Approach to the Energy Policies and Investments in Renewable Energy Resources via the Fuzzy Hybrid Models. *Energies* **2019**, *12*, 4536. [CrossRef]
6. Fuchs, C.; Marquardt, K.; Kasten, J.; Skau, K. Wind Turbines on German Farms—An Economic Analysis. *Energies* **2019**, *12*, 1587. [CrossRef]
7. Kryzia, D.; Kopacz, M.; Kryzia, K. The Valuation of the Operational Flexibility of the Energy Investment Project Based on a Gas-Fired Power Plant. *Energies* **2020**, *13*, 1567. [CrossRef]
8. Wang, C.; Yan, Y.; Li, J.; Quan, Y.; Li, S.; Wu, G. Integrated Environmental Effect Assessment on Large Coal-Electricity Production Base and Sustainability Strategy: A Case in Xilin Gol, China. *Sustainability* **2020**, *12*, 5943. [CrossRef]
9. Lu, Y.; Khan, Z.A.; Alvarez-Alvarado, M.S.; Zhang, Y.; Huang, Z.; Imran, M. A Critical Review of Sustainable Energy Policies for the Promotion of Renewable Energy Sources. *Sustainability* **2020**, *12*, 5078. [CrossRef]
10. Tian, J.; Fang, C.; Qiu, J.; Wang, J. Analysis of Pollution Characteristics and Influencing Factors of Main Pollutants in the Atmosphere of Shenyang City. *Atmosphere* **2020**, *11*, 766. [CrossRef]
11. Dettori, M.; Pittaluga, P.; Busonera, G.; Gugliotta, C.; Azara, A.; Piana, A.; Arghittu, A.; Castiglia, P. Environmental Risks Perception Among Citizens Living Near Industrial Plants: A Cross-Sectional Study. *Int. J. Environ. Res. Public Health* **2020**, *17*, 4870. [CrossRef] [PubMed]
12. Halkos, G.E.; Gkampoura, E.-C. Reviewing Usage, Potentials, and Limitations of Renewable Energy Sources. *Energies* **2020**, *13*, 2906. [CrossRef]
13. Bielecka, A.; Kulczycka, J. Coal Combustion Products Management toward a Circular Economy—A Case Study of the Coal Power Plant Sector in Poland. *Energies* **2020**, *13*, 3603. [CrossRef]
14. Waluś, K.J.; Warguła, Ł.; Krawiec, P.; Adamiec, J.M. Legal regulations of restrictions of air pollution made by non-road mobile machinery—The case study for Europe: A review. *Environ. Sci. Pollut. Res.* **2018**, *25*, 3243–3259. [CrossRef]
15. Kuklinska, K.; Wolska, L.; Namiesnik, J. Air quality policy in the U.S. and the EU—A review. *Atmos. Pollut. Res.* **2015**, *6*, 129–137. [CrossRef]
16. Elkamhawy, A.; Jang, C.-M. Performance Evaluation of Hybrid Air Purification System with Vegetation Soil and Electrostatic Precipitator Filters. *Sustainability* **2020**, *12*, 5428. [CrossRef]
17. Karvelas, E.; Liosis, C.; Benos, L.; Karakasidis, T.; Sarris, I. Micromixing Efficiency of Particles in Heavy Metal Removal Processes under Various Inlet Conditions. *Water* **2019**, *11*, 1135. [CrossRef]
18. Onrubia-Calvo, J.A.; Pereda-Ayo, B.; González-Velasco, J.R. Perovskite-Based Catalysts as Efficient, Durable, and Economical NO_x Storage and Reduction Systems. *Catalysts* **2020**, *10*, 208. [CrossRef]
19. Hamadeh, H.; Toor, S.Y.; Douglas, P.L.; Sarathy, S.M.; Dibble, R.W.; Croiset, E. Techno-Economic Analysis of Pressurized Oxy-Fuel Combustion of Petroleum Coke. *Energies* **2020**, *13*, 3463. [CrossRef]

20. Gonzales-Zurita, Ó.; Clairand, J.-M.; Peñalvo-López, E.; Escrivá-Escrivá, G. Review on Multi-Objective Control Strategies for Distributed Generation on Inverter-Based Microgrids. *Energies* **2020**, *13*, 3483. [CrossRef]
21. Gram-Hanssen, K.; Hansen, A.R.; Mechlenborg, M. Danish PV Prosumers' Time-Shifting of Energy-Consuming Everyday Practices. *Sustainability* **2020**, *12*, 4121. [CrossRef]
22. Allen, J.; Halberstadt, A.; Powers, J.; El-Farra, N.H. An Optimization-Based Supervisory Control and Coordination Approach for Solar-Load Balancing in Building Energy Management. *Mathematics* **2020**, *8*, 1215. [CrossRef]
23. Perea-Moreno, A.-J.; Manzano-Agugliaro, F. Energy Saving at Cities. *Energies* **2020**, *13*, 3758. [CrossRef]
24. Felseghi, R.-A.; Carcadea, E.; Raboaca, M.S.; TRUFIN, C.N.; Filote, C. Hydrogen Fuel Cell Technology for the Sustainable Future of Stationary Applications. *Energies* **2019**, *12*, 4593. [CrossRef]
25. Achillas, C.H.; Vlachokostas, C.H.; Moussiopoulos, N.; Banias, G.; Kafetzopoulos, G.; Karagiannidis, A. Social acceptance for the development of a waste-to-energy plant in an urban area. *Resour. Conserv. Recycl.* **2011**, *55*, 857–863. [CrossRef]
26. Yap, K.Y.; Sarimuthu, C.R.; Lim, J.-Y. Virtual Inertia-Based Inverters for Mitigating Frequency Instability in Grid-Connected Renewable Energy System: A Review. *Appl. Sci.* **2019**, *9*, 5300. [CrossRef]
27. Vadi, S.; Padmanaban, S.; Bayindir, R.; Blaabjerg, F.; Mihet-Popa, L. A Review on Optimization and Control Methods Used to Provide Transient Stability in Microgrids. *Energies* **2019**, *12*, 3582. [CrossRef]
28. Sikorski, T.; Jasiński, M.; Ropuszyńska-Surma, E.; Węglarz, M.; Kaczorowska, D.; Kostyla, P.; Leonowicz, Z.; Lis, R.; Rezmer, J.; Rojewski, W.; et al. A Case Study on Distributed Energy Resources and Energy-Storage Systems in a Virtual Power Plant Concept: Technical Aspects. *Energies* **2020**, *13*, 3086. [CrossRef]

energies

Article

New HFC/HFO Blends as Refrigerants for the Vapor-Compression Refrigeration System (VCRS)

Bartosz Gil *, Anna Szczepanowska and Sabina Rosiek

Department of Thermodynamics and Renewable Energy Sources, Faculty of Mechanical and Power Engineering, Wrocław University of Science and Technology, Wybrzeże Wyspiańskiego 27, 50-370 Wrocław, Poland; aniaszczepanowska@gmail.com (A.S.); sabina.rosiek@pwr.edu.pl (S.R.)
* Correspondence: bartosz.gil@pwr.edu.pl; Tel.: +48-71-320-4826

Abstract: In this work, which is related to the current European Parliament Regulation on restrictions affecting refrigeration, four new three-component refrigerants have been proposed; all were created using low Global Warming Potential(GWP) synthetic and natural refrigerants. The considered mixtures consisted of R32, R41, R161, R152a, R1234ze (E), R1234yf, R1243zf, and RE170. These mixtures were theoretically tested with a 10% step in mass fraction using a triangular design. The analysis covered two theoretical cooling cycles at evaporating temperatures of 0 and −30 °C, and a 30 °C constant condensing temperature. The final stage of the work was the determination of the best mixture compositions by thermodynamic and operational parameters. R1234yf–R152a–RE170 with a weight share of 0.1/0.5/0.4 was determined to be the optimal mixture for potentially replacing the existing refrigerants.

Keywords: hydrofluoroolefins; hydrofluorocarbons; refrigerants; low GWP

check for updates

Citation: Gil, B.; Szczepanowska, A.; Rosiek, S. New HFC/HFO Blends as Refrigerants for the Vapor-Compression Refrigeration System (VCRS). *Energies* **2021**, *14*, 946. https://doi.org/10.3390/en14040946

Received: 17 December 2020
Accepted: 8 February 2021
Published: 11 February 2021

Publisher's Note: MDPI stays neutral with regard to jurisdictional claims in published maps and institutional affiliations.

Copyright: © 2021 by the authors. Licensee MDPI, Basel, Switzerland. This article is an open access article distributed under the terms and conditions of the Creative Commons Attribution (CC BY) license (https://creativecommons.org/licenses/by/4.0/).

1. Introduction

It was only a few years ago that most refrigeration appliances used refrigerants containing chlorine; these had a destructive effect on the ozone layer. The use of chlorofluorocarbons (CFCs) and hydrochlorofluorocarbons (HCFCs), which actually possess very good thermodynamic properties, is now prohibited. Apart from the destructive impact on the ozone layer, refrigerants also affect the natural environment by contributing to the greenhouse effect. Bearing in mind the changes taking place in the environment, the European Parliament approved Regulation No. 517/2014 [1] on fluorinated greenhouse gases, which significantly limits the possibility of using the refrigerants currently available on the market, especially those with a high Global Warming Potential (GWP). The purpose of legal regulations is to limit global climate change in accordance with the Paris Agreements and to prevent the adverse effects of this change. In response, greater use of natural refrigerants and hydrocarbon derivatives is necessary, as they have a lower environmental impact. The extensive use of carbon dioxide or hydrocarbons, as well as hydrofluoroolefins (HFOs) and hydrofluorocarbons (HFCs) with a GWP not exceeding 150, is a key challenge in transforming the refrigerant market.

HFO substances quickly decompose in the lower atmosphere due to the double carbon bond in the molecule; this guarantees very low GWP, but also results in flammability [2]. The most popular refrigerant in this group is R1234yf, which is used as a replacement for R134a in automotive air-conditioning systems. The second leading representative of the new generation of refrigerants is R1234ze(E), which has zero Ozone Depletion Potential (ODP) and a low GWP. It has been proposed as a replacement for R32, R410A, and R134a. However, R1234ze(E) is not an ideal substitute for the phased-out refrigerants, as it has a lower specific cooling capacity [3], a lower coefficient of performance (COP) [4], and a lower heat transfer coefficient [5,6]. According to EN 378-1 [7], both HFO refrigerants have a very low level of GWP, at only 4 and 7, respectively. In addition to compressor cycles,

HFOs are also considered for operation in other types of refrigeration equipment, such as ejector devices [8,9] or combined systems [10,11].

Combining HFO refrigerants with HFCs has become popular as a way to improve their properties, especially low cooling capacity, due to low latent heat of vaporization and specific refrigerating effect. In scientific research, it has often been combined with R32, a high-pressure refrigerant with satisfactory thermodynamic properties, high latent heat, and a relatively low GWP of 675 [12]. The R32–R1234ze(E) mixture was tested in the proportions of 0.5/0.5 and 0.2/0.8 to determine the heat transfer coefficient obtained in a horizontal tube [13] and the isobaric heat capacity (a mole fraction of R32 from 0.226 to 0.946) [2]. Another proposed combination of HFO/HFC refrigerants is the R32–R1234yf mixture, at different weight proportions, tested to determine a truncated virial equation of state [14].

Akasaka [15] presented models of thermodynamic properties for the mixtures R32–R1234ze(E) and R32/R1234yf, the uncertainties of which are 1% for the bubble point pressure and 0.25% for the liquid density. The author states that while the models show slightly greater uncertainties than the typical Helmholtz energy state equations for pure fluids, they are applicable to preliminary analysis of refrigeration equipment and heat pumps. R1234ze(E), like R1234yf, was also combined with R134a to determine the vapor–liquid balance of the mixtures [2] and to test the possibility of replacing pure R134a with new substances in home refrigerators.

Aprea et al. [16] proposed a mixture of R1234yf–R134a with a 10% HFC weight share, defined by the GWP limit of 150. This mixture achieved a 17% lower life-cycle climate performance (LCCP) index than pure R134a.

Other experimental results have shown that for R134a–R1234ze(E) mixture, despite the larger required refrigerant charge, a shorter daily operating time was obtained while maintaining the set temperature level, which resulted in a reduction of energy consumption by 14% compared to R134a [17]. Another combination is the R152a–R1234ze(E) mixture, for which the equilibrium (vapor + liquid) was tested; this is one of the most important parameters used to calculate and optimize cooling cycle efficiency, as well as the organic Rankine cycle and other chemical processes [18].

Ternary mixtures are another way to combine the new generation of refrigerants with HFCs. Various mixtures containing HFO are currently commercially available. These include, among others:

- R447A, which has a favorable GWP of 572, and, in terms of heat exchange, can be treated as a potential alternative to R410A (GWP = 2088) [19];
- R457A and R459B, with a low GWP (less than 150), which can provide a smaller refrigerant charge and improve energy efficiency in relation to R404A [20];

It is not only synthetic refrigerants that are combined together. Natural substances, such as carbon dioxide and hydrocarbons, are also used to create refrigeration blends. Kondou et al. [21] examined the heat transfer coefficient during evaporation and condensation of the ternary mixture R32–R1234yf–R744 with different weight shares (0.29/0.62/0.09, 0.43/0.53/0.04 and 0.06/0.34/0.6) and compared them with the binary R32–R1234ze(E) mixture (with weight shares of 0.4/0.6 and 0.73/0.27). Saengsikhiao et al. [22] considered using R463A as a replacement for R404A. It has been shown that R463A may operate at a higher ambient temperature, achieving higher COP in a low-temperature application, and lower GWP compared to R404A.

A further example of combining HFCs with HFOs and natural substances is R134a–R1234yf–R600a. Isobutane has good latent heat and excellent thermodynamic performance but is highly flammable and explosive. R1234yf is also slightly flammable, so non-flammable R134a has been added to reduce this flammability. The vapor–liquid equilibrium data were collected for the mixture [23]. Research has also been carried out on implementing a model for forecasting vapor + liquid balance data in a binary mixture of CO2, HFC, or HC, with two low GWP refrigerants (R1234yf and R1234ze (E)) [24]. Another flammable natural refrigerant that has attracted interest in recent years is dimethyl ether

(DME or RE170), which is widely used in the chemical industry, medicine, etc., and is therefore considered an alternative refrigerant. It has a low boiling point, high latent heat, and is non-toxic, slightly corrosive, and more environmentally friendly, because its ODP and GWP are 0 and 1, respectively. The study investigated the flammability of the RE170–R1234yf–R134a mixture (with weight fractions of 0.1/0.8/0.1) [25].

There are many possible refrigerant combinations; HFO is usually combined with well-known substances from the HFC group due to their excellent environmental parameters. As shown above, it is mainly R1234yf and R1234ze(E) that are selected for analysis. Others are not as popular for a variety of reasons, including a normal boiling points above 0 °C or even non-zero ODP. In this study, along with R1234ze(E) and R1234yf, it was decided to use one of the rarer hydrofluoroolefins as a substrate for the newly defined mixture, namely R1243zf (a GWP of approx. 1, an ODP of 0).

Finding the perfect refrigerant, which has favorable thermodynamic properties and can work efficiently in the refrigeration cycle, while also being cheap, easily accessible, and safe for the environment, is a very difficult task, if only because of the numerous legal and technical changes taking place in the refrigeration sector. There are many possibilities for combining and mixing pure refrigerants together to achieve the desired properties and to have a negligible effect on the atmosphere. In this work, four ternary mixtures were proposed and theoretically tested for their cooling capacities in theoretical cooling circuits to determine the optimal composition for the new blends. Significant effort has gone into developing and testing different refrigerant blends. However, there are still many unresolved issues and opportunities that need to be investigated. In light of the state of the art in this field, combining and mixing pure refrigerants together to achieve the desired properties while having a negligible effect on the atmosphere is still an open problem and a meaningful endeavor. This combining and mixing took several aspects into account, such as the use of the rarer hydrofluoroolefins (instead of commonly used refrigerants) as a substrate for the newly defined mixture, the triangular design, which involves the use of three refrigerant components, and the thermodynamic and operational parameters. Such an approach is hardly visible in the literature, which confirms the novelty of the presented research. Its realization required the application of new approaches that employed advanced analysis of the four ternary mixtures, along with their experimental verification. This allowed for the advanced testing of their cooling capacities in theoretical cooling circuits to determine the optimal composition for the new blends. It should also be highlighted that the weight fraction of the presented blends was determined (and not fixed, as a frequent goal of scientific articles) to find the optimal mixture for potentially replacing the existing refrigerants. Consequently, this article contributes to the state of the art in the new HFC/HFO blend selection by developing appropriate methods that can be feasibly implemented.

2. Materials and Methods

2.1. Methodology and Prerequisites for the Mixture Components

The selection of the optimal zeotropic or azeotropic mixture components was based on a triangular-basis plan, which involves the use of three refrigerant components. The sum of the weight shares of the selected substances must always be equal to 1 (100%), and the single share must be zero or positive. Thus, each blend will be defined by three independent variables. The shares of individual components (pure refrigerants) in the mixture change from 0 to 1 (from 0% to 100%) in steps of 0.1 (10%). The basic properties and operating parameters of each mixture, depending on their composition, are presented on ternary charts. In the corners of the graph are the pure substances selected as mixture components. The edges of the triangle depict binary mixtures, while the ternary mixtures are inside the triangle.

The basic criteria for selecting the constituent substances were the environmental parameters of a zero ODP and a low GWP. A condition was also set regarding the normal boiling point, the value of which had to be at least −25 °C to avoid under-pressure conditions. Given the final evaporation temperature levels obtained, all the proposed mixtures were assigned to the implementation of air-conditioning or freezing cycles and compared to the corresponding reference refrigerants. An additional criterion was the small distribution of individual components in already existing and commercially available mixtures. It was decided that the newly created working fluid should contain at least one alternative refrigerant (natural or from the HFO group), the rest being HFCs. The criteria were ranked according to their weight and checked in accordance with the diagram presented in Figure 1. The main initial consideration when creating mixtures and selecting their components was to limit the GWP value to 750 for air-conditioning systems (or 150 if possible) and 150 for low-temperature refrigeration devices. Obtaining low temperatures for the assumed operating conditions of the condenser may be associated with obtaining correspondingly higher pressure ratios in the system (compression ratio). In this case, the operation of the system may be disadvantageous due to a number of phenomena, such as low volumetric efficiency of the compressor and decrease in refrigerant mass flow due to an increase in specific vapor volume, increase in the compressor power consumption, high discharge temperature, and deterioration of lubricating properties of oils. In order to avoid the above-mentioned problems, it was assumed that the maximum value of compression ratio of the analyzed mixtures should not exceed 8.0.

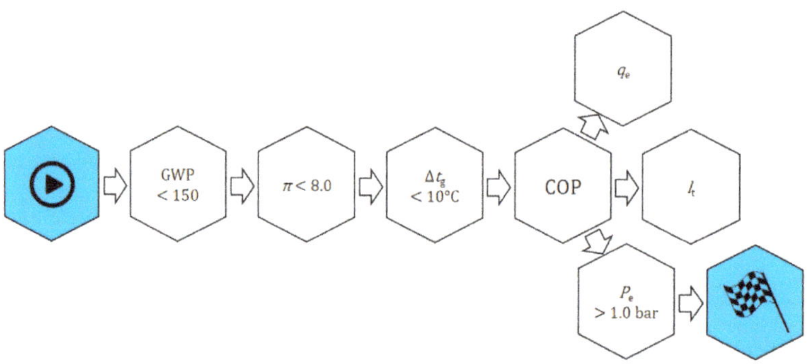

Figure 1. Selection criteria flow chart.

An equally important issue in the operation of real cooling cycles is the temperature glide, which may cause evaporator malfunctions. Components with extremely different vapour pressures can cause frosting to the initial sections of the evaporator due to the evaporation of low-boiling components. On the other hand, components with a high boiling point may not completely evaporate, which can lead to fractionation of the refrigerant inside the system and change its operating parameters. In case of extreme temperature glides, it is necessary to increase the vapour superheat set point to prevent the compressor from sucking in liquid refrigerant. In this analysis, the temperature glide of the mixtures was limited to 10 K. After meeting the initial criteria, the working parameters of the mixtures were analyzed.

Four mixture types have been presented and tested in this paper:

- R32–R41–R1234ze(E)—a mixture combining R32 that is currently gaining popularity, the R41, which has a low normal boiling point and a very low GWP, and R1234ze(E), the second most-studied HFO refrigerant;
- R32–R161–R1234ze(E)—a mixture similar to the previous one. However, the second component has been changed to R161, which has a slightly higher boiling point at 1 bar pressure but an almost eight-times lower GWP;
- R1234yf–R152a–RE170—this mixture combines the most popular refrigerant from the HFO group—R1234yf, with R152a often used in HFC/HFO mixtures, and a natural

substance RE170, which has been increasingly used in refrigeration mixtures, since the early 2000s;

- R1243zf–R152a–RE170—a mixture related to the previous one. However, the HFO group has been replaced by another, R1243zf, which thermodynamic properties are the most similar to widely used R134a, and which has a higher heat of vaporization than R1234yf.

All have been included in the European Patent (EP3309233A1) [26], which relates to compositions for use in refrigeration and air-conditioning. However, we would like to highlight that this does not negate the possibility of studying them from the scientific standpoint, with no intention of using them commercially. Basic information on the pure components used is presented in Table 1.

Table 1. Basic properties of pure components [27].

Refrigerant	Unit	R1234yf	R1234ze(E)	R1243zf	R152a	R161	R32	R41	RE170
Name	-	2,3,3,3-Tetrafluoroprop-1-ene	trans-1,3,3,3-Tetrafluoroprop-1-ene	3,3,3-Trifluoropropene	1,1-Difluoroethane	Fluoroethane	Difluoro-methane	Fluoro-methane	Dimethyl ether
CAS No.	-	754-12-1	29118-24-9	677-21-4	75-37-6	353-36-6	75-10-5	593-53-3	115-10-6
GWP [1]	-	4	7	1	124	12	675	92	1
Critical temperature	°C	94.7	109.4	103.8	113.3	102.1	78.11	44.13	127.2
Critical pressure	bar	33.82	36.35	35.18	45.17	50.46	57.82	58.97	53.37
Normal boiling point	°C	−29.48	−18.97	−25.42	−24.02	−37.54	−51.65	−78.31	−24.78
Molar mass	g/mol	114.0	114.0	96.05	66.05	48.06	52.02	34.03	46.07
Flamability class [7]	-	2 L	2 L	2 L	2	3 *	2 L	3 *	3

* flammable compounds, not included in EN 378; determined based on NFPA 704.

2.2. Theoretical Refrigeration Cycle—Assumptions

Two theoretical single-stage refrigeration cycles were determined for comparing the mixtures. In the first cycle, the evaporating temperature (t_e) was set at 0 °C, with the condensing temperature (t_c) equal to 30 °C. This cycle corresponds to the work done by air-conditioning systems. The evaporating temperature in the second cycle was set to −30 °C, with the condensing temperature unchanged. This allowed to examine the behavior of mixtures in low-temperature systems. In both cases, it was assumed that the compression isentropic efficiency was η = 0.7. For the purposes of theoretical analysis, it was assumed that the liquid subcooling in the condenser and the vapour superheating in the evaporator were equal to zero, as presented in Figure 2. Pressure drops in the heat exchangers and in the pipeline flow were also omitted.

The theoretical analysis of comparative cycles allowed us to determine the basic operating parameters of the proposed mixtures. With the help of enthalpy at the characteristic operating points, the specific cooling capacity (q_e), the specific work of the cycle (l_t), the volumetric cooling capacity (defined as $q_e/v_{suction}$), and the COP of each of the proposed mixtures were defined in terms of the assumed variability in the weight shares of the individual components. On the basis of the evaporation and condensation pressures obtained, the compression ratio and the temperature glide were determined. The specific refrigeration system parameters were determined for both the high-temperature and the low-temperature cycles. In addition to the operating parameters, the basic properties were also determined, such as the GWP, the critical point temperature and pressure, the normal boiling point (1 bar), and the molar mass.

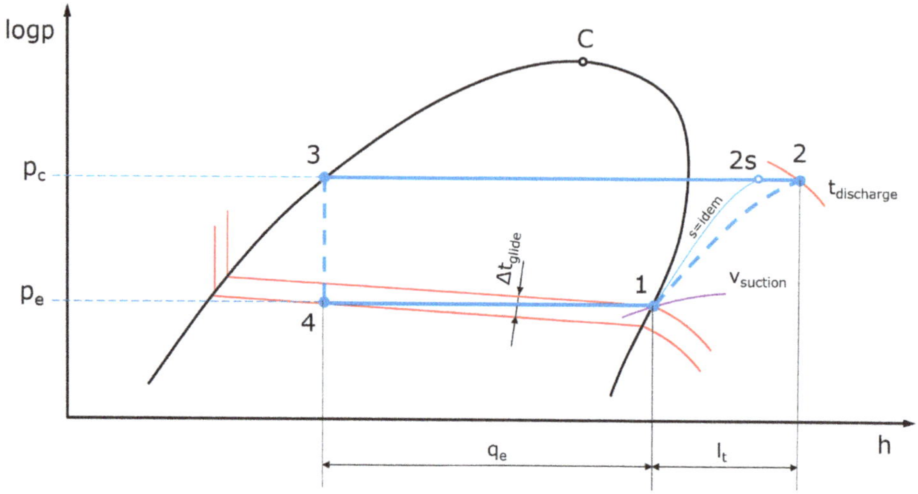

Figure 2. Single stage refrigeration cycle analyzed.

The theoretical analysis of the newly defined zeotropic or azeotropic mixtures was carried out using the REFPROP 10.0 program [27], from which the thermodynamic and transport properties of the fluids relevant for the study and their mixtures were taken, such as the critical temperature and pressure, the normal boiling point, and the molar mass. The thermodynamic properties of the mixtures were determined by employing a model that applies mixing rules to the Helmholtz energy of the mixture components, along with a departure function to account for the departure from ideal mixing. The same "XR0" mixing rule was used in REFPROP for all analyzed mixtures. It should be emphasized that in order to determine the properties of mixtures, it is necessary to have properties for binary subsystems, and among the mixtures under consideration, not all subsystems were tested experimentally. Therefore, it should be borne in mind that despite the generally good accuracy of determining thermal-flow properties, some of the presented values may differ from the actual ones, which requires confirmation in the field of further experimental studies.

The newly defined zeotropic or azeotropic mixtures will be compared (at work) to the refrigerants being withdrawn from cooling equipment as a result of the European Parliament regulation. It is important to find the parameters that will be favored by the proposed compositions. Taking into account the scope of application, the benchmark refrigerants for low temperature circuits will be R404A and R507A, and for high temperature circuits, R410A, R134a, R32, and R429A. The R429A mixture was used as a benchmark due to its similar composition to the two newly defined blends tested. This refrigerant consists of RE170, R152a, and R600a in the corresponding weight proportions of 60%, 10%, and 30%.

3. Results

Presented paper analyzes the results, which aimed to define the optimal composition of a new refrigerant mixture. Four lists of refrigerants were considered.

3.1. *R32–R41–R1234ze(E)*

The first of the mixtures contains the now popular R32 refrigerant, which is treated as a substitute for the R410A refrigerant and has very good thermodynamic properties, as well as a relatively low GWP; nevertheless, it is in the A2L flammability class. The second HFC component is R41, which has a very low global warming potential. The representative from the HFO group is the well-known R1234ze(E), which is freely available on the market.

Analyzing the composition of the discussed mixture, it is most likely to belong to the A2L flammability class.

The paper presents the value gradients determined for the R32–R41–R1234ze(E) mixture. This started from the basic values: the GWP and the normal boiling point, presented in Figure 3.

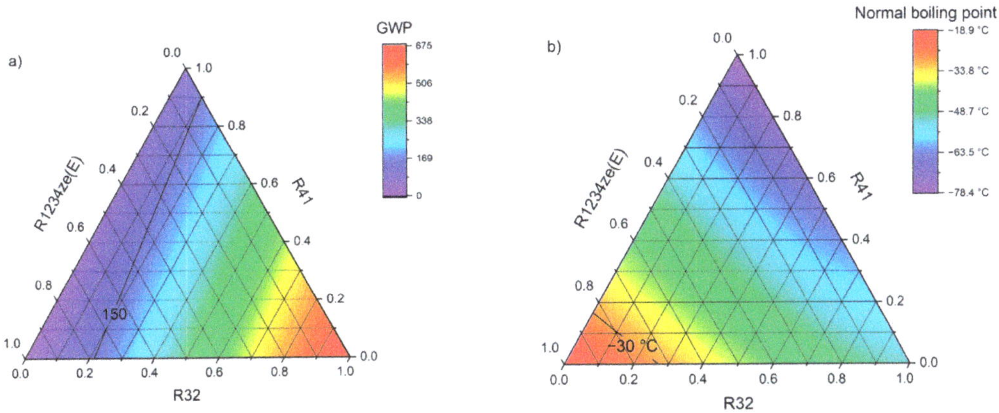

Figure 3. Summary of the basic properties for the R32–R41–R1234ze(E) mixture: (**a**) GWP and (**b**) normal boiling point.

The mixture, regardless of the shares of individual components, has a GWP below 750, as none of the components exceed this threshold. However, in order to meet the criterion for low-temperature cycles, it is necessary to limit the share of R32 to a maximum of 20%. Thus, only the left side of the triangle shown in Figure 3a is potentially usable. Attention should also be paid to the lower left corner of the normal boiling point diagram (Figure 3b). The boiling points at normal pressure in this area are higher than −30 °C, which means that the system will work under pressure. In the case of failure and leakage, it may lead to the appearance of air inside the system, which in the case of flammable refrigerants may pose a real risk of ignition of the installation due to the possible exceeding of the lower flammability limit of the mixture. Moreover, analyzing Figure 4b, it can be seen that the area at the left edge of the triangle is almost entirely covered by high temperature glides, reaching even over 25 K. In practice, only the mixtures located in the upper corner and in the right corner should be considered as suitable for low-temperature cycles.

For air conditioning cycles, the GWP limit according to [1] is 750, which is not exceeded at any point in the graph. The only limit here is the temperature glide, the value of which should not exceed the assumed threshold of 10 K (compare Figure 4a). Figure 5a,b show that it is impossible to select the mixture in such a way as to ensure both the possible high COP and the high volumetric cooling capacity of the mixture. Optimizing the mixture for high volumetric cooling capacity will provide benefits in terms of reducing the compressor (piston) displacement, which will translate into the number of pistons, compressor weight, dimensions, and price. Optimization for the possible high COP will reduce operating costs due to the minimization of the energy needed to compress the refrigerant. Regardless of the chosen direction of optimization, only a few mixtures are possible to use. Taking into account that GWP should be limited to 150, only two ternary mixture are available with mass fractions 0.1/0.8/0.1 and 0.1/0.7/0.2. Unfortunately, both of these mixtures contain a high proportion of R41, a refrigerant that is currently hardly available on the market due to the lack of widespread use in refrigeration. At the same time, it can be seen that there is no other R41-free mixture that meets these criteria. Extending the analysis to the threshold of 750 adopted by Regulation (EU) 517/2014, it is possible to select two binary mixtures R32–R1234ze(E) with the weight shares of 0.9/0.1 and 0.8/0.2. Of the two mixtures, the first is the more promising, as it is characterized by an increase in volumetric cooling capacity by 8%, with an almost identical COP. It should be remembered that the GWP threshold of 750 applies only to single split air-conditioning systems containing less than 3 kg of F-gases and only in this group of devices can the mixture proposed above be used.

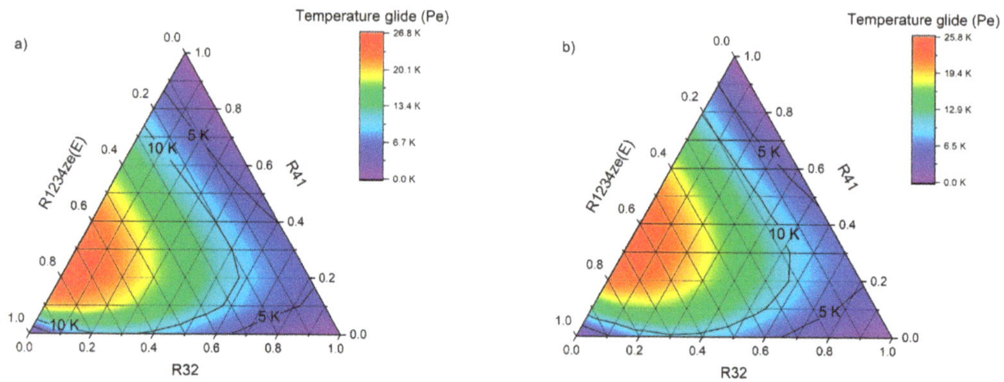

Figure 4. Temperature glide at the evaporation pressure for cooling cycles with R32–R41–R1234ze(E) mixture as refrigerant: (**a**) AC system (t_e/t_c = 0/30 °C); (**b**) low-temperature system (t_e/t_c = −30/30 °C).

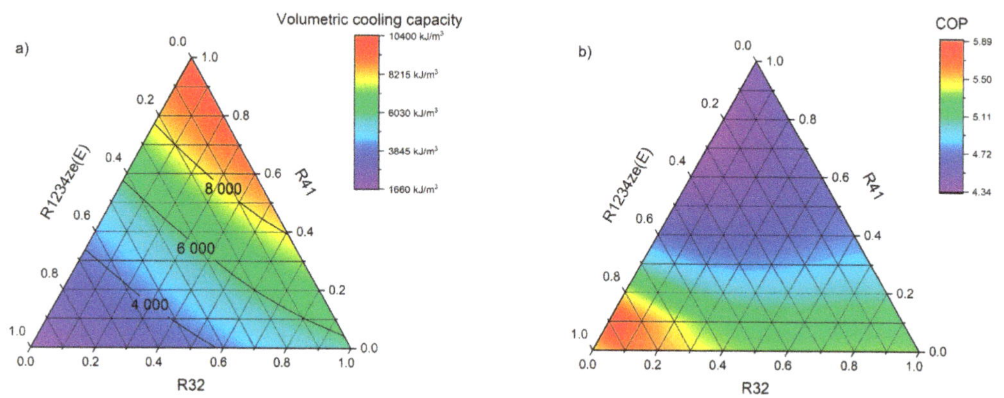

Figure 5. Volumetric cooling capacity (**a**) and (**b**) the COP for air-conditioning system (t_e/t_c = 0/30 °C) with R32–R41–R1234ze(E) mixture as refrigerant.

In the case of low-temperature cycles, additional problems in finding the proper mixture are the areas of high pressure ratio and high discharge temperature (Figure 6a,b). The use of a mixture with a mass fraction of R1234ze(E) greater than or equal to 0.4 leads to exceeding the assumed permissible level of the compression ratio for a single-stage cycle. In addition, a fraction greater than 0.8 largely leads to an evaporation pressure below 1 bar. For freezing circuits, it is not possible to select a mixture that does not contain R41, as the normal boiling point and GWP limitations effectively exclude the use of the R32–R1234ze(E) binary mixture. It is necessary to introduce a third component that will lower both these parameters at the same time. Similarly to the case of air-conditioning cycles for the considered ternary mixture, there is only one composition that meets all the assumptions—0.1/0.8/0.1. However, further analysis of Figure 6a–d shows that by abandoning the R1234ze(E) component, all the operating parameters of the circuit are improved. The use of the binary mixture R32–R41 (0.1/0.9) allow the increase of both the COP and the volumetric cooling capacity, by 3% and 20.6%, respectively. A 6 K reduction in discharge temperature is also achieved, while the pressure ratio value drops below 5.5.

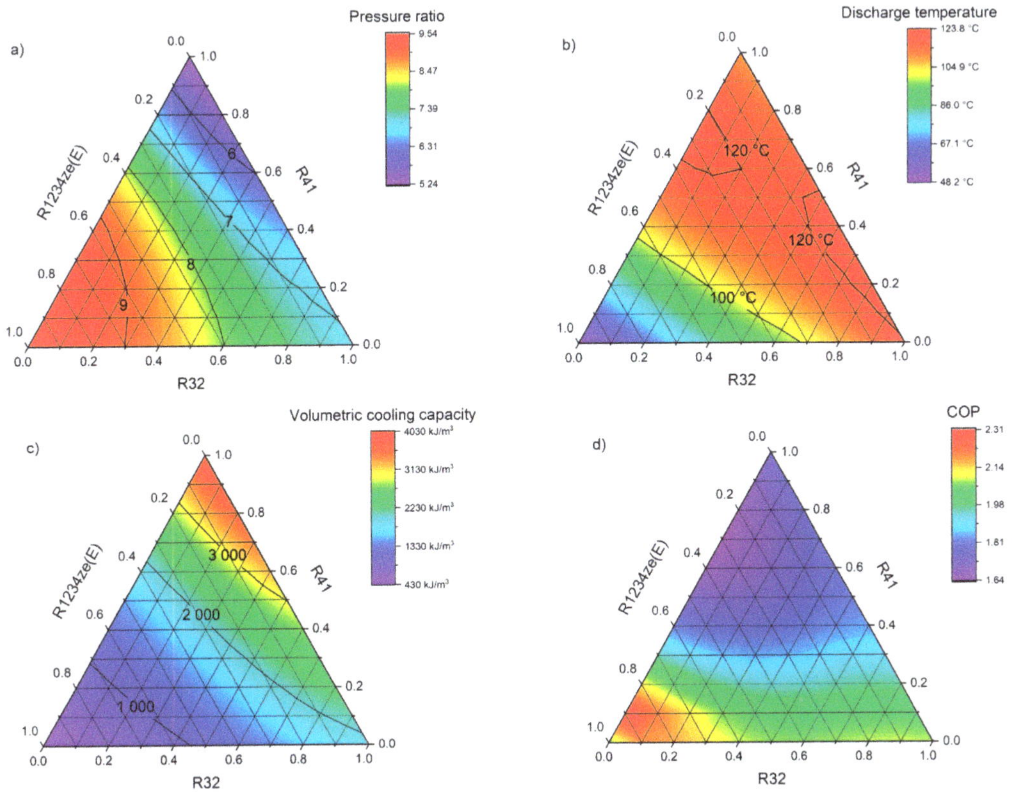

Figure 6. Low-temperature system parameters ($t_e/t_c = -30/30$ °C) of the R32–R41–R1234ze(E) mixture: (**a**) pressure ratio; (**b**) temperature of the refrigerant vapour at the compressor discharge; (**c**) volumetric cooling capacity; and (**d**) COP.

3.2. R161–R41–R1234ze(E)

The second mixture analyzed is a combination of R161, R41, and R1234ze(E). This composition is similar to the first mixture, but the R32 refrigerant has been replaced with R161, which also belongs to the HFC group. It has a very low GWP (equal to 12). In this mixture, R161 and R41 are highly flammable, while R1234ze(E) belongs to the 2 L class; therefore, it can be assumed that ternary mixtures will also belong to the highest flammability class. Each mixture component has a GWP below 150 (Figure 7a), so none of the points exceeded the strictest limit. Similar to the previously considered mixture, the normal boiling point of mixtures with high proportions of R1234ze(E) precludes the use of part of the composition in low temperature systems (Figure 7b). The heterogeneity of the mixtures is a serious problem in both the high- and low-temperature systems. Due to the very high temperature glides of this mixture for both analyzed cycles, only the narrow range of compositions meet the criterion of the maximum Δt_{glide} of 10 K, as presented in Figure 8.

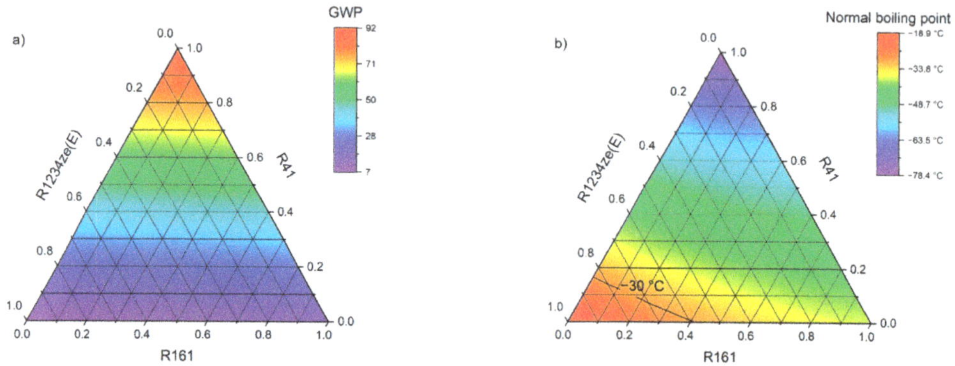

Figure 7. Summary of the basic properties for the R161–R41–R1234ze(E) mixture: (**a**) GWP and (**b**) normal boiling point.

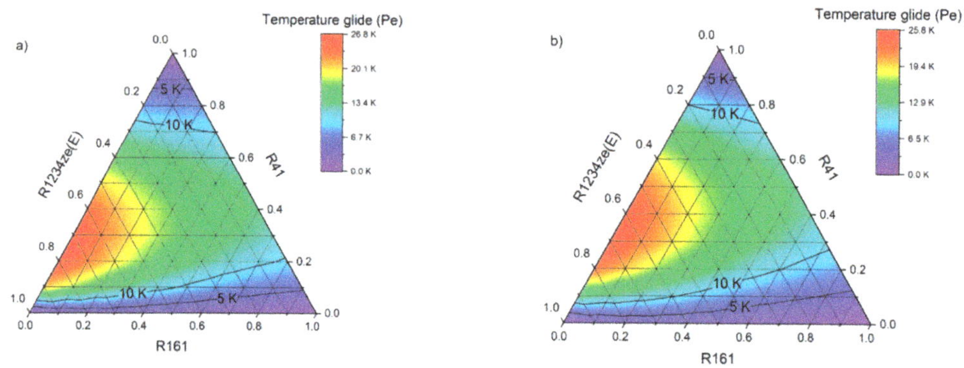

Figure 8. Temperature glide at the evaporation pressure for cooling cycles with R161–R41–R1234ze(E) mixture as refrigerant: (**a**) AC system (t_e/t_c = 0/30 °C); (**b**) low-temperature system (t_e/t_c = −30/30 °C).

The similarity to the previous mixture can also be seen in the COP and volumetric capacity charts (Figure 9). The areas of the highest values of these parameters are mutually exclusive, so it is necessary to consider which of the values will be more important for the end user. Considering the low cooling capacity of split air-conditioning devices, it can be concluded that higher energy efficiency will be more beneficial. Choosing a mixture with mass fractions of 0.8/0.1/0.1 will result in a much higher COP compared to a mixture with mass fractions of 0.1/0.8/0.1. An additional advantage is also a lower temperature glide.

In the case of mixtures predestined for operation in low-temperature circuits, the choice of a mixture with the 0.8/0.1/0.1 composition gives an additional advantage resulting from a significantly lower temperature of the medium after the compression process (see Figure 10b). Lowering this temperature by more than 30 K will be crucial for the operation of the system, especially in the summer. Discharge temperature drop will result in the lack of restrictions in terms of thermal stability of oils or the need for additional cooling of the compressor working elements. Compared to the 0.1/0.8/0.1 composition, the pressure ratio does increase, but its value remains at an acceptable level of 7.55 (Figure 10a). The mixture R161–R41–R1234ze(E) with mass fractions of 0.8/0.1/0.1 seems to be optimal also for low-temperature cycles, where for the assumed operating parameters it obtains COP = 2.2, with a volumetric cooling capacity of 1129 kJ/m^3 (Figure 10c,d).

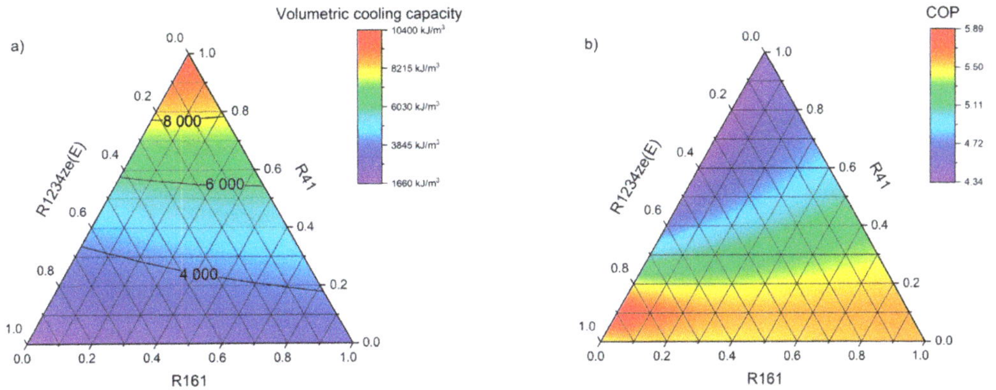

Figure 9. Volumetric cooling capacity (**a**) and (**b**) the COP for air-conditioning system (t_e/t_c = 0/30 °C) with R161–R41–R1234ze(E) mixture as refrigerant.

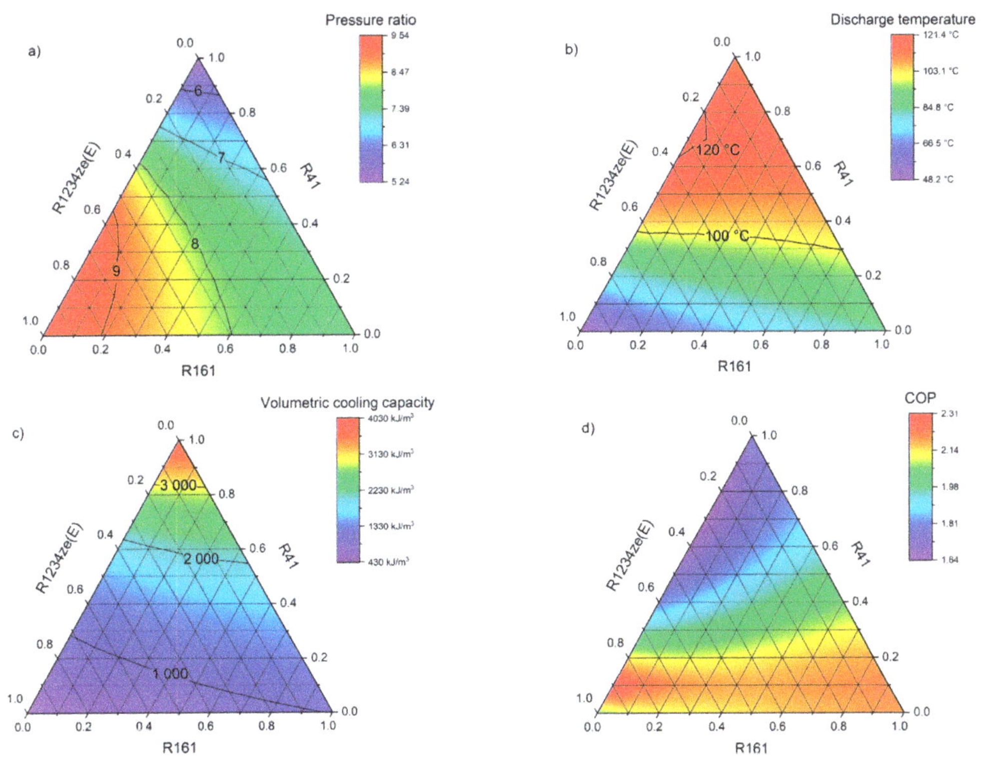

Figure 10. Low-temperature system parameters (t_e/t_c = −30/30 °C) of the R161–R41–R1234ze(E) mixture: (**a**) pressure ratio; (**b**) temperature of the refrigerant vapour at the compressor discharge; (**c**) volumetric cooling capacity; and (**d**) COP.

3.3. R1234yf–R152a–RE170

The third mixture has a completely different composition to the previous ones. It was decided to check the working parameters of the mixture, which in addition to fluorinated HFC/HFO refrigerants also contains dimethyl ether, known as RE170. The proportion of dimethyl ether in the mixture is likely to influence the toxicity and flammability class of the new refrigerant. The mixture's flammability class is assumed to be 3, as to be expected from a mixture with a high proportion of RE170. The representative from the HFC group is R152a. This is often used in existing mixtures because of its low GWP and its relatively high normal boiling point, thus prompting its use in medium to high temperature circuits. The last mixture component is the most common and most easily available fluid from the HFO group—R1234yf. As in the R161–R41–R1234ze(E) mixture, all the constituents have a GWP lower than 150; therefore, the GWP of the mixture itself does not exceed this value (Figure 11a). The flaws of this blend is that normal boiling point never reached −30 °C, which proves third mixture is not predisposed for use in low-temperature units (Figure 11b). The device's operation under negative pressure is associated with certain dangers and requires the use of a larger compressor. Considering all of the above, in the case of the R1234yf–R152a–RE170 mixture, it was decided to analyze only the high-temperature cycle.

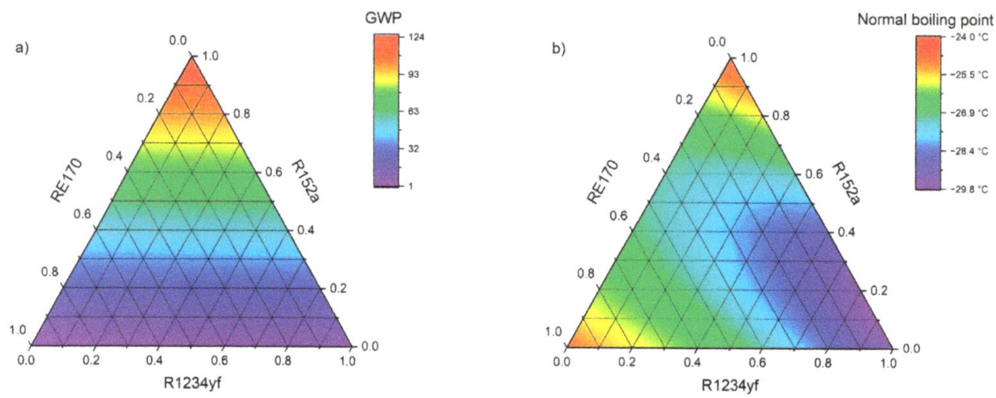

Figure 11. Summary of the basic properties for the R1234yf–R152a–RE170 mixture: (**a**) GWP and (**b**) normal boiling point.

When analyzing the third mixture (R1234yf–R152a–RE170) and the temperature glides obtained (Figure 12), the conclusion is that the temperature difference during the phase change does not exceed 1 K for the assumed evaporating pressure level regardless of the composition; thus, it can be considered at least as near-azeotropic. The achieved volumetric cooling capacity and COP play the greatest role in the assessment (Figure 13a,b). The highest COP value determined was 5.67, but it was for pure RE170. Given the importance of the fluid share from the A3 group being as small as possible, this point was not analyzed. It is worth noting that, with a binary and ternary mixture, as many as 54 points out of 63 have a COP > 5.48, which is the reference value obtained by R134a. There are two ternary mixtures that meet all the assumptions of the work and appear to be good potential substitutes for phased-out fluorinated greenhouse gases. The first is the 0.1/0.1/0.8 mixture, which has a COP of 5.65. The second mixture with a very similar COP has a mass fraction of 0.1/0.5/0.4, but is additionally characterized by a lower normal boiling point and temperature glide well below 0.1 K. In this term, it can be considered as azeotropic. The second composition also has a volumetric cooling capacity that is 124 kJ/m^3 higher than that of the first mixture.

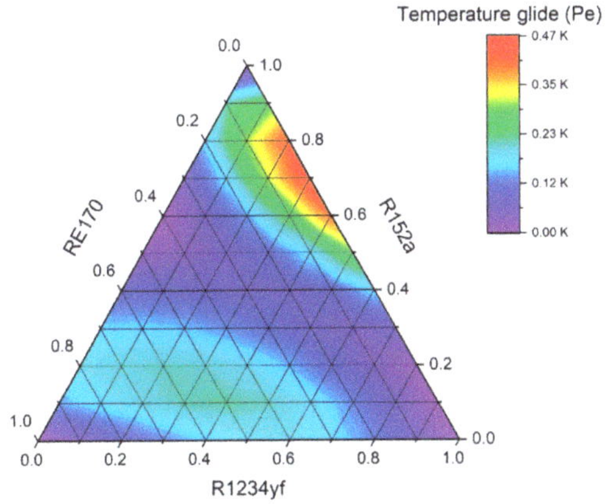

Figure 12. Temperature glide at the evaporation pressure for air-conditioning system (t_e/t_c = 0/30 °C) with R1234yf–R152a–RE170 mixture as refrigerant.

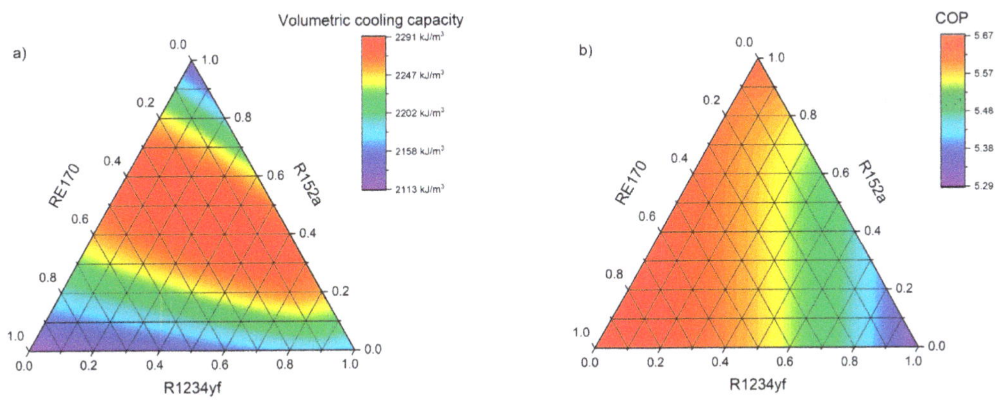

Figure 13. Volumetric cooling capacity (**a**) and (**b**) the COP for air-conditioning system (t_e/t_c = 0/30 °C) with R1234yf–R152a–RE170 mixture as refrigerant.

3.4. R1243zf–R152a–RE170

The last analyzed mixture is similar in composition to the third mixture, because it combines the two previously used refrigerants, RE170 and R152a. The HFO group refrigerant has been changed to a less popular R1243zf, as yet unused in any of the encountered mixtures. The compositions discussed, like those in the previous point, will most likely belong to the higher flammability group due to the presence of dimethyl ether. As with the previous combination of fluids, the GWP does not exceed that of R152a, and due to the normal boiling point, the mixture is considered only in high-temperature circuits (Figure 14).

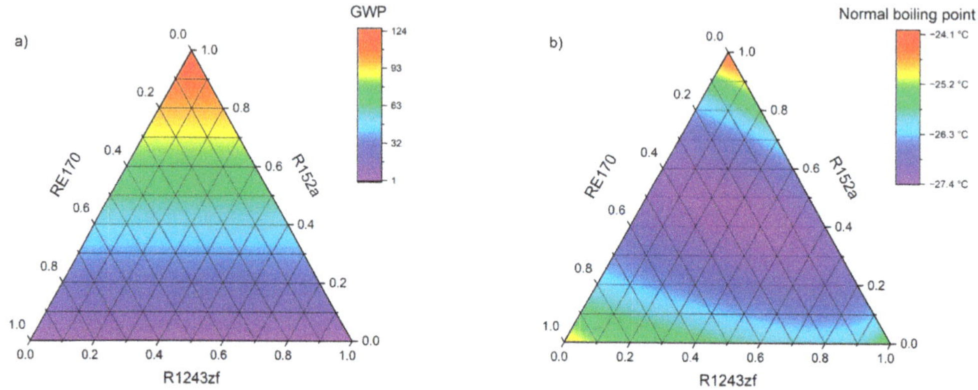

Figure 14. Summary of the basic properties for the R1243zf–R152a–RE170 mixture: (**a**) GWP and (**b**) normal boiling point.

The mixture R1243zf–R152a–RE170, regardless the composition, fulfils the requirements for the Δt_{glide} and π limits. For this mixture, an interesting parameter is the normal boiling point (Figure 14b), which reaches its minimum in the center of the triangle, not at the corners or sides. The lowest temperature achieved was −27.4 °C for the composition of 0.4/0.4/0.2. This phenomenon did not occur in the previously tested mixtures. Although all the points reached a negative normal boiling point, it is not low enough to use this mixture safely in freezing equipment. The temperature glide presented in Figure 15 is definitely below 1 K and has the lowest values in the center of the graph.

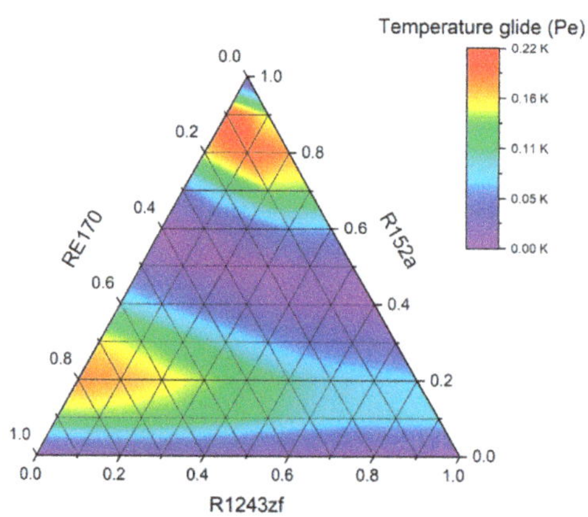

Figure 15. Temperature glide at the evaporation pressure for air-conditioning system (t_e/t_c = 0/30 °C) with R1243zf–R152a–RE170 mixture as refrigerant.

Moreover, a significant number of mixtures with a weight fraction of R152a in the range of 40–60% show a temperature glide significantly below 0.1 K, which makes them azeotropic. Considering the volumetric cooling capacity and the COP (Figure 16), as well as the previously mentioned normal boiling point and temperature glide, the best composition turns out to be 0.2/0.5/0.3. This is a compromise between a moderately high COP and a nearly zero temperature glide. Furthermore, it is a ternary mixture, and the RE170 belonging to the A3 flammability class does not make up the largest share. The second optimal substance, with a slightly higher COP, but also Δt_{glide}, is the mixture with a weight composition of 0.1/0.5/0.4. At the same time, it is a composition that achieves almost the highest volumetric cooling capacity of 2284 kJ/m^3.

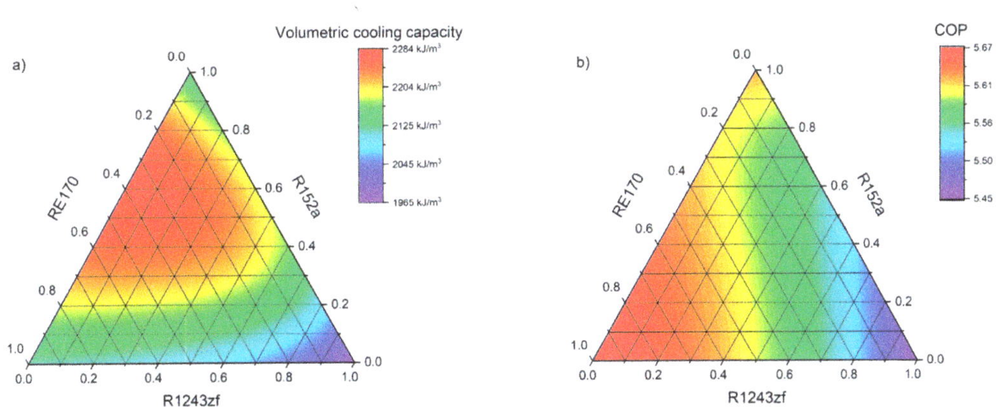

Figure 16. Volumetric cooling capacity (**a**) and (**b**) the COP for air-conditioning system (t_e/t_c = 0/30 °C) with R1243zf–R152a–RE170 mixture as refrigerant.

4. Discussion

Summarizing the results presented in this study, seven optimal mixture compositions were selected, of which six are ternary mixtures and two are binary:

- R32–R41–R1234ze(E) 0.9/0/0.1 air-conditioning cycle
- R32–R41–R1234ze(E) 0.1/0.9/0 low-temperature cycle
- R161–R41–R1234ze(E) 0.8/0.1/0.1 both cycles
- R1234yf–R152a–RE170 0.1/0.1/0.8 air-conditioning cycle
- R1234yf–R152a–RE170 0.1/0.5/0.4 air-conditioning cycle
- R1243zf–R152a–RE170 0.2/0.5/0.3 air-conditioning cycle
- R1243zf–R152a–RE170 0.1/0.5/0.4 air-conditioning cycle

Their distribution on the experiment plan is shown in Figure 17.

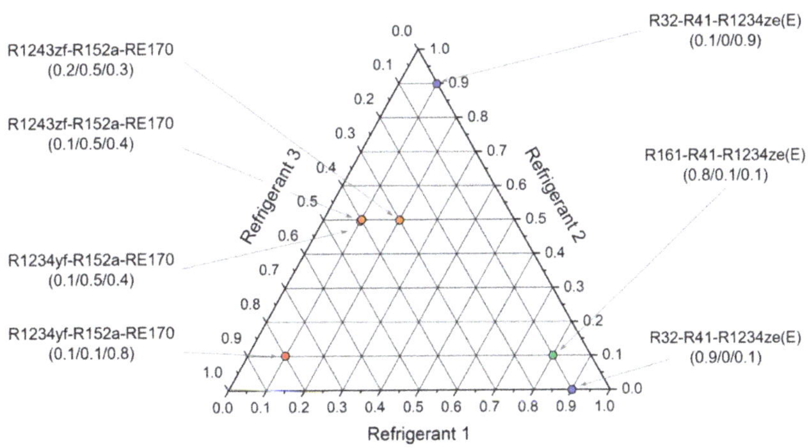

Figure 17. Marking the optimal mixtures in a triangular pattern.

There were many criteria for proving the applicability of a given refrigerant. The main ones were the temperature glide values achieved at p_e and within the whole range of working pressures. Based on the temperature differences obtained for the evaporation process, for which the limit value was set as 10 K, a significant proportion of the points for the first two mixtures were rejected. The temperature drop in the evaporation process for the assumed cooling cycle and variable evaporation temperatures for selected mixtures are presented in Figure 18. The figure shows that only for the mixture R161–R41–R1234ze(E), significant changes in the evaporation temperature are obtained, and therefore this mixture must be classified as zeotropic (ZEO). At the same time, it is clearly visible that with the increase in the evaporation temperature, the temperature glide increases significantly.

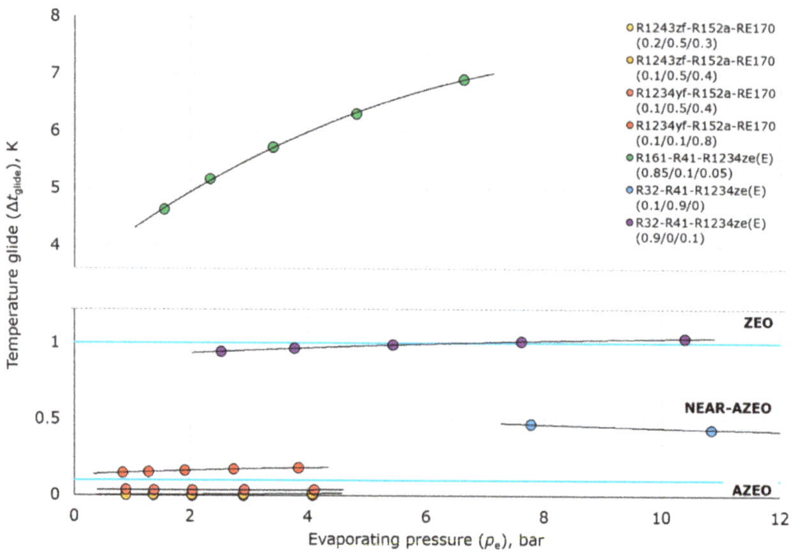

Figure 18. Temperature glide of the new mixtures depending on the evaporating pressure.

On the other hand, all the mixtures containing the R152a and RE170 fluids are characterized by a small temperature glide, the value of which does not exceed 0.5 K. The binary mixture R32–R1234ze(E) shows a temperature glide of almost exactly 1 K (limit for near-azeotropic mixtures). Figure 18 also shows that among the selected mixtures, R32–R41 achieves much higher evaporating pressures.

Table 2 shows a comparison of the newly defined mixtures with the reference refrigerants dedicated to low-temperature installations. The parameters that are favorable to the new mixtures are the GWP and specific cooling capacity. Compared to currently used refrigerants, it was found that the most optimal is a R32–R41 binary mixture with a mass fraction of 0.9/0.1, for which the specific cooling capacity is 243 kJ/kg and is more than twice as high as for the R404A or R507A. The great advantage of the R32–R41 binary mixture is the increase in volumetric cooling capacity by over 2700 kJ/m^3. The disadvantages are the slightly higher temperature glide and high working pressures. The biggest problem seems to be the much higher discharge temperature exceeding 110 °C, which is typical for the currently implemented and used R404A substitutes. Conversely, R161–R41–R1234ze(E) obtains a lower volumetric cooling capacity, which is dictated by a much higher specific vapor volume, at the same time showing a much higher temperature glide, which can eliminate this refrigerant from systems requiring precise evaporator temperature. The use of this mixture, however, allows to significantly improve the COP; compared to R404A, the increase is as much as 15.8%. However, the use of these mixtures requires a profound change in the refrigerant market, as currently R41 is not widely available.

Table 2. Comparison of the newly defined mixtures to the reference refrigerants suitable for low-temperature systems; properties based on [27].

Refrigerant	Unit	R404A (R125–R143a–R134a)	R507A (R125/R143a)	R32–R41-R1234ze(E)	R161–R41-R1234ze(E)
Weight share	-	0.44/0.52/0.04	0.5/0.5	0.1/0.9/0	0.8/0.1/0.1
GWP [1]	-	3922	3985	150	20
Critical temperature	°C	72.12	70.62	45.32	95.99
Critical pressure	bar	37.35	37.05	58.93	53.10
Normal boiling point	°C	−45.74	−47.01	−76.80	−39.77
Molar mass	g/mol	97.60	98.86	35.25	48.87
Evaporating temperature	°C	−30	−30	−30	−30
Evaporating pressure	bar	2.02	2.13	7.77	1.55
Condensing temperature	°C	30	30	30	30
Condensing pressure	bar	14.14	14.59	41.66	11.73
Pressure ratio	-	6.99	6.85	5.36	7.55
Compressor Discharge temperature	°C	52.86	51.75	110.63	87.43
Specific suction vapour volume	m^3/kg	0.095	0.089	0.062	0.252
Liquid density	kg/m^3	1019.4	1022.6	957.6	708.6
Temperature glide for p_e	K	0.48	0.003	0.47	4.63
Volumetric cooling capacity	kJ/m^3	1118.2	1151.6	3894.1	1128.9
Specific cooling capacity	kJ/kg	106.01	102.09	243.28	284.94
Specific heating capacity	kJ/kg	161.86	156.24	381.77	414.37
Specific work of the cycle	kJ/kg	55.85	54.15	138.49	129.43
COP	-	1.90	1.88	1.76	2.20

Table 3 shows the properties of the mixtures considered as refrigerants in air conditioning cycle. R410A, R134a, R32, and R429A were used as reference. An indicator that definitely favors the new refrigerants is the GWP, the value of which for R32-free mixtures does not exceed 63 and is less than half the permissible limit. Of the reference substances, only R429A has the same low GWP. The COP for the new mixtures are also favorable, since they all exceed a value of 5.34. When comparing the temperature glide, it can be observed that it is almost zero for the mixtures containing RE170, which is a slight advantage over R429A. These mixtures achieve a volumetric cooling capacity almost identical to that of R134a and approximately 6–12% higher than that of R429A. However, they are not in competition with R32 or mixtures containing it, so a higher amount of refrigerant in the system will be required. In terms of the obtained volumetric cooling capacity, the mixtures R32–R1234ze(E) and R161–R41–R1234ze (E) may be an interesting proposition. For both mixtures, the obtained values are much higher than for R134a, and in the case of a binary mixture R32–R1234ze(E) also at a similar level as for R32 and R410A. By analyzing all the variables, it can be summarized that the mixture R1234yf–R152a–RE170 with the weight shares of 0.1/0.5/0.4 seems to be the most promising for the implementation in air-conditioning cycles.

The conclusions drawn from the theoretical analysis should be confirmed by means of experimental studies of the various evaporation and condensation temperatures. To do this, mixtures with similar compositions should be tested, with a smaller jump in the weight shares of the individual components. However, this is future research work, as the scope of this work only included theoretical considerations regarding the new mixtures.

Table 3. Comparison of the newly defined mixtures to the reference air-conditioning refrigerants; properties based on [27].

Refrigerant	Unit	R134a	R32	R410A (R32–R125)	R429A (RE170–R152a–R600a)	R1234yf-R152a–RE170	R1234yf-R152a–RE170	R1243zf-R152a–RE170	R1243zf-R152a–RE170	R32–R41-R1234ze(E)	R161–R41-R1234ze(E)
Weight share/name	-	1,1,1,2-Tetrafluoro-ethane	Difluoro-methane	0.5/0.5	0.6/0.1/0.3	0.1/0.1/0.8	0.1/0.5/0.4	0.2/0.5/0.3	0.1/0.5/0.4	0.9/0/0.1	0.8/0.1/0.1
GWP [1]	-	1430	675	2088	14	14	63	63	63	608	20
Critical temperature	°C	101.06	78.10	71.34	121.95	123.28	114.90	113.04	115.22	80.58	95.99
Critical pressure	bar	40.59	57.82	49.01	47.30	51.41	47.82	46.07	47.77	58.07	53.10
Normal boiling point	°C	−26.07	−51.65	−51.62	−25.37	−25.88	−27.44	−27.33	−27.27	−50.23	−39.77
Molar mass	g/mol	102.03	52.02	72.58	50.76	50.62	58.38	61.86	57.82	55.02	48.87
Evaporating temperature	°C	0	0	0	0	0	0	0	0	0	0
Evaporating pressure	bar	2.93	8.13	7.98	2.64	2.75	2.92	2.86	2.89	7.61	4.82
Condensing temperature	°C	30	30	30	30	30	30	30	30	30	30
Condensing pressure	bar	7.70	19.28	18.84	6.64	6.99	7.42	7.30	7.36	18.16	11.73
Pressure ratio	-	2.63	2.37	2.36	2.52	2.54	2.54	2.55	2.54	2.39	2.43
Compressor Discharge temperature	°C	41.97	64.99	51.26	44.62	48.78	48.56	47.94	48.84	62.75	53.36
Specific suction vapor volume	m^3/kg	0.069	0.045	0.033	0.157	0.151	0.122	0.118	0.125	0.046	0.086
Liquid density	kg/m^3	1187.5	939.6	1032.7	633.9	697.8	780.6	804.9	773.4	957.6	688.5
Temperature glide for p_e	K	0.00	0.00	0.08	0.64	0.15	0.00	0.00	0.02	1.01	6.30
Volumetric cooling capacity	kJ/m^3	2263.5	5743.2	5292.9	2026.9	2158.4	2271.7	2226.5	2255.3	5393.2	3572.2
Specific cooling capacity	kJ/kg	156.88	259.98	173.11	318.32	326.24	277.84	261.77	281.34	248.74	307.00
Specific heating capacity	kJ/kg	185.53	308.70	206.27	374.80	384.02	327.34	308.48	331.43	295.30	362.11
Specific work of the cycle	kJ/kg	28.65	48.72	33.17	56.48	57.78	49.51	46.71	50.08	46.55	55.11
COP	-	5.48	5.34	5.22	5.64	5.65	5.61	5.60	5.62	5.34	5.57

5. Conclusions

Referring to the applicable environmental parameter limits set on refrigerants, four new refrigerant mixtures have been proposed in this work. The optimal weight shares of the individual components were estimated by analyzing the GWP, thermodynamic, and operational parameters.

Theoretical tests were performed using the REFPROP 10.0 program, and the collected data allowed for a preliminary estimation to be made. On the basis of theoretical analyzes, it was shown that all the proposed compositions, except for the R161–R41–R1234ze(E) mixture, can be classified as near-azeotropes or even azeotropes, because their temperature glide in a wide range of evaporation pressure does not exceed 1K. At optimal compositions, the share of HFOs in all mixtures does not exceed 20%. After considering the advantages and disadvantages of the refrigerants proposed, it was determined that the most optimal composition in high-temperature (air conditioning) systems was the R1234yf–R152a–RE170 mixture with a weight share of 0.1/0.5/0.4. This is argued by its low GWP, equal to 63, the relatively high COP of 5.61, the relatively low normal boiling point of -27.27 °C, and a lower weight share of the most flammable components (flammability class 3) than in the case of R429A.

The analyzes also show that it is extremely difficult to find a blend with a negligible impact on the greenhouse effect and at the same time good thermodynamic properties, which could be used as a replacement for R404A or R507A in low-temperature systems. Both of the proposed mixtures have disadvantages compared to currently used refrigerants. Although the R32–R41 achieves high volumetric cooling capacity, it is also characterized by high evaporating and condensing pressures and high discharge temperatures, which will result in higher thermal and force loads of the compressor working elements and may lead to their shorter life span. On the other hand, the R161–R41–R1234ze(E) mixture, despite the high coefficient of performance, shows nearly ten times higher temperature glide than R404A, which may cause evaporator malfunctions. Components with extremely different vapor pressures can cause excessive frosting to the initial sections of the evaporator due to the evaporation of low-boiling components. On the other hand, components with a high boiling point may not completely evaporate, which can lead to fractionation of the refrigerant inside the system and change its operating parameters. In case of extreme temperature glides, it is necessary to increase the vapor superheat set point to prevent the compressor from sucking in liquid refrigerant, which obviously affects the efficiency of the system. Therefore, further research should be directed towards this application. It should be emphasized that the presented analyzes do not explore the issue of using these mixtures in cooling cycles completely. Above all, further studies of the flammability and safe use of the presented mixtures are required, as all the components used are flammable, and a significant part of them belong to the highest flammability class.

Author Contributions: Conceptualization: B.G. and A.S.; methodology: B.G.; validation: B.G.; investigation: A.S. and B.G.; writing—original draft preparation: B.G., A.S. and S.R.; writing—review and editing: S.R. and B.G.; supervision: B.G. and S.R.; project administration: B.G. and S.R.; funding acquisition: B.G. and S.R. All authors have read and agreed to the published version of the manuscript.

Funding: Financial support for this work was provided by the Polish National Science Center under the MINIATURA3 project (2019/03/X/ST8/01192) as well as by the Polish National Agency for Academic Exchange under the RadMAT project (PPN/PPO/2018/1/00042/U/00001).

Conflicts of Interest: The authors declare no conflict of interest.

References

1. Regulation (EU) No 517/2014 of the European Parliament and of the Council of 16 April 2014 on fluorinated greenhouse gases and repealing Regulation (EC) No 842/2006. *Off. J. Eur. Union* **2014**, L150/195–230.
2. Gao, N.; Chen, G.; Wang, Y.; Tang, L. Experimental isobaric heat capacity of liquid HFC-32 + HFO-1234ze(E) mixture and extension of a predictive corresponding state equation to HFC mixtures. *Int. J. Refrig.* **2018**, *88*, 318–323. [CrossRef]

3. Janković, Z.; Sieres Atienza, J.; Martínez Suárez, J.A. Thermodynamic and heat transfer analyses for R1234yf and R1234ze(E) as drop-in replacements for R134a in a small power refrigerating system. *Appl. Therm. Eng.* **2015**, *80*, 42–54. [CrossRef]
4. Mota-Babiloni, A.; Navarro-Esbrí, J.; Barragán, Á.; Molés, F.; Peris, B. Drop-in energy performance evaluation of R1234yf and R1234ze(E) in a vapor compression system as R134a replacements. *Appl. Therm. Eng.* **2014**, *71*, 259–265. [CrossRef]
5. Diani, A.; Mancin, S.; Rossetto, L. Flow boiling heat transfer of R1234yf inside a 3.4 mm ID microfin tube. *Exp. Therm. Fluid Sci.* **2015**, *66*, 127–136. [CrossRef]
6. Mancin, S.; Diani, A.; Doretti, L.; Rossetto, L. R134a and R1234ze(E) liquid and flow boiling heat transfer in a high porosity copper foam. *Int. J. Heat Mass Transf.* **2014**, *74*, 77–87. [CrossRef]
7. *Refrigerating Systems and Heat Pumps–Safety and Environmental Requirements–Part 1: Basic Requirements, Definitions, Classification and Selection Criteria. Annex E Safety Classification and Information about Refrigerants* BS Standard EN 378-1:2016+A1:2020, 2020. Available online: https://shop.bsigroup.com/ProductDetail?pid=000000000030396148 (accessed on 31 December 2016).
8. Gil, B.; Kasperski, J. Efficiency Evaluation of the Ejector Cooling Cycle using a New Generation of HFO/HCFO Refrigerant as a R134a Replacement. *Energies* **2018**, *11*, 2136. [CrossRef]
9. Nguyen, V.V.; Varga, S.; Dvorak, V. HFO1234ze(e) as an alternative refrigerant for ejector cooling technology. *Energies* **2019**, *12*, 4045. [CrossRef]
10. Emmi, G.; Bordignon, S.; Carnieletto, L.; De Carli, M.; Poletto, F.; Tarabotti, A.; Poletto, D.; Galgaro, A.; Mezzasalma, G.; Bernardi, A. A Novel Ground-Source Heat Pump with R744 and R1234ze as Refrigerants. *Energies* **2020**, *13*, 5654. [CrossRef]
11. Sleiti, A.K.; Al-Khawaja, M.; Al-Ammari, W.A. A combined thermo-mechanical refrigeration system with isobaric expander-compressor unit powered by low grade heat-Design and analysis. *Int. J. Refrig.* **2020**, *120*, 39–49. [CrossRef]
12. Hu, X.; Meng, X.; Wu, J. Isothermal vapor liquid equilibrium measurements for difluoromethane (R32) + trans-1,3,3,3-tetrafluoropropene (R1234ze(E)). *Fluid Phase Equilib.* **2017**. [CrossRef]
13. Kondou, C.; Baba, D.; Mishima, F.; Koyama, S. Flow boiling of non-azeotropic mixture R32/R1234ze(E) in horizontal microfin tubes. *Int. J. Refrig.* **2013**, *36*, 2366–2378. [CrossRef]
14. Yang, J.; Jia, X.; Wu, J. Vapor phase pvTx measurements of binary mixtures of difluoromethane (R32) and 2,3,3,3-tetrafluoroprop-1-ene (R1234yf). *J. Chem. Thermodyn.* **2019**, *134*, 41–51. [CrossRef]
15. Akasaka, R. Thermodynamic property models for the difluoromethane (R-32)+trans-1,3,3,3-tetrafluoropropene (R-1234ze(E)) and difluoromethane+2,3,3,3-tetrafluoropropene (R-1234yf) mixtures. *Fluid Phase Equilib.* **2013**, *358*, 98–104. [CrossRef]
16. Aprea, C.; Greco, A.; Maiorino, A. HFOs and their binary mixtures with HFC134a working as drop-in refrigerant in a household refrigerator: Energy analysis and environmental impact assessment. *Appl. Therm. Eng.* **2018**. [CrossRef]
17. Aprea, C.; Greco, A.; Maiorino, A. Comparative performance analysis of HFO1234ze/HFC134a binary mixtures working as a drop-in of HFC134a in a domestic refrigerator. *Int. J. Refrig.* **2017**, *82*, 71–82. [CrossRef]
18. Yang, Z.; Gong, M.; Guo, H.; Dong, X.; Wu, J. Phase equilibrium for the binary mixture of {1,1-difluoroethane (R152a)+trans-1,3,3,3-tetrafluoropropene (R1234ze (E))} at various temperatures from 258.150 to 288.150K. *Fluid Phase Equilib.* **2013**, *355*, 99–103. [CrossRef]
19. Li, M.; Guo, Q.; Lv, J.; Li, D. Research on condensation heat transfer characteristics of R447A, R1234ze, R134a and R32 in multi-port micro-channel tubes. *Int. J. Heat Mass Transf.* **2018**, *118*, 637–650. [CrossRef]
20. Llopis, R.; Calleja-Anta, D.; Sánchez, D.; Nebot-Andrés, L.; Catalán-Gil, J.; Cabello, R. R-454C, R-459B, R-457A and R-455A as low-GWP replacements of R-404A: Experimental evaluation and optimization. *Int. J. Refrig.* **2019**, *106*, 133–143. [CrossRef]
21. Kondou, C.; Mishima, F.; Koyama, S. Condensation and evaporation of R32/R1234ze(E) and R744/R32/R1234ze(E) flow in horizontal microfin tubes. *Sci. Technol. Built Environ.* **2015**, *21*, 564–577. [CrossRef]
22. Saengsikhiao, P.; Taweekun, J.; Maliwan, K.; Sae-Ung, S.; Theppaya, T. Investigation and analysis of R463A as an alternative refrigerant to R404A with lower global warming potential. *Energies* **2020**, *13*, 1514. [CrossRef]
23. Hu, P.; Zhu, W.-B.; Chen, L.-X.; Cai, X.-D.; Chen, Z.-S. Vapor–liquid equilibria measurements of 1,1,1,2-tetrafluoroethane (HFC-134a) + 2,3,3,3-tetrafluoroprop-1-ene (HFO-1234yf) + isobutane (HC-600a) ternary system. *Fluid Ph. Equilibria* **2016**, *3*, 3. [CrossRef]
24. Barati-Harooni, A.; Najafi-Marghmaleki, A. Prediction of vapor–liquid equilibrium for binary mixtures containing R1234yf or R1234ze (E). *Int. J. Refrig.* **2018**, *88*, 239–247. [CrossRef]
25. Chen, Q.; Yan, J.W.; Chen, G.; Zhao, Y.; Shi, Y.; Zeng, Z.; Pan, Q. Experimental studies on the flammability of mixtures of dimethyl ether. *J. Fluor. Chem.* **2015**, *176*, 40–43. [CrossRef]
26. Rao, V.N.M.; Minor, B.H. Compositions comprising HFC-1234yf and HFC-32. *Eur. Pat. Off.* **2018**, *39*, EP3378919A.
27. Lemmon, E.W.; Bell, I.H.; Huber, M.L.; McLinden, M.O. *NIST Standard Reference Database: REFPROP Reference Fluid Thermodynamic and Transport Properties (Version 10)*; National Institute of Standards and Technology: Gaithersburg, MD, USA, 2018.

energies

MDPI

Article

The Use of Capsuled Paraffin Wax in Low-Temperature Thermal Energy Storage Applications: An Experimental and Numerical Investigation

Agnieszka Ochman, Wei-Qin Chen, Przemysław Błasiak, Michał Pomorski and Sławomir Pietrowicz *

Department of Thermodynamics and Renewable Energy Sources, Faculty of Mechanical and Power Engineering, Wrocław University of Science and Technology, 27 Wybrzeże Wyspiańskiego Street, 50-370 Wrocław, Poland; agnieszka.ochman@pwr.edu.pl (A.O.); weiqin.chen7@gmail.com (W.-Q.C.); przemyslaw.blasiak@pwr.edu.pl (P.B.); michal.pomorski@pwr.edu.pl (M.P.)
* Correspondence: slawomir.pietrowicz@pwr.edu.pl; Tel.: +48-71-320-36-17

Abstract: The article deals with the experimental and numerical thermal-flow behaviours of a low-temperature Phase Change Material (PCM) used in Thermal Energy Storage (TES) industrial applications. The investigated PCM is a composition that consists of a mixture of paraffin wax capsuled in a melamine-formaldehyde membrane and water, for which a phase change process occurs within the temperature range of 4 °C to 6 °C and the maximum heat storage capacity is equal to 72 kJ/kg. To test the TES capabilities of the PCM for operating conditions close to real ones, a series of experimental tests were performed on cylindrical modules with fixed heights of 250 mm and different outer diameters of 15, 22, and 28 mm, respectively. The module was tested in a specially designed wind tunnel where the Reynolds numbers of between 15,250 to 52,750 were achieved. In addition, a mathematical model of the analysed processes, based on the enthalpy porosity method, was proposed and validated. The temperature changes during the phase transitions that were obtained from the numerical analyses in comparison with the experimental results have not exceeded 20% of the relative error for the phase change region and no more than 10% for the rest. Additionally, the PCM was examined while using a Scanning Electron Microscope (SEM), which indicated no changes in the internal structure during phase transitions and a homogeneous structure, regardless of the tested temperature ranges.

Keywords: low-temperature phase change material; paraffin wax; thermal energy storage; numerical modelling; scanning electron microscope

check for updates

Citation: Ochman, A.; Chen, W.-Q.; Błasiak, P.; Pomorski, M.; Pietrowicz, S. The Use of Capsuled Paraffin Wax in Low-Temperature Thermal Energy Storage Applications: An Experimental and Numerical Investigation. *Energies* **2021**, *14*, 538. https://doi.org/10.3390/en14030538

Received: 21 December 2020
Accepted: 17 January 2021
Published: 21 January 2021

Publisher's Note: MDPI stays neutral with regard to jurisdictional claims in published maps and institutional affiliations.

Copyright: © 2021 by the authors. Licensee MDPI, Basel, Switzerland. This article is an open access article distributed under the terms and conditions of the Creative Commons Attribution (CC BY) license (https://creativecommons.org/licenses/by/4.0/).

1. Introduction

Thermal Energy Storage (TES) technology has gained increasing worldwide attention, because it, among others, has been regarded as an effective way to compensate for the intermittence of renewable sources [1,2]. Among various TES technologies, Latent Heat Thermal Energy Storage, (LHTES), utilising Phase Change Materials, (PCMs), is one of the most attractive forms, with a relatively high storage density and small temperature changes from storage to retrieval [3]. PCMs are substances with the property of heat absorption when they undergo a phase change from solid to liquid, liquid to gas, or vice versa [4–6]. These PCMs are widely applicable in a broad range of industrial areas. For instance, they can be encapsulated in building materials, e.g., gypsum plasterboard, cubicle, and wall board in order to enhance the thermal storage capacity [7,8]. PCMs are also considered to improve the frosting/defrosting operating performance of air source heat pumps [9]. Likewise, PCMs are frequently used in order to produce thermoregulated textiles, where they are generally entrapped in micro/nano-capsules to prevent leakage [10]. PCMs-assisted packaging is an innovative technology, which can plenarily control temperature-sensitive food products under different conditions [11]. In addition, PCMs also play an important role in a large number of fields, like temperature-adjustable greenhouses [12,13], waste heat recovery [14], and building air-conditioning [15]. High-temperature PCMs have attracted considerable interests over the past decade, which is extraordinarily promising in the concentrated solar thermal field and other high-temperature-required domains [16–18]. On the other hand, Amin et al. [19] considered a different temperature interval with a focus on low-temperature PCMs that are encapsulated in spheres.

PCMs can be generally classified into two types: organic and inorganic. Most inorganic PCMs are barely applicable to the TES system due to their toxicity, corrosivity, and supercooling properties. In contrast, organic PCMs are relatively safe, chemically inert and ecologically friendly [20]. Among diverse organic PCMs, paraffin wax is one of the most common materials, with several remarkable properties, e.g., high energy storage density, relatively low cost in commerce, and a small super-cooling trend. Paraffin wax (usually extracted from ozokerite, petroleum, natural gas, etc.) has a wide range of phase change temperatures and it has a general chemical formula $C_n H_{2n+2}$ ($n \geq 4$, the higher the value of n the higher the melting temperature) [21]. Consequently, an ideal scheme seems to be to use capsuled paraffin wax as a storage material and proposals for such a method have already been tackled by groups of researchers [22,23].

Plenty of investigations on the mathematical models and numerical analyses with PCMs that are capsuled in different configurations have been performed and reported in the literature. Shamsundar and Sparrow [24] resolved the enthalpy equation using the finite difference approach and performed an analysis of the multidimensional transient solidification process with the change of density and an increasing shrinkage cavity. The authors noted that the highest impact of the density ratio and the Stefan number on the heat transfer occurred at almost the end of solidification. In a horizontal tube with unfixed solid PCMs inside, the thermal behaviours of the PCMs (heat flux densities, geometric shape, melting rates, etc.) were obtained by Bareiss and Beer [25] through neglecting the inertial force. The authors pointed out that the gravity of the solid PCMs and pressure forces in a thin liquid layer jointly formed a force balance. Bilir and Ilken [26] investigated PCMs capsuled in a spherical/cylindrical container employing the third kind of boundary condition. The authors derived correlations that utilise the Biot number, the Stefan number and the dimensionless surface temperature to present the dimensionless total solidification time of PCMs. Verma et al. [27] studied the mathematical models that are based on the first and second law of thermodynamics regarding the LHTES system with PCMs inside. The authors indicated that the model based on the first law of thermodynamics had been experimentally validated, and could be employed to model PCMs. However, the model based on the second law of thermodynamics required additional work related to its experimental verification. A numerical model regarding the solidification process of PCMs in a triplex tube with external and internal fins was proposed by Al-Abidi et al. [28]. The authors noted that factors, including the fin length, the fin thickness, and the numbers of fins, had a considerable impact on the heat transfer. However, the effect of the fin thickness was assessed to be less than that of the fin length. Similarly, Li et al. [29] explored the enhancement effect of aluminium oxide on phase change heat transfer in the triplex tube with fins. The conducted research revealed that an extra alumina contributed to a stronger conduction and the best discharging rate was determined in the case of dp = 40 nm. The entropy optimization method was applied in order to study the solidification behaviour of nanoparticle-enhanced PCMs in the LHTES system affected by a magnetic field by Shah et al. [30]. The authors indicated that the Lorentz force, caused by the Hartmann number, and buoyancy forces had positive and negative impacts on the solidification rate of nanoparticle-enhanced PCMs, respectively. Jourabian et al. [31] utilised the enthalpy-based Lattice Boltzmann method and double distribution function to explore the melting process of the ice within a semicircle enclosure. The authors noticed that the concentration of nanoparticles had a positive and an adverse effect on the thermal conductivity and the latent heat of PCMs, respectively, but a negligible impact on the average Nusselt number. A novel PCM-air tubular heat exchanger and the corresponding analytical solution were proposed by Dubovsky et al. [32]. The authors successfully predicted the results of separate tubes and verified the applicability of the analytical solution to the practical heat exchangers. Darzi et al. [33] presented several simulations of the symmetric melting process between two cylinders in an eccentric and concentric position using N-eicosane as the PCM. The authors pointed out that the downward movement of the inner cylinder caused a significant increase in the melting rate due to the dominance of convective heat transfer in most areas of the PCM. Mahdaoui et al. [34] proposed a numerical model involving the natural convection phenomenon in the PCM-melted region around a horizontal cylinder. As a result of the conducted research, the authors assessed that regardless of the assumed boundary conditions, (constant temperature of the cylinder walls, constant heat flux), the melting of the PCM in the lower part was ineffective since the energy was transferred mainly by convection to the top of the cylinder. Regin et al. [35] focused on the cylindrical PCM-melting model integrated into the LHTES system, which was combined with a solar water heating collector. The conducted analyses indicated that the melting of PCMs was primarily dependent on the magnitude of the temperature range of the phase transformation, the Stefan number, and the capsule radius. Although these papers have successfully studied diverse mathematical and numerical PCM melting/solidification models employing different constraints. Additionally, it was considered various influencing factors and configurations, none of them pays attention to the changes in the internal structure of the PCMs during the phase change process. Furthermore, most of them only deal with the PCMs capsuled in horizontal cylinders and heat transfer is uniformly along the circumference of the circular cylinder. Accordingly, we have made an attempt to investigate the case of PCMs capsuled in the vertical position of a cylinder, which will undergo several heat-flow processes of the cross-flow of air with different velocities. We are highly expecting to test and compare the capabilities of the mixture of paraffin wax and water in different heat flux environments and to more explicitly describe the characteristic parameters of five-phase change regions of the mixed PCMs. In addition, a Scanning Electron Microscope (SEM) will be expedient in confirming the specific changes in the internal structure of the mixed PCMs during the phase transition.

The motivation to undertake the analyses carried out in the article is to design a container for storing/accumulating cold "energy" cooperating with the heat pump. The presented literature analysis shows that one of the most promising candidates that can be directly applied in the analysed device and its ranges of the temperature is the PCM-capsuled paraffin wax. This container unit filled with PCM has to operate with a system named a Flower Shape Oscillating Heat Pipe (FSOHP), which was described in detail by Czajkowski et al. in [36] and it has also been patented in [37] by Pietrowicz et al. Thus, an integral element in the innovative system for cooling mixed substances is a special exchanger that contains the PCM described and studied in this paper. The operational parameters of the exchanger have been defined and described by Ochman and Pietrowicz in [38], and have also been patented by Pietrowicz et al. [39]. What is important, due to the required technological process, PCMs are expected to store the "cold" energy in the range of 4 °C to 6 °C. The paraffin wax, due to its thermodynamic and functional properties, is a pertinent candidate for this novel system, as it was mentioned.

Thus, when preparing the design procedures that are dedicated for a storage tank, the authors of the article needed to have a complete, validated mathematical model of the thermal-flow processes occurring in the tested phase change material and to have knowledge of the impact of the operating conditions of the designed storage tank on the temperature change in the PCMs, depending on the total applied mass of the PCM. It was also important to determine the time that is needed for the phase changes and to compare them with different values of the supplied heat flux. The authors of the article also believe that the developed numerical procedures together with the conducted research will help in the future during the optimization process of the construction of a cold accumulator cooperating with a heat pump and a mixing/dissolver device.

In the presented article, the thermal-flow processes occurring in the low-temperature PCM were experimentally tested and then numerically investigated. The experiments were carried out for fully turbulent flow (15,250 < Re < 52,750) and for three cylindrical modules filled with a PCM with fixed heights of 250 mm and with outer diameters of 15, 22, and 28 mm, respectively. For this purpose, a special set-up was designed and constructed, in which a wind tunnel is the main element. Additionally, special test procedures were developed and adapted. Subsequently, a mathematical model of thermal-flow processes existing in the phase change material, based on the enthalpy porosity method, was proposed and validated. Finally, the numerical calculations, during the transient processes, were carried out for various boundary conditions that are close to those expected during the real operation of the device.

The structure of the article consisted of the following elements: Section 2 describes the tested phase change material, with a description of the thermophysical properties and the analysis of the internal structure while using a SEM. The test stand, tested modules filled with PCM and measurement procedures are described in Section 3. Section 4 presents the mathematical model and numerical procedures, numerical domains with the applied boundary conditions and applied thermal properties. The experimental studies are detailed and discussed in Section 5. Additionally, this Section compares the results of the numerical studies with the experimental data and summarizes them by the relative error analysis. Section 6 concludes the work, where the most important results and observations from the conducted research are presented.

2. Phase Change Material (PCM)

2.1. General Description of the PCM

To meet the thermal requirements, i.e., the phase change temperature range and usability, the research was conducted on a commercial phase-change material, which was paraffin wax capsules in a melamine-formaldehyde membrane and water mixture, produced by the *MikroCaps* company [40]. Table 1 shows the selected thermophysical properties describing the tested PCM.

Table 1. The thermophysical properties of the Phase Change Material (PCM) used during the experiments [40].

Property	Unit	Minimum	Maximum
PCM content in the dispersion	%	25	30
PCM content in dry capsule	%	75	80
Dry content in the dispersion	%	35	38
PCM melting range	°C	4	6
Heat storage capacity (of slurry)	kJ/kg	60	72
Heat storage capacity (of dried microcapsules)	kJ/kg	180	192
pH	–	7.0	9.0
Density	kg/m^3	900	970
Dynamic viscosity (at 25 °C)	kg/(m s)	0.1	0.5
Average particles size	µm	10	30

2.2. Analysis of the PCM's Internal Structure

2.2.1. Test Conditions and Testing Procedures

The purpose of the analysis described in this section was to observe whether and how the internal structure of the PCM changes at the characteristic temperature points. The structure of the material was studied while using a Scanning Electron Microscope (SEM). A *Prisma* E scanning electron microscope from Thermo Fisher Scientific was used for this purpose. A special table with a Peltier system was installed in the test chamber in order to regulate the sample temperature in the phase transition range, i.e., from 0 °C to 16 °C. Figure 1 shows the system with the studied sample along with the details of the test chamber.

(**a**) Test chamber with mounted the Peltier system and sample. (**b**) Prepared sample.

Figure 1. Description of the Scanning Electron Microscope (SEM) chamber and the prepared sample.

During the measurements, the focusing beam power was set from 20 to 25 kV. Those values were experimentally selected to ensure the right quality of contrast and sharpness of an image. It should be mentioned that the power of the selected high-energy electron beam allowed for the analysis of the sample without its visible degradation of the sample.

Efforts were made to keep the relative humidity value in the measuring chamber at 17 ÷ 42% in order to maintain the emulsion structure of the sample during the analyses of the PCM structure. Maintaining these parameters ensured that the evaporation of water from the material was minimized by achieving thermal conditions far from saturation conditions. Figure 2 presents the change in sample temperature and humidity during the experiment.

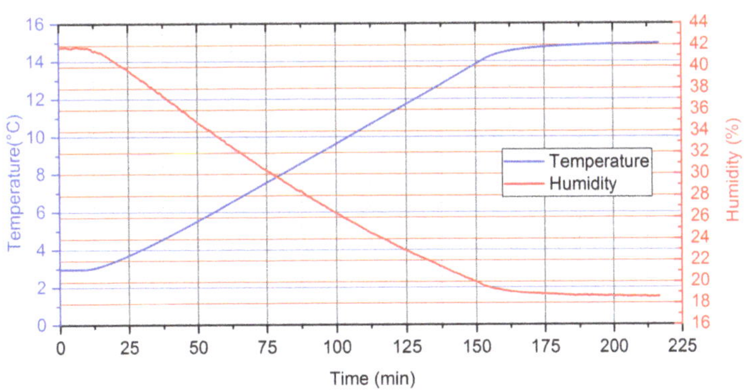

Figure 2. Temperature and relative humidity profiles maintained during measurement.

2.2.2. SEM Analysing

Figure 3 presents the results of the SEM's investigation of the PCM structure for selected temperature points that were defined at 0 °C, 5.5 °C, and 15 °C.

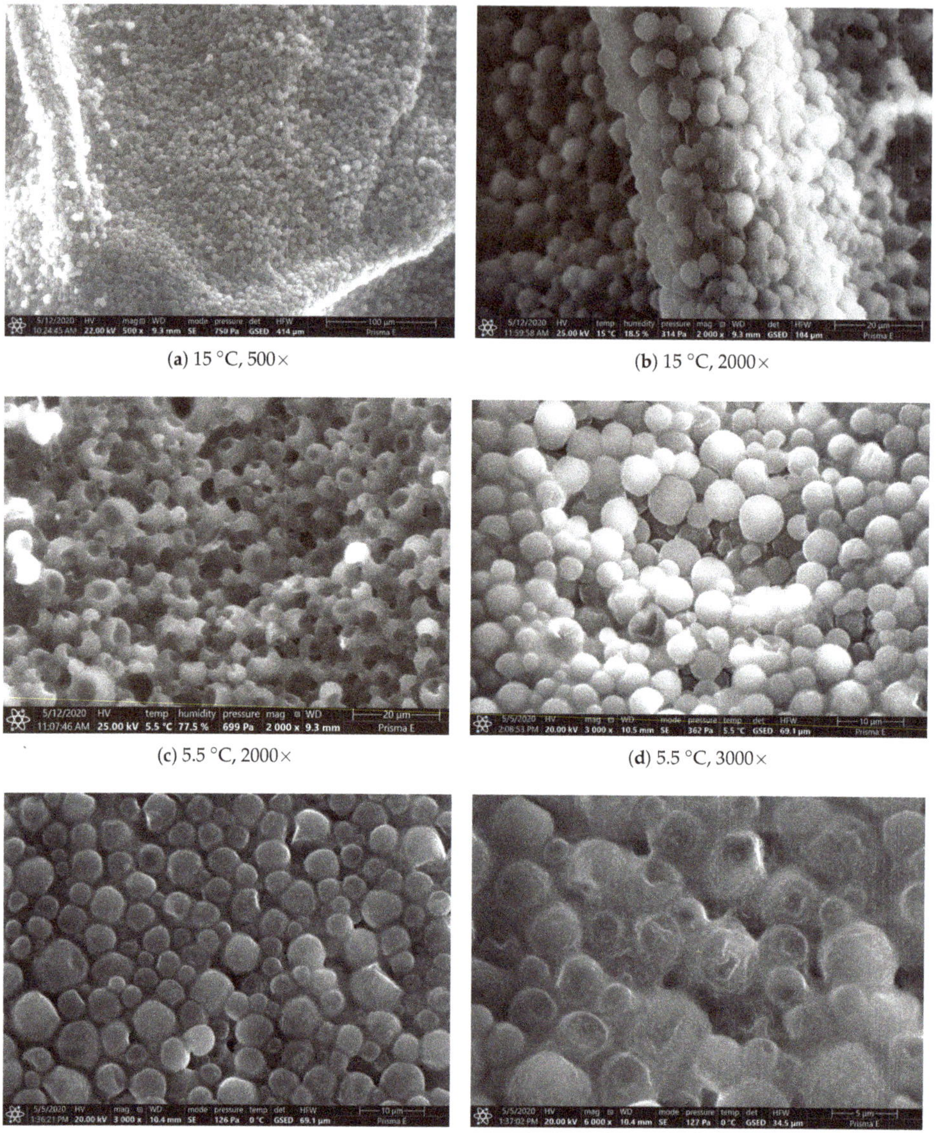

(**a**) 15 °C, 500×　(**b**) 15 °C, 2000×

(**c**) 5.5 °C, 2000×　(**d**) 5.5 °C, 3000×

(**e**) 0 °C, 3000×　(**f**) 0 °C, 6000×

Figure 3. Analysis of the PCM internal structure, made for different temperatures and magnifications obtained while using the SEM.

Those temperatures have been selected, because the potential changes in the structure were expected. Thus, the temperature of 15 °C (Figure 3a,b) is significantly separated from the phase transition temperature, which was estimated at between 4 °C and 6 °C (Figure 3c,d). Because the analysed substance is an emulsion of two components-paraffin wax and water, analysis at 0 °C (Figure 3e,f) was also potentially considered, due to the phase change of the water.

The investigation showed that the diameter of the analysed paraffin capsules is between 3 and 8 µm. This differs from the values declared by the manufacturer and described in Table 1. Additionally, it was observed from Figure 3e,f, that, for 0 °C, the distances between microcapsules increase in comparison with other tested temperatures, which are a consequence of the presence of water in the form of ice and a higher value of the specific volume compared to other temperature tests. Generally, it can be concluded that, in the studied temperature ranges, no significant changes in the internal structure were noticed. This is an important fact that was used later during numerical simulations. The substance that is taken into account is homogeneous and the separation of components, i.e., sedimentation, is not later considered.

3. Experimental Set-Up and Measuring Procedure

3.1. General Description

In Section 1, it was mentioned that the tested PCM is dedicated to working in a special storage heat exchanger/tank [38,39]. It was decided that the PCM, due to the design of the heat exchanger, should be studied in a cylindrical system for the conditions most similar to those in which it will be operated later. For this purpose, a wind tunnel, as presented in Figure 4, was used that met the operational and functional parameters. The obtained range of the Reynolds numbers (15,260–52,767) is fully turbulent and the heat transfer conditions (temperature and the Nusselt numbers) are similar to the considered storage tank.

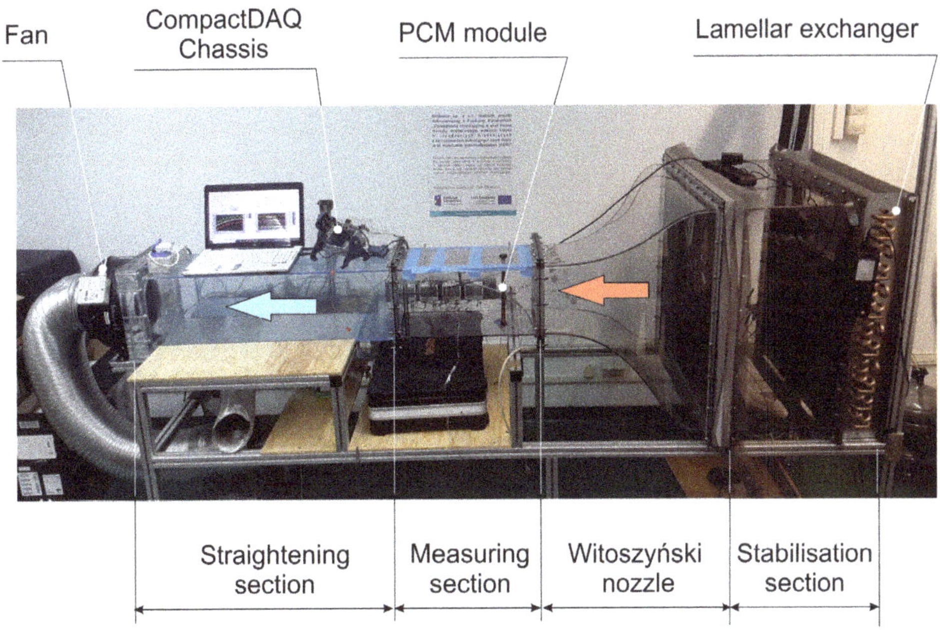

Figure 4. General view of the wind tunnel with installed PCM module.

The experimental set-up consists of four basic sections, operated in an open circuit suction mode. The first section is related to the air inlet. It is composed of a lamellar exchanger with a cross-section of 0.75 m × 0.74 mm in which it was possible to control the temperature in the range of −15 °C to 35 °C while using a chiller with a maximum cooling capacity of 4 kW. Subsequently, a stabilisation flow section was installed behind the lamellar exchanger. The stabilisation section is made of three net layers with a mesh size of approximately 1.5 mm × 1.5 mm. Such a system ensured a stable flow, without recirculation zones while also eliminating potential turbulence behind the lamellar exchanger. The next section is the so-called the Witoszyński nozzle, whose task is to shape a flat velocity profile in the measuring section. This special velocity profile allowed for homogeneous thermal-flow parameters along the entire length of the tested module to be obtained. The measuring section had a square cross section with dimensions of 0.25 m × 0.25 m and a length of 0.52 m. In order to eliminate the impact of the installed fan, between the measuring section and the fan a 0.9 m long channel with an identical cross-section as the measuring section was installed. The purpose of this channel was to stabilise the fan working conditions and eliminate any possible flow disorders that are caused by its operation. The fan that is installed in the experimental set-up provides the opportunity to achieve three mean values of velocity in the measuring section equal to 0.92, 2.27, and 3.18 m/s, respectively, which correspond to the Reynolds numbers 15,260, 37,688, and 52,767, based on the channel height of the test section. In order to achieve different mean value of air velocities, the fan power was regulated by a capacitor system. Generally, the maximum volume air flow rate was 1000 m^3/h, which was generated for the maximum electrical power of the fan engine, equal to 150 W.

The parameters of inlet air, such as ambient temperature, pressure, and humidity, were measured just before the lamellar exchanger by a high-precision digital temperature, humidity, and airflow meter Testo 480 with Testo Robust Humidity Probe.

3.2. Tested Module

The PCM was experimentally tested in a special cylindrical module made of copper, as presented in Figure 5. This module was characterised by a fixed height and a wall thickness of 250 mm and 1 mm, respectively. Three different outer diameters of 15, 22, and 28 mm were studied. At the ends of the module, two pipe caps that were made of the same material with a height of 15 mm were mounted. Additionally, three T-type thermocouples have been installed in situ in the axis of the module, at half height, and one-third above and below, as is shown in Figure 5. The module was then placed and tested in the measuring section, as depicted in Figures 4 and 6.

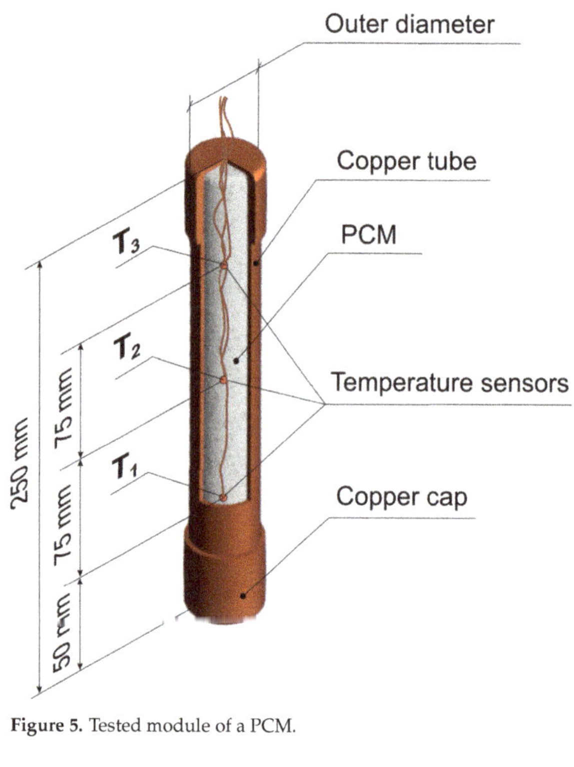

Figure 5. Tested module of a PCM.

The PCM mass contribution in the module varied from 25.23% to 37.82%, (see Table 2). The theoretical heat transfer from the air to the modules were estimated and presented in Table 2, according to the producer information, as described in Table 1 and assumptions concerning paraffin and copper properties used during the numerical calculations, as summarised in Table 4. For paraffin, it varies from 2.87 kJ to 10.19 kJ for sensible heat and from 2.46 kJ to 8.74 kJ for latent heat. For copper the heat transfer varies from 1.62 kJ to 3.71 kJ. The heat storage in copper is 23.31% for d = 15 mm, 20.78% for d = 22 mm and 16.37% for d = 28 mm of the total system, respectively. Those estimations show that PCM phase change plays an important role during the heating of the module that is filled with the PCM.

Table 2. Contributions of the module components in heat transfer.

Outer Diameter of Module	mm	15	22	28
PCM/module mass fraction	%	25.23	31.53	37.82
Copper/module mass fraction	%	74.77	68.47	62.18
Sensible heat of copper	kJ	1.62	2.66	3.71
Sensible heat of PCM	kJ	2.87	5.46	10.19
Latent heat of PCM	kJ	2.46	4.68	8.74
PCM/module heat fraction	%	76.69	79.22	83.63
Copper/module heat fraction	%	23.31	20.78	16.37

3.3. Preparation of the Module to the Test

The module has been specially prepared in order to test the thermal behaviour of the module filled with the PCM for various thermal-flow conditions. First, before filling with the PCM, the module was degreased with propyl alcohol and thoroughly dried. Next, three T-type thermocouples with an accuracy of ±0.5 °C were placed in the prepared arrangement in the configuration that is shown in Figure 5. Subsequently, the module was filled with the phase change material. Before testing, the module was placed in a freezer for about 4 h to cool to a temperature of about −18 °C. The freezing process was completed when the material obtained the same temperature in the entire volume, which was monitored by thermocouples that were placed inside the module. The module prepared in this way was then mounted in the wind tunnel measuring section. Figure 6 presents the measuring section with the installed PCM module.

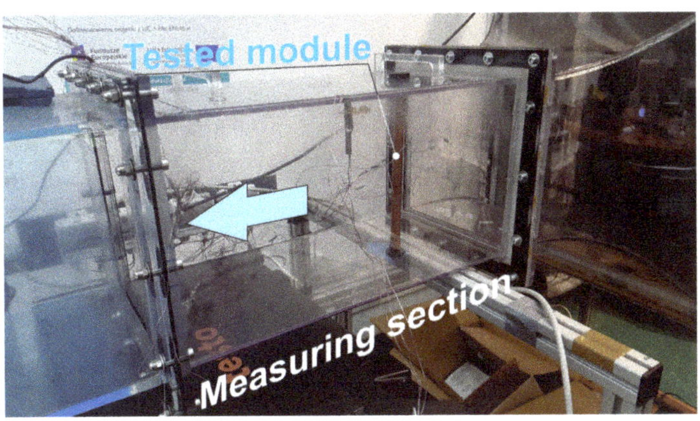

Figure 6. PCM module placed in the wind tunnel.

3.4. Measurement and Control Systems

The experimental set-up was equipped with two types of measurement system: one was based on Testo devices and the second one on the National Instruments module

4-Slot USB CompactDAQ Chassis (cDAQ 9174) working under the LabView environment. The high-precision digital temperature, humidity, and airflow meter Testo 480 had a build-in absolute pressure meter with an accuracy of ±3 hPa and allowed for up to three probes to be connected. The first Testo device was a Robust Humidity Probe, which was used to measure humidity and temperature of the suction air with an accuracy of temperature ±0.5 °C (from −20 °C to 0 °C) and ±0.4 °C (from 0.1 °C to 50 °C), relative humidity: ±2% RH (from 2.1 to 98% RH). The second installed device was a Thermal Flow Velocity Probe, Testo 480, which was used to measure the velocity profile. The device is characterised by the accuracy of measured velocity profile amounting to ±0.03 m/s + 5% of the measured velocity.

A special thermocouples net system was installed in order to measure inlet and outlet air temperatures at the measured section. At the inlet to the section, eight T-type thermocouples were installed and used to determine the average inlet temperature. At the outlet, one T-type thermocouples type of temperature sensor was mounted. The accuracy of installed T-type thermocouples was ±0.5 °C.

The combination of the National Instruments equipment and LabView software allowed for the measurement of temperature with a high resolution and frequency. During the experiments, it was sufficient that the measurements were carried out with a frequency of 0.2 Hz (every 5 s).

3.5. Velocity Profiles at the Inlet and Outlet of the Measurement Section

A special construction of the inlet section with the installed Witoszyński nozzle contributes to the formation of quasi-uniform thermal and velocity profiles in the measurement section, as mentioned in Section 3.

Before the measurement campaign, the velocity profiles at the inlet and outlet were determined. The measurements of the velocity profile were carried out at a distance of 20 mm from the inlet and outlet edges of the measuring section. The velocity profiles were determined at measuring points 33, 73, and 105 mm from the channel's symmetry axis.

In total, seven holes have been made in the top of the measuring section cover for measurement purposes. The thermal flow velocity probe that was connected to the Testo 480 module has been inserted into these holes, mounted on a special measuring instrument. A measuring device, similar to a caliper allowed for precise measurement of the velocity profile along with the channel height with an accuracy of ±0.1 mm. In total, one profile was determined while using 19 measurements, concentrated in the central part of the channel. The results of the obtained profiles at the inlet and outlet of the measuring section for different powers of the fan are presented in Figure 7, where a larger number represents higher air velocity.

The measurements that are presented in Figure 7 showed that, for all the tested ranges of fan power, the velocity profiles are characterised by a flat profile. The obtained flat velocity profile is a consequence of the use of a specially shaped inlet contraction section designed according to the Witoszynski curve equation [41]. The analysis points out that the differences between the average values determined do not exceed 7.4% of the local value of velocity. Table 3 summarises the test results. In addition, for further analysis, each analysed value of average speed: 0.0, 0.92, 2.18, and 3.18 m/s was defined as Case 0, Case I, Case II, and Case III, respectively.

Table 3. The summarised results that were obtained from profile measurements.

Parameter	Unit	Case 0	Case I	Case II	Case III
Power of the fan	W	0	60	80	150
Mean velocity	m/s	0.00	0.92	2.27	3.18
Maximum deviation from the average value	%	–	7.36	7.15	5.19

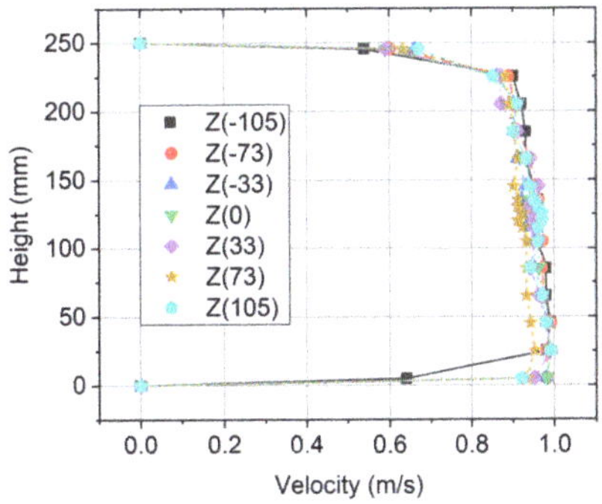

(**a**) Inlet velocity profile-Case I.

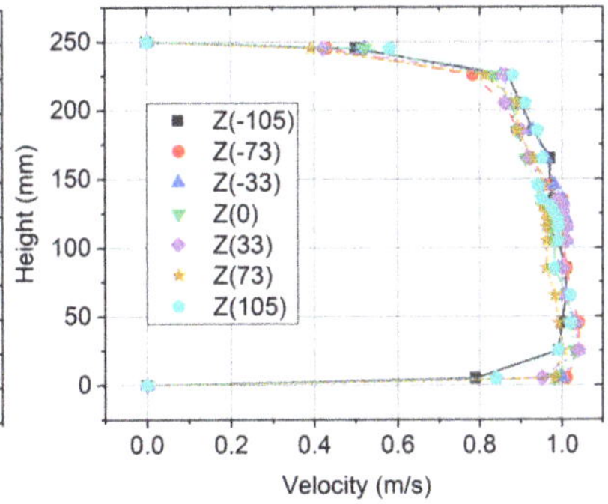

(**b**) Outlet velocity profile-Case I.

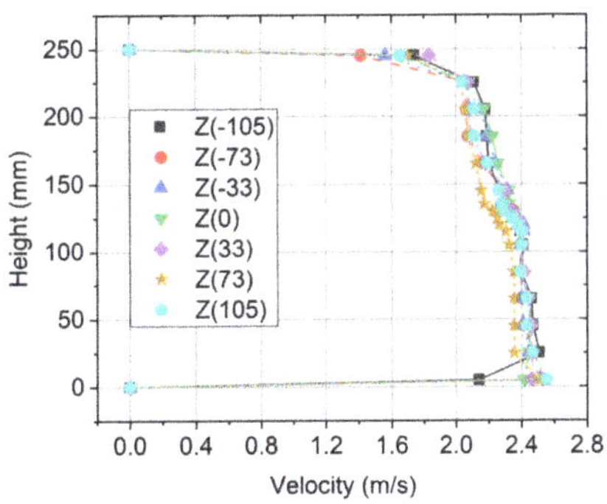

(**c**) Inlet velocity profile-Case II.

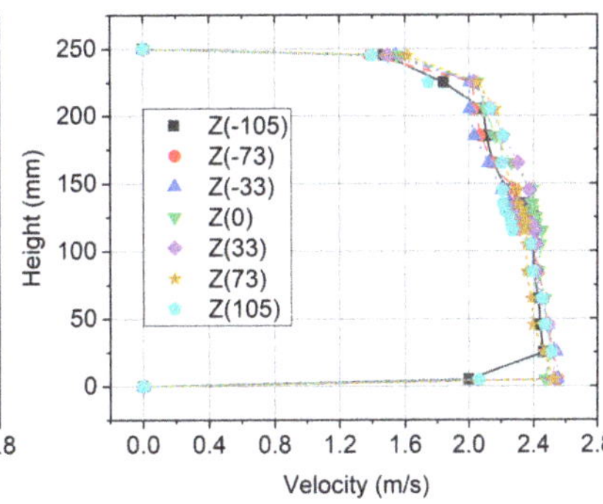

(**d**) Outlet velocity profile-Case II.

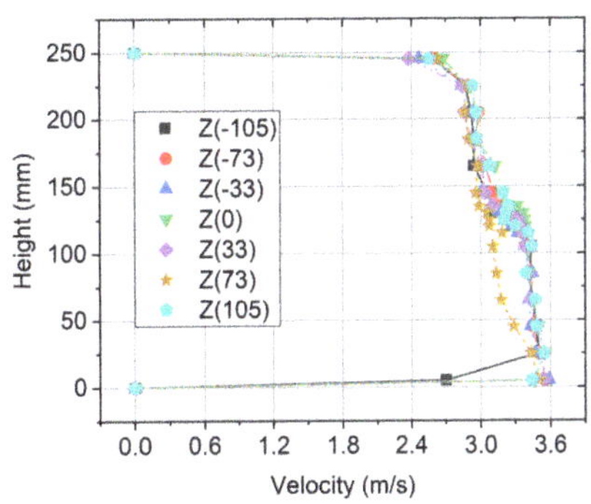

(**e**) Inlet velocity profile-Case III.

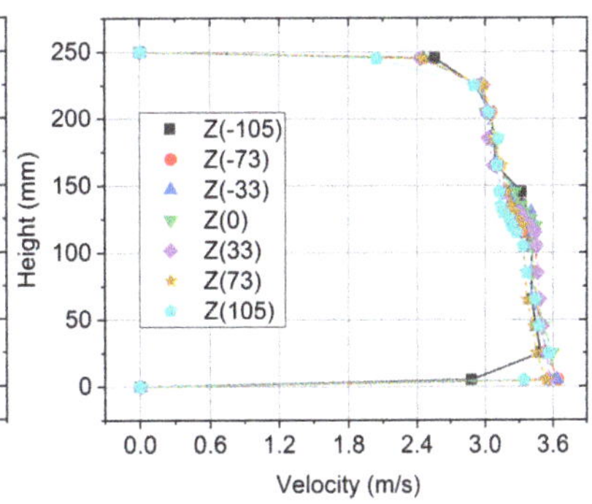

(**f**) Outlet velocity profile-Case III.

Figure 7. Velocity profiles of cross section at different power.

4. Numerical Modeling of the Analyzed System with PCM

4.1. Process Governing Equations

The thermal-flow behaviour in the PCM and flowing air can be described by a standard set of equations [42,43] and for the three-dimensional case, comes down to solutions of Partial Differential Equations, (PDEs), which include a mass, a momentum, and an energy equation [44]. In this context, for solving the set of equations, one of the most frequently used numerical methods, called the Finite Volume Method (FVM), [45,46] was applied.

In order to solve transport equations of mass, momentum and energy, the commercial CFD tool-Ansys Fluent has been utilised [47]. To take the phase change of the PCM into account, an enthalpy-porosity technique [48] has been used, which is already implemented in Ansys Fluent software module. Using this technique, the interface is not explicitly tracked, but the liquid fraction for each domain cell is solved in each iteration. The cells that contain a liquid fraction β between 0 and 1 create an interface region, called the mushy zone. This region is treated as a porous medium with porosity ranging from 1 for a liquid to 0 for the solid. The porosity is equal to the liquid fraction.

The liquid fraction is calculated knowing the balance of the enthalpy H. The enthalpy of a material is computed as a sum of sensible enthalpy h and the latent heat content ΔH

$$H = h + \Delta H \tag{1}$$

where sensible enthalpy h is defined as follows

$$h = h_{ref} + \int_{T_{ref}}^{T} c_p \mathrm{d}T \tag{2}$$

h_{ref} and T_{ref} are reference to enthalpy and temperature, respectively, and c_p is specific heat at constant pressure. Regarding the liquid fraction, it is defined in the following manner:

$$\beta(T) = \begin{cases} 0 & \text{if } T < T_{solidus} \\ \dfrac{T - T_{solidus}}{T_{liquidus} - T_{solidus}} & \text{if } T_{solidus} < T < T_{liquidus} \\ 1 & \text{if } T > T_{liquidus} \end{cases} \tag{3}$$

Knowing the latent heat L of the material used, the latent heat content in Equation (1) is calculated as

$$\Delta H = \beta\, L \tag{4}$$

where ΔH varies between 0 for a solid and L for a liquid. Based on the liquid fraction distribution, the enthalpy is calculated and then inserted into the energy equation, where the temperature distribution is solved via an iterative manner.

4.2. Geometry, Boundary and Initial Conditions and Numerical Schemes

The geometry used during numerical calculation is based on the experiments that are described in Section 3. For the purpose of simplifying the calculations, only the measuring section is modelled. Figure 8 presents a scheme of the studied case. During the simulation, to the rectangular duct, made of plexiglass, hot air of a temperature 24 °C inflows according to the experimentally obtained uniform velocity changing in a range of 0.92 ÷ 3.18 m/s. The duct is 0.7 m long and the inlet surface is a square, with a side length of 0.25 m. The wall thickness of the duct is 3 mm. In the middle of the duct, the copper cylinder with an outer diameter of 15 ÷ 28 mm and 250 mm long is set. To close the module, pipe caps that were made of copper, 15 mm long and 15 ÷ 28 mm in diameter, were mounted at the ends of the test element. The cylinder and bases are hollowed and the wall thickness is 1 mm. The inside of the cylinder is filled with the tested PCM. The air outflow has an atmospherical pressure of 1 bar.

For the air, plexiglass and copper materials constant thermophysical properties are assumed. For the real PCM, physical properties, such as heat capacity or thermal conductivity, vary with temperature [49]. Thermal conductivity k is higher in the solid than in the liquid phase. In the case of the PCM, the density, thermal conductivity, and dynamic viscosity are regarded as temperature-dependent and calculated via Equations (5)–(7), respectively. Table 4 describes the thermophysical properties of all materials that are used during the numerical simulations.

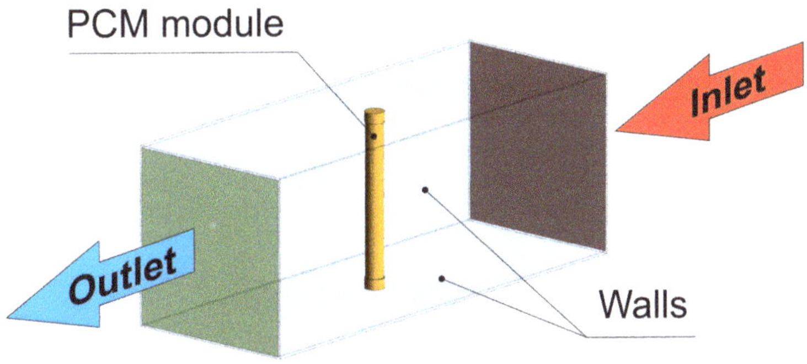

Figure 8. Three-dimensional geometrical model used during numerical calculation.

Density of the PCM:

$$\begin{aligned} for\ T < T_{solidus} &\Rightarrow \rho = 910\,\frac{\text{kg}}{\text{m}^3} \\ for\ T \geq T_{solidus} &\Rightarrow \rho = 790\,\frac{\text{kg}}{\text{m}^3} \end{aligned} \tag{5}$$

Thermal conductivity of the PCM:

$$\begin{aligned} for\ T < T_{liquidus} &\Rightarrow k = 0.5\,\frac{\text{W}}{\text{m K}} \\ for\ T \geq T_{liquidus} &\Rightarrow k = 0.4\,\frac{\text{W}}{\text{m K}} \end{aligned} \tag{6}$$

Dynamic viscosity of the PCM [50]:

$$\eta = 0.001 \exp\left(-4.25 + \frac{1700}{T}\right) \text{Pa s} \tag{7}$$

For all analysed cases, the axis of the cylinder is parallel to the gravity vector. The ratio of the module height to the inside diameter is large and it varies from 8.93 to 16.67, in the PCM volume, natural convection current occurs, which has an important impact on heat transfer and fluid flow behaviour in PCM. To take this phenomenon into account, the Boussinesq approximations that are shown in Equation (8) have been made in the PCM domain using UDF procedure. The following term has been added to the y-momentum equation for cells with liquid fraction higher than 0 within PCM domain:

$$S_v = -\rho_{liquidus}\, g\, \beta (T - T_{solidus}) \tag{8}$$

where: the thermal expansion coefficient of the paraffin equalled $\beta = 0.000778$ 1/K has been included in the numerical calculation procedures [51].

Table 4. Thermophysical properties of materials used during calculations.

Material	Density kg/m^3	Specific Heat J/(kg K)	Thermal Conductivity W/(m K)	Dynamic Viscosity kg/(m s)	Melting Heat J/kg	Solidus Temperature °C	Liquidus Temperature °C
Paraffin wax	Equation (5)	2500	Equation (6)	Equation (7)	72,000	4	6
Air	1.125	1006.43	0.0242	1.7894×10^{-5}	-	-	-
Plexi-glass	1000	1000	0.2	-	-	-	-
Copper	8978	381	387.6	-	-	-	-

The geometrical model that is presented in Figure 8 was discretized in space with the use of Ansys Meshing software. A tetragonal mesh was used within the air and PCM space. Solid elements were discretized while using hexahedral elements. Additionally, on the outer and inner surfaces of the cylinder 12 inflation layers have been applied. The thickness of the first element adjacent to the wall was set to 0.1 mm. In order to ensure energy transfer between solid and fluid regions, the applied meshes were non-conformal and were connected with the use of coupled wall boundary condition [47]. Figure 9 presents the employed mesh with details.

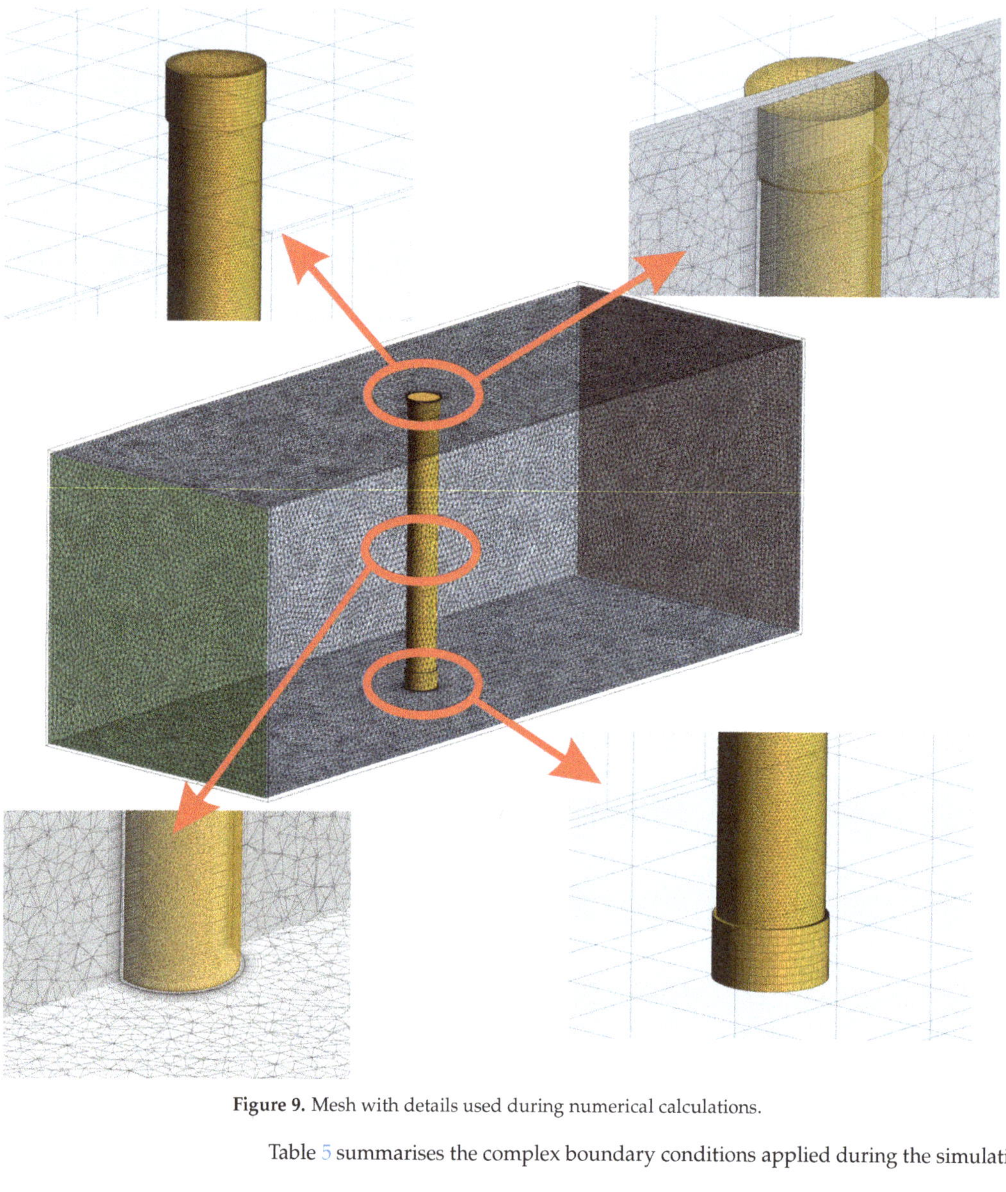

Figure 9. Mesh with details used during numerical calculations.

Table 5 summarises the complex boundary conditions applied during the simulations.

Table 5. Boundary conditions employed during calculations.

Domain/Material	Location	Boundary	Unit	Value
Duct/plexi-glass	External walls	Wall-temperature	°C	24
	Internal walls	Interface	–	Coupled wall [47]
Air/air	Duct wet walls	Interface	–	Coupled wall [47]
	Upper base wet walls	Interface	–	Coupled wall [47]
	Lower base wet walls	Interface	–	Coupled wall [47]
	Cylinder base wet walls	Interface	–	Coupled wall [47]
	Inlet	Inlet-velocity	m/s	0.92, 2.27, 3.18
		Inlet-temperature	°C	24
	Outlet	Outlet-pressure	bar	1
Upper base/copper	External and internal walls	Interface	–	Coupled wall [47]
Lower base/copper	External and internal walls	Interface	–	Coupled wall [47]
Cylinder/copper	External and internal walls	Interface	–	Coupled wall [47]
PCM module/paraffin wax	External walls	Interface	–	Coupled wall [47]

Air and PCM were both modelled as incompressible fluids. In order to account for turbulence, in the air region a standard $k-\varepsilon$ turbulence model with enhanced wall treatment was employed. In the PCM region, laminar flow is assumed. The boundary conditions used in the model are depicted in Figure 8. A no-slip boundary condition was utilised on all walls. The initial temperature of the tunnel walls and the air was set to 24 °C, while the PCM, cylinder, and lower and upper bases to 2 °C.

Calculations have been performed in a transient mode with a time-step of 0.01 s with the first-order implicit treatment. The total time of the simulation was 1440 s. The SIMPLE algorithm with the second-order scheme was used for pressure-velocity coupling. Transport equations of momentum, turbulent kinetic energy, turbulent dissipation, and energy were discretized while using the second order upwind scheme. Gradients were calculated with the use of a least-square cell-based scheme [47]. The solution in each time-step was considered as converged if the residuals were less than 10^{-6}; however, a maximum of 30 iterations per time-step was done. For energy, liquid fraction, and turbulence equations the under-relaxation of 0.5 was applied. For other equations, the default values were set. The calculations were conducted in a parallel mode on a cluster with the use of 16 Intel Xeon E5-2670 v3 2.3 GHz (Haswell) processors. The time of the calculations was approximately two weeks.

5. Results

5.1. Experimental Results

The experimental tests were carried out for the outer diameters of 15, 22, and 28 mm, and the thermal-flow conditions obtained in the wind tunnel, as defined in Table 3. This combination of the outer diameters and air tunnel conditions necessitated 12 measurements. Figure 10 shows a representative example of the heating process for a module with an outer diameter of 28 mm and an air flow rate of 0.92 m/s (Case I).

Generally, from the performed experiments, it can be concluded that the heating process of a module filled with the PCM can be divided into five phases. This solution is unique in comparison to homogeneous PCM systems, where only three phases can be observed [49,52,53]. In the first phase, starting with a uniform initial temperature of −18 °C throughout, the mixture of paraffin wax and water coexisted in solid form is heated. This phase that is presented in Figure 10 as Phase I is characterised by a linear temperature change with a high value of the inclination angle of the characteristic. There are two reasons why temperature increases rapidly. The first reason is related to the fact that in this temperature range there is the highest temperature gradient between the tested module and the air flowing outside; therefore, the highest temperature driving force occurred. The second reason may also be that ice has half of the specific heat lower than water.

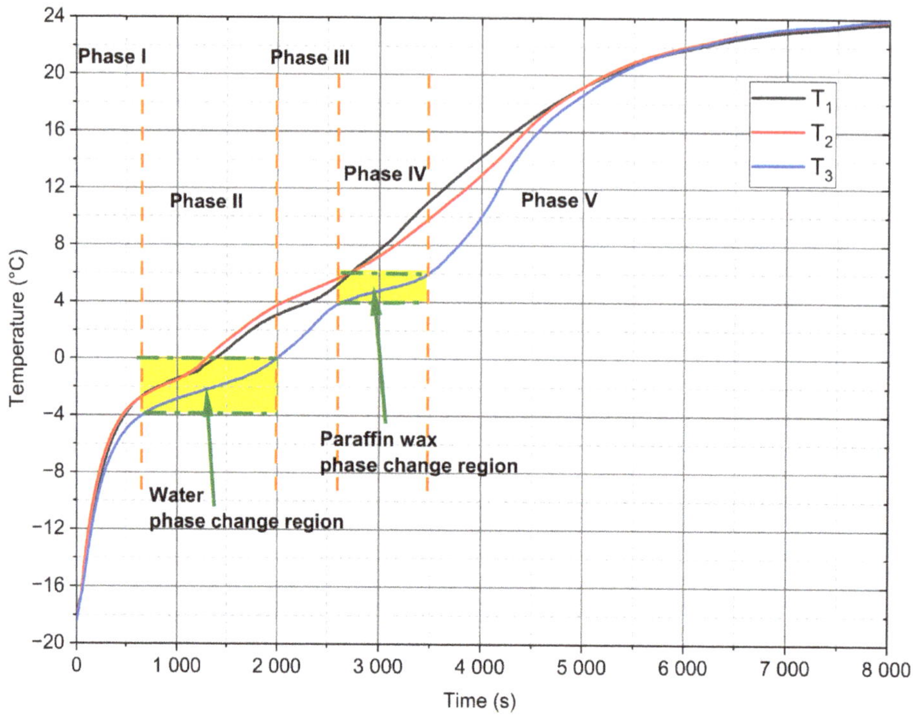

Figure 10. Measured temperatures evolution for d = 28 mm PCM module and 0.92 m/s value of the velocity.

Additionally in this temperature range, mention should be made of the condensation of water vapour that is contained in the humid air. This phenomenon also affects the thermal processes that take place during the heating of the module. Subsequently, starting from approximately −3 °C to 0 °C, flattened temperature curves can be observed (the first area marked in yellow called Phase II), and the temperature stratification process in the module begins. This phase is directly related to the process of changing the water phase from ice to water. The stratification of temperature curves also indicates the convective flow term in the module itself. Figure 10 shows the third phase as Phase III between the phase transitions of water and paraffin wax. It is characterised by an increase to a temperature of about 4 °C. From this temperature the fourth phase starts (Phase IV indicated in Figure 10 as the second area marked as the yellow). The curves reach a plateau and can be treated as a process associated with the phase transformation of the paraffin wax contained in the emulsion. This phase ends when the temperature reaches 6 °C. Subsequently, during the rest of the process, described as the fifth phase (Phase V), the temperature of the mixture steadily increases to an ambient condition.

The phase analysis performed above is mainly conducted for the temperature evolution recorded by the T_3 thermocouple. However, during the measurements, some differences between the individual thermocouples were observed. Despite the fact that the design of the wind tunnel ensures the achievement of homogeneous thermal and flow conditions on the outer surface of the cylinder, in the upper part of the measuring section, as shown in Figure 7, the velocity values are decreased and, thus, the heat flux to PCM is also decreasing. A second major reason for the temperature evolution discrepancy may be that the thermocouples are not positioned in the module axis. This applies to the two bottom thermocouples, which were the most difficult to install in the module and during the cooling process of the module, could have moved closer to the inside of the module wall. However, regardless of the differences, all of the curves have similar trends and phases.

The identification of the aforementioned phase transitions for ice-water and paraffin wax can also be observed in Figure 11. If a derivative of the temperature function $dT/d\tau$ versus time is determined, then the first two "valleys" of the function with values smaller than 0.005 °C/s can be interpreted as the phase transitions (areas marked in yellow in Figure 11). The rest of the derivative close to zero starting around 6000 s describes the so-called steady-state process, i.e., the temperature in the module reached about 90% of the ambient value.

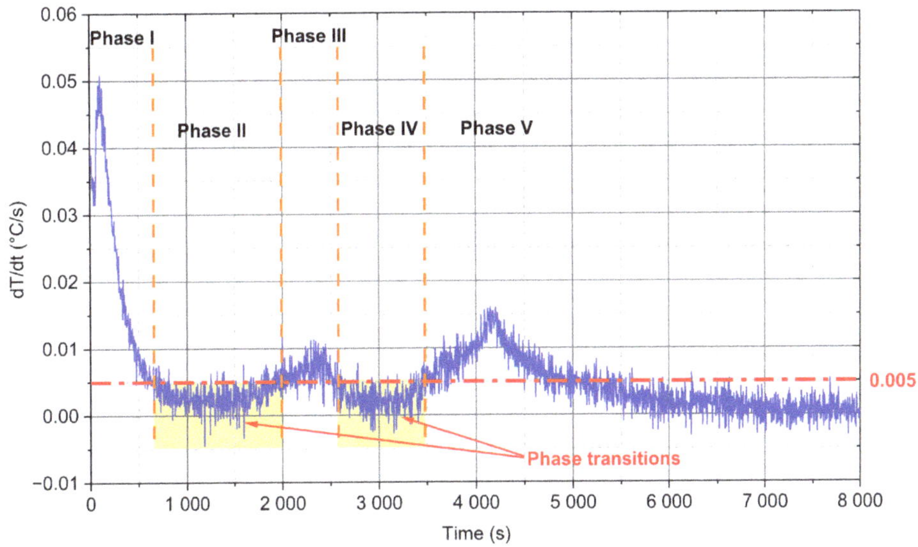

Figure 11. Derivation evolution of T_3 thermocouple for d = 28 mm PCM module and 0.92 m/s value of the velocity.

Figure 12 shows the comparison of the changes in the recorded values of T_3 thermocouple for different velocity and external diameters of the module. As the airflow rate flowing around the module increases, the heat transfer to the PCM also increases. This is evident in the dynamics of temperature increase and the time of phase transitions, and it can be seen that, for higher velocity values, the temperature increase is faster. However, as the outside diameter of the module increases, the amount of PCM used increases and, hence, the amount of cold accumulated. For this reason, the increase in temperature is less dynamic.

From a practical point of view, it is important to assess how fast the process is going through the characteristic phase. In the case of the tested module, two characteristic temperature ranges were selected: the first temperature range is the range from −4 °C to 0 °C (phase marked in Figures 10 and 11 as Phase II-ice–water phase transition region), the second is the phase change range of the tested PCM contained in temperatures from 4 °C to 6 °C, showed in Figures 10 and 11 as Phase IV-paraffin wax phase transition region. Table 6 presents the measured time interval for the various velocity and module diameter values that are defined in Table 3 as Case 0, Case I, Case II, and Case III for the water in the mixture, while Table 7 specifies this for paraffin wax.

Table 6. Water melting time.

Case	PCM 15 mm	PCM 22 mm	PCM 28 mm
Case 0	05 min 45 s	18 min 55 s	35 min 30 s
Case I	03 min 00 s	11 min 00 s	17 min 30 s
Case II	02 min 20 s	07 min 15 s	12 min 30 s
Case III	01 min 55 s	05 min 15 s	11 min 05 s

Table 7. Paraffin melting time.

Case	PCM 15 mm	PCM 22 mm	PCM 28 mm
Case 0	05 min 25 s	24 min 05 s	36 min 5 s
Case I	02 min 25 s	10 min 50 s	13 min 50 s
Case II	02 min 00 s	04 min 55 s	07 min 00 s
Case III	00 min 50 s	03 min 00 s	04 min 40 s

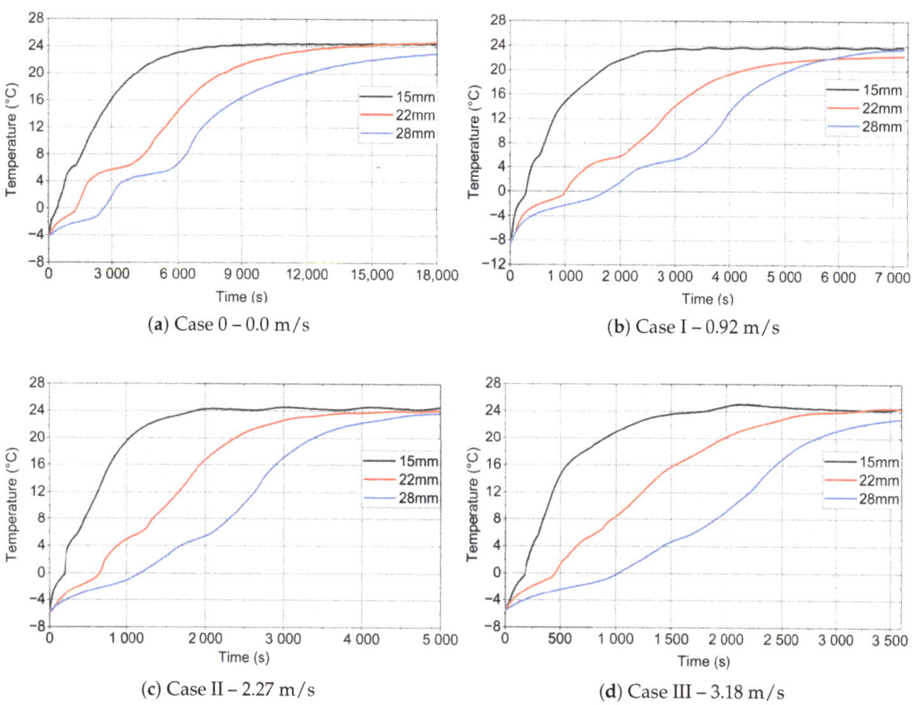

(**a**) Case 0 – 0.0 m/s (**b**) Case I – 0.92 m/s (**c**) Case II – 2.27 m/s (**d**) Case III – 3.18 m/s

Figure 12. T_3 thermocouple evolution for different diameter of PCM module and velocity: (**a**) Case 0–0.0 m/s, (**b**) Case I–0.92 m/s, (**c**) Case II–2.27 m/s, and (**d**) Case III–3.18 m/s.

It is obvious that the longest time of phase transitions will occur for processes in which the velocity value is zero (Case 0) and it will increase with increasing diameter, i.e., mass of PCM. In order to show this process in a graphical form, it would be advisable to refer to the longest process (Case 0) and when a system with forced convection is used (Case I-yellow, Case II-cyan, and Case III-grey). In the article, a dimensionless parameter called Relative Time Increase (RTI) was introduced and defined by the following formula:

$$RTI = \frac{\tau_0 - \tau}{\tau_0} 100\%, \tag{9}$$

where τ_0-time of phase change for Case 0 and τ-time of phase change.

In the case of the phase transitions of water that are shown in Figure 13a, the smallest time increments occur for the lowest velocities and amount from 42% to 50%. With an increase in the value of the velocity of the flowing air, the time needed to achieve the phase changes is reduced from 66% to 71%, respectively.

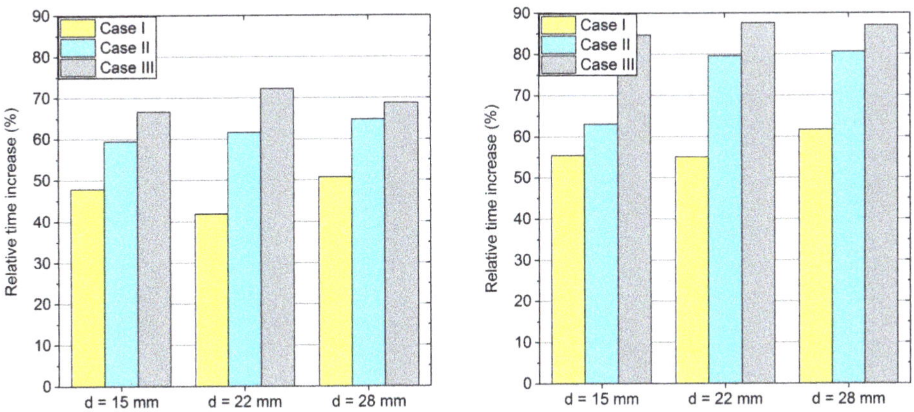

(a) Water's RTI. (b) Paraffin wax's RTI.

Figure 13. Relative time increase (RTI) for analysed Cases: (**a**) water and (**b**) paraffin wax.

For paraffin wax, as presented in Figure 13b, the situation is very similar, only the values of the RTI are different. For the smallest air velocities, the RTI varies from 55% to 61%, and for the highest velocities from 85% to 87%.

Figure 13 also shows how effectively the time of the melting process changes. The analysis presents that for paraffin it is more "sensitive" than for water to alterations in external conditions, i.e., heat transfer coefficients. Interestingly, the effect of changing the diameter is less effective and it only causes a change in the accumulated heat value, not in the phase transition time.

5.2. Numerical Results

In Figures 14–16, the comparison of the experimental and numerical results for the analysed values of air velocity 0.92 (Figure 14), 2.27 (Figure 15), and 3.18 m/s (Figure 16) and module diameters (a) 15, (b) 22, and (c) 28 mm are presented. All of the comparisons were carried out up to 1440 s, which allowed for the analysis of mainly phase changes taking place in the paraffin wax, due to the long computing time. Additionally, the initial temperature for all cases was 2 °C, because of the fact that the considered model did not take the ice-water phase change into account.

It is visible that, especially for low diameter (d = 15 mm) of the cylinder (see Figures 14a, 15a, and 16a), the numerical model agrees quite well with the experimental data. At the beginning of the simulation, an inaccurate prediction can be caused by the initial conditions for the temperature and velocity contours. In the experiment, the flow is fully developed while in the model, uniform distributions of velocity and temperature are assumed. For larger diameters, the time of heating is higher and the initial influence of the conditions is slightly larger. For $d = 28$ mm (Figures 14c, 15c, and 16c) numerical errors are marginally higher than in the case of $d = 22$ mm (Figures 14b, 15b, and 16b).

For the higher velocity of $v = 2.27$ m/s and diameter of $d = 15$ mm (Figure 15a), the accuracy of the model is similar to the case of $v = 0.92$ m/s (Figure 14a). The numerical errors slightly increase with the increase of the cylinder diameter.

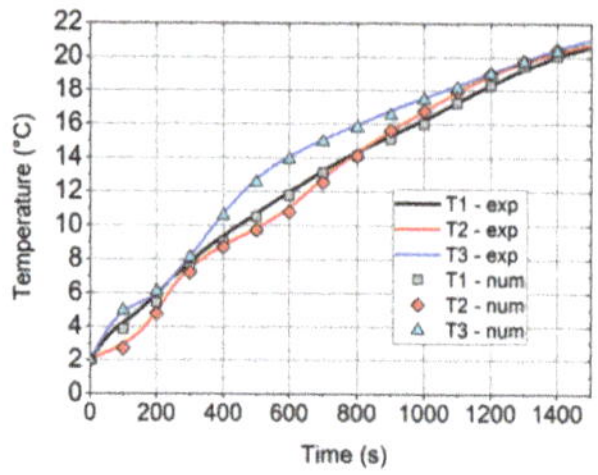

(**a**) d = 15 mm.

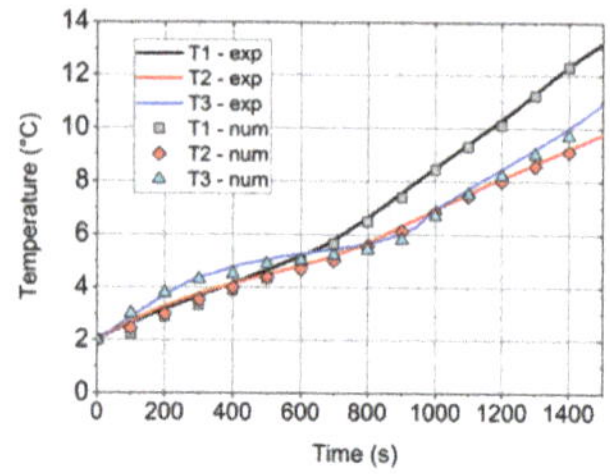

(**b**) d = 22 mm.

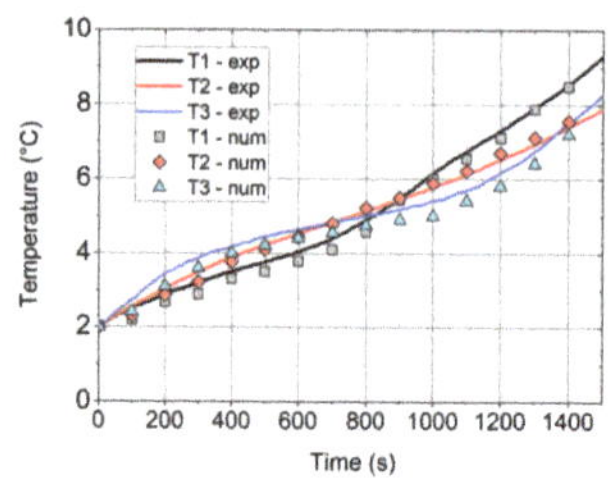

(**c**) d = 28 mm.

Figure 14. Comparison of the experimental and numerical results for the inlet velocity of 0.92 m/s and different cylinder diameters.

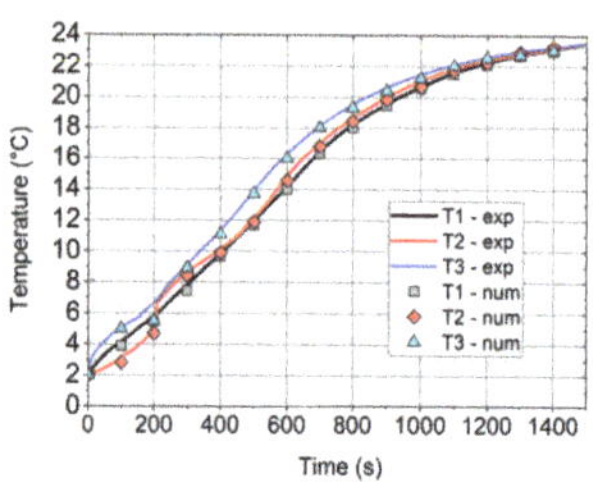

(**a**) d = 15 mm.

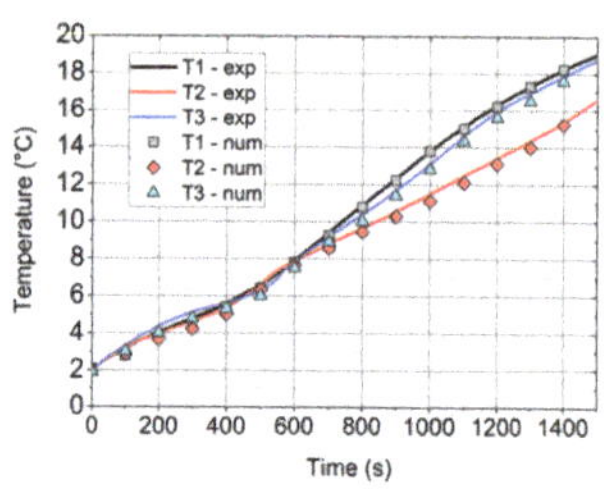

(**b**) d = 22 mm.

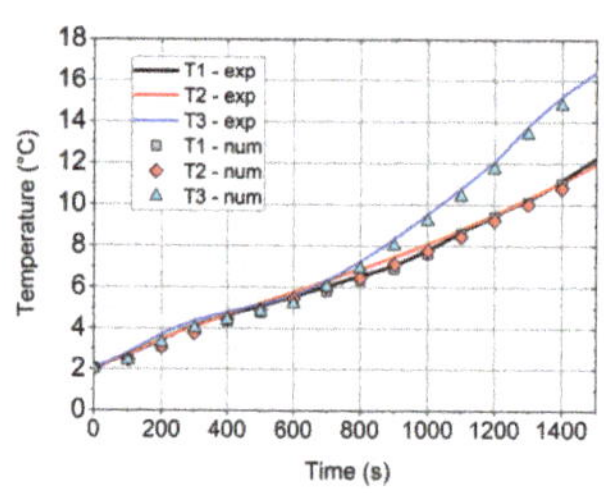

(**c**) d = 28 mm.

Figure 15. Comparison of the experimental and numerical results for the inlet velocity of 2.27 m/s and different cylinder diameters.

With the increase of the inlet velocity to 3.18 m/s (see Figure 16), higher discrepancies between the experimental and model results were recorded. In the case of a small diameter ($d = 15$ mm, Figure 16a), the model agrees fairly well with the experiment. For larger diameters, i.e., $d = 22$ (Figure 16b) and $d = 28$ mm (Figure 16c), the accuracy is still satisfactory and the model predicts the temperature evolution well; however, the errors are more visible. For higher velocities, the turbulence model may also have a greater impact on the results. On the other hand, probably, the influence of the turbulence model was not the goal of this work and further studies should be conducted.

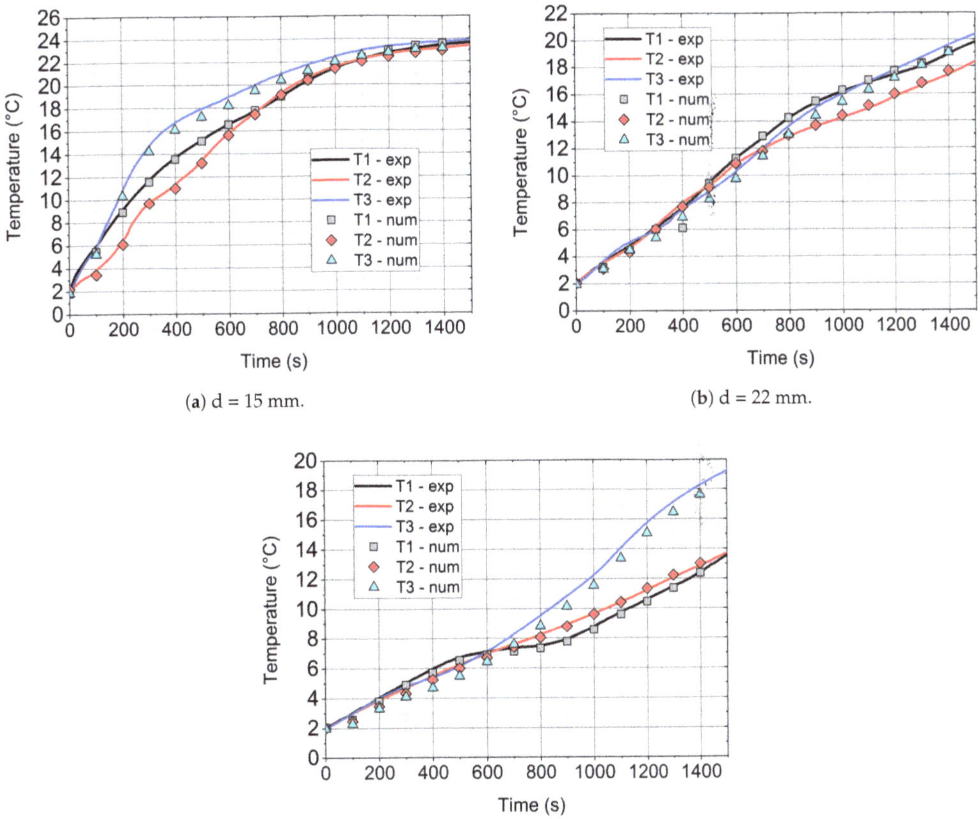

(a) d = 15 mm.

(b) d = 22 mm.

(c) d = 28 mm.

Figure 16. Comparison of the experimental and numerical results for the inlet velocity of 3.18 m/s and different cylinder diameters.

Figures 17 and 18 show the exemplary numerical results of temperature distributions in the vertical and horizontal middle plane, respectively, for different times. It is visible from Figure 17a–d that the heating of the PCM firstly takes place in the top and bottom space of the cylinder. This is caused by the conduction heat flux at the contact surfaces of the cylinder and direct contact with the tunnel walls. Moreover, more mass is melted in the upper part of the cylinder when compared to the bottom one. The reason of this is natural convection that intensifies thermal-flow processes in the upper part of the solid [54]. Similar results and observations are also presented in the literature [55–58].

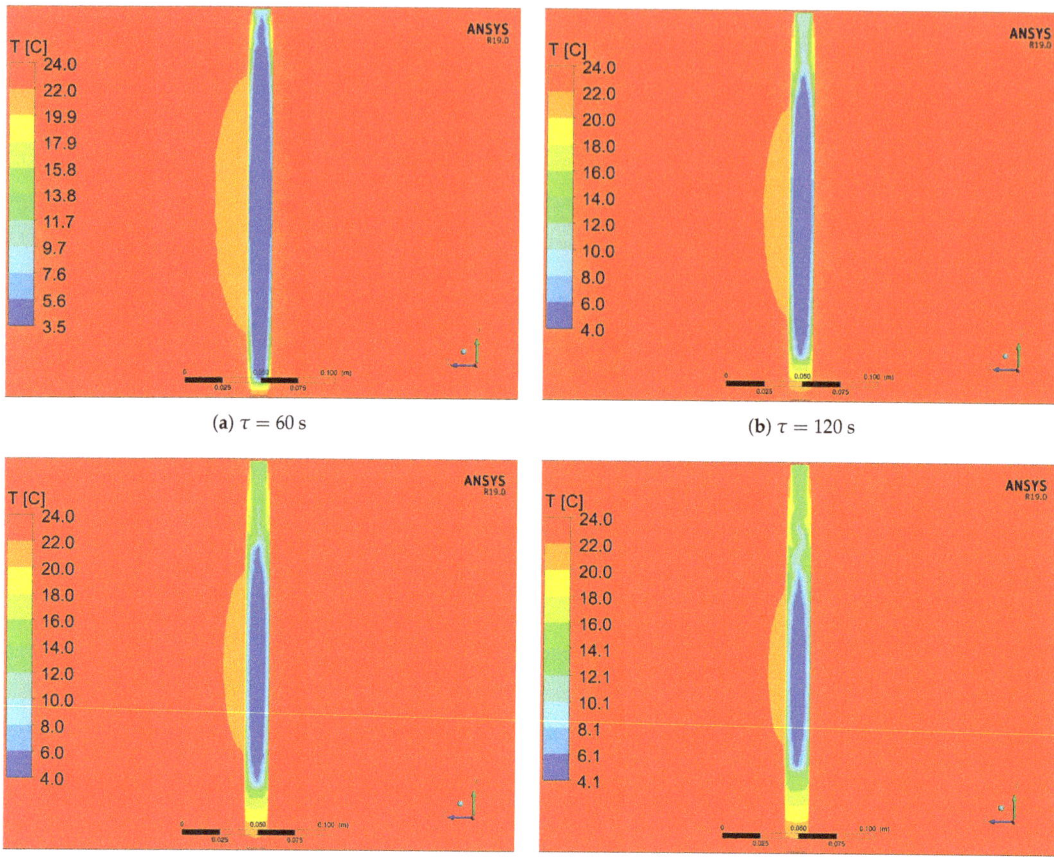

(**a**) $\tau = 60$ s (**b**) $\tau = 120$ s

(**c**) $\tau = 180$ s (**d**) $\tau = 240$ s

Figure 17. Numerical results of temperature distribution in the middle vertical plane for the inlet velocity of 0.92 m/s and cylinder diameter of 15 mm and different times of the simulation.

Over time, the isotherms of high temperature propagate into the interior of the PCM. Behind the cylinder, a wake is developing, which gives rise to change temperature distribution in the front of, as well as in the back, of the cylinder (see Figure 18). The differences in temperature distribution around cylinder are caused by increasing local heat transfer coefficient, which was described by Cengel [59] and Incropera et al. [60]. At the beginning (Figure 18a,b), behind the cylinder, one large area of lower temperature exists. For greater times (Figure 18c,d), this area slightly shrinks and its length behind the cylinder is shortened. This process is caused by changes in temperature difference between the cylinder and air. With an increase of the PCM temperature, the temperature difference decrease; therefore, the air faster reach surrounding temperature.

As it results from the comparison of the results that were experimentally obtained with the results obtained from the proposed numerical model, presented in Figure 19, the relative error in a wide temperature range does not exceed 10%. However, it has been observed that at the beginning of the heating process and in the temperature region where the phase change occurs, this error can reach even 20% (see enlarged analysis region in Figure 19 in the upper left corner). A few factors could cause the observed differences between the numerical and experimental results, in particular, in the initial heating phase and in the phase change region. During the numerical calculations, only pure paraffin wax was analysed. In the experiment, paraffin wax that was capsuled in the melamine-formaldehyde membrane micro-capsules was used. Such an interior structure is difficult to model and it was not considered in the simulations. Moreover, during the preparing of the experimental set-up, unintentional human errors could occur, e.g., imperfect placement of the thermocouples in the module axis or its movement due to a melted PCM, which can result, in particular, in the region of phase change. In the numerical investigations, boundary conditions were averaged and stable during the heat transfer processes. On the other hand, in the experiment, chiller hysteresis caused unstable inlet air temperature and, therefore, unstable boundary conditions. Another thing could be the air humidity impact on the heat transfer when the air dew point temperature was higher than the module temperature. During this situation, vapour from the air condensed on the module walls, which probably caused an intensification of heat transfer.

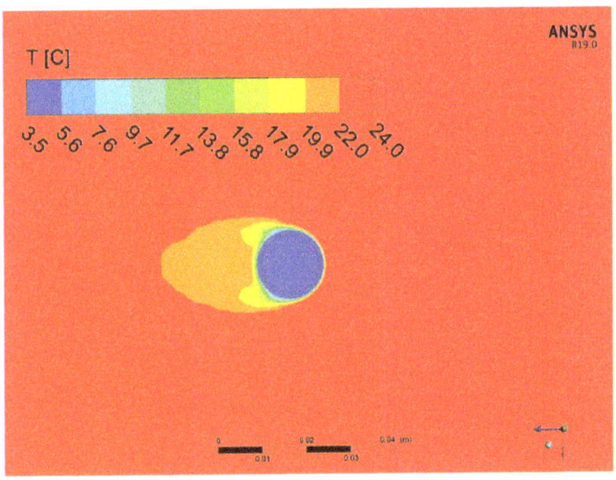

(**a**) $\tau = 60$ s.

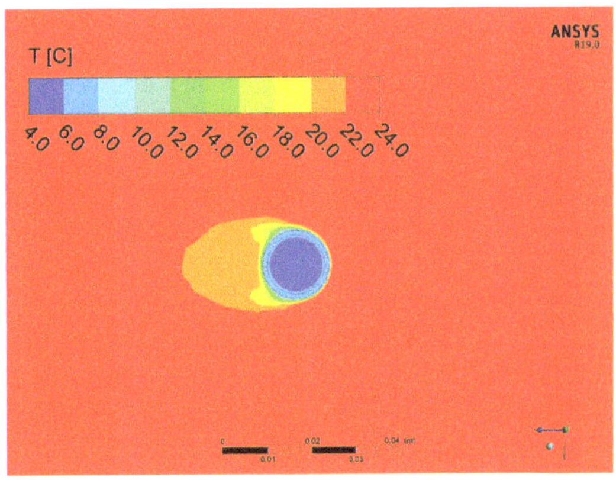

(**b**) $\tau = 120$ s.

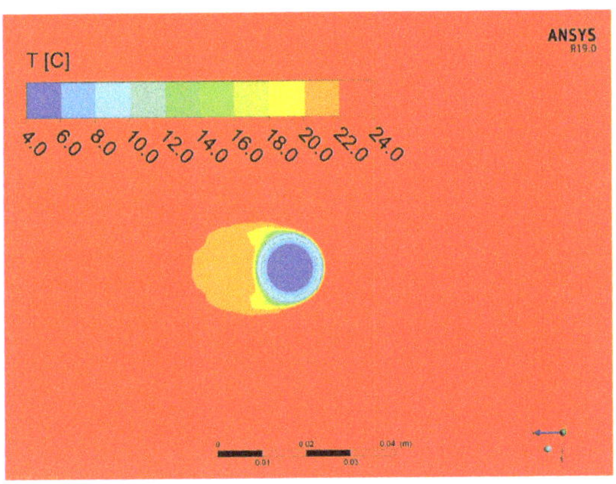

(**c**) $\tau = 180$ s.

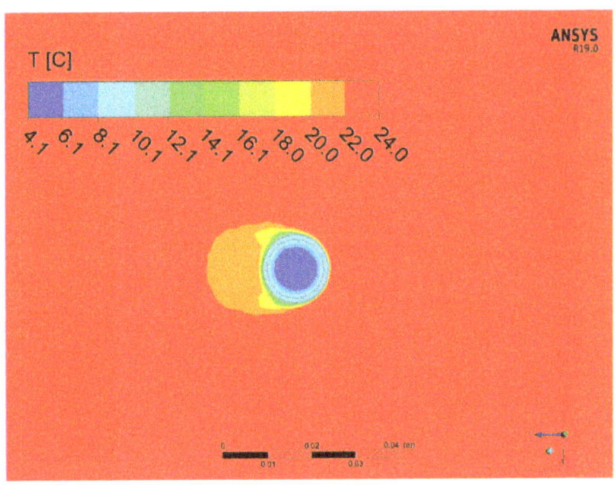

(**d**) $\tau = 240$ s.

Figure 18. Numerical results of temperature distribution in the middle horizontal plane for the inlet velocity of 0.92 m/s and cylinder diameter of 15 mm and different times of the simulation.

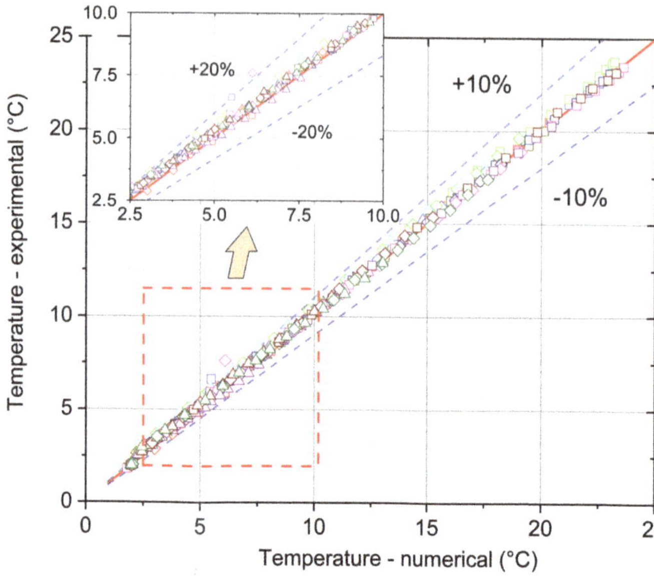

Figure 19. The comparison of numerical and experimental temperatures and obtained relative error.

6. Conclusions

The thermal-flow processes taking place in the low-temperature phase-change material, which was a mixture of paraffin wax that was capsuled in a melamine-formaldehyde membrane and water, were presented and analysed in the article. The experimental and numerical investigations during transient processes were carried out for the copper modules filled with a PCM for three different outer diameters (15, 22, and 28 mm) and a fixed height of 250 mm. A total of four cases i.e., for stagnant condition (0.0 m/s) and for three different mean velocity values of 0.92, 2.27, and 3.18 m/s for each module were tested. The experiment was performed in a wind tunnel, especially designed for this purpose. In addition, the internal structure of the tested PCMs was also analysed while using a SEM. The following conclusions can be drawn from the research carried out:

- For analysed PCM, five distinct phases during the heating process can be observed (Figure 10). Two of these phases are a phenomenon, in which phase change processes can be singled out. The first transformation is characteristic for water in the temperature range from −4 °C to 0 °C and the second is identified for paraffin from 4 °C to 6 °C.
- Because of the use of forced convection, the time of the melting process can be reduced by even up to 87% (*RTI*) as compared to stagnation conditions, as shown in Tables 6 and 7 and Figure 13. It is also worth emphasising that, when changing the outer diameter from the module from 22 to 28 mm, these changes are less noticeable than for the diameters of 18 to 22 mm.
- The mathematical model that is based on the enthalpy porosity method (Equations (1)–(4)) reproduces the conditions in the phase change material during the heating process with a maximum error of up to 20%. (Figures 14–16 and 19). It is very important that the proposed model should be implemented in the numerical code, together with variable thermophysical properties (Equations (5)–(7)).
- The SEM tests confirmed the homogeneous structure of the mixture and the lack of changes in the internal structure during phase transformations at characteristic temperatures (Figure 3).

Author Contributions: Conceptualisation, A.O., W.-Q.C., P.B., M.P. and S.P.; methodology, A.O., W.-Q.C., P.B. and S.P.; software, W.-Q.C. and P.B.; validation, A.O., W.-Q.C., P.B. and S.P.; formal analysis, A.O., W.-Q.C., P.B., M.P. and S.P.; investigation, A.O. and W.-Q.C.; resources, S.P.; writing—original draft preparation, A.O., W.-Q.C., P.B., M.P. and S.P.; writing—review and editing, A.O., W.-Q.C., P.B., M.P. and S.P.; visualisation, A.O., W.-Q.C., P.B. and S.P.; supervision, S.P.; project administration, S.P., funding acquisition, S.P. All authors have read and agreed to the published version of the manuscript.

Funding: This research was partly funded by National Centre for Research and Development of Poland grant number POIR.04.01.04-00-0037/15 and partly supported by the internal research funds of the Department of Thermodynamics and Renewable Energy Sources at Wrocław University of Science and Technology, Poland, No. 8201003902 (MPK 9090750000). Calculations have been carried out using resources provided by Wrocław Center for Networking and Supercomputing (http://wcss.pl), grant Nos. 202 and 309.

Institutional Review Board Statement: Not applicable.

Informed Consent Statement: Not applicable.

Acknowledgments: The authors would like to thank Halina Pawlak-Kruczek and Michał Czerep from Wrocław University of Science and Technology for providing and assistance in conducting tests using a Scanning Electron Microscope (SEM) for PCM analysis.

Conflicts of Interest: The authors declare no conflict of interest.

Abbreviations

The following abbreviations are used in this manuscript:

CFD	Computational Fluid Dynamics
FSOHP	Flower Shape Oscillating Heat Pipe
FVM	Finite Volume Method
LHTES	Latent Heat Thermal Energy Storage
PCM	Phase Change Material
PDE	Partial Differential Equation
RTI	Relative Time Increase
SEM	Scanning Electron Microscope
TES	Thermal Energy Storage

References

1. Edenhofer, O.; Pichs-Madruga, R.; Sokona, Y.; Seyboth, K.; Kadner, S.; Zwickel, T.; Eickemeier, P.; Hansen, G.; Schlömer, S.; von Stechow, C.; et al. *Renewable Energy Sources and Climate Change Mitigation: Special Report of the Intergovernmental Panel on Climate Change*; Cambridge University Press: Cambridge, UK, 2011.
2. Paksoy, H.Ö. *Thermal Energy Storage for Sustainable Energy Consumption, Fundamentals, Case Studies and Design*; Springer: Dordrecht, The Netherlands, 2007.
3. Zalba, B.; Marín, B.; Cabeza, J.; Mehling, H. Review on thermal energy storage with phase change: Materials, heat transfer analysis and applications. *Appl. Therm. Eng.* **2003**, *23*, 251–283. [CrossRef]
4. Jiji, L.M.; Gaye, S. Analysis of solidification and melting of PCM with energy generation. *Appl. Therm. Eng.* **2006**, *26*, 568–575. [CrossRef]
5. An, C.; Su, J. Lumped parameter model for one-dimensional melting in a slab with volumetric heat generation. *Appl. Therm. Eng.* **2013**, *60*, 387–396. [CrossRef]
6. Kalaiselvam, S.; Veerappan, M.; Arul Aaron, A.; Iniyan, S. Experimental and analytical investigation of solidification and melting characteristics of PCMs inside cylindrical encapsulation. *Int. J. Therm. Sci.* **2008**, *47*, 858–874. [CrossRef]
7. Cheng, R.; Wang, X.; Zhang, Y.P. Energy-Efficient Building Envelopes with Phase-Change Materials: New Understanding and Related Research. *Heat Transf. Eng.* **2014**, *35*, 970–984. [CrossRef]
8. Shi, X.; Memon, S.A.; Tang, W.C.; Cui, H.Z.; Xing, F. Experimental assessment of position of macro encapsulated phase change material in concrete walls on indoor temperatures and humidity levels. *Energy Build.* **2014**, *71*, 80–87. [CrossRef]
9. Qu, M.L.; Xia, L.; Deng, S.M.; Jiang, Y.Q. Improved indoor thermal comfort during defrost with a novel reverse-cycle defrosting method for air source heat pumps. *Build. Environ.* **2010**, *45*, 2354–2361. [CrossRef]
10. Mondal, S. Phase change materials for smart textiles—An overview. *Appl. Therm. Eng.* **2008**, *28*, 1536–1550. [CrossRef]
11. Alehosseini, E.; Jafari, S.M. Micro/nano-encapsulated phase change materials (PCMs) as emerging materials for the food industry. *Trends Food Sci. Technol.* **2019**, *91*, 116–128. [CrossRef]

12. Najjar, A.; Hasan, A. Modeling of greenhouse with PCM energy storage. *Energy Convers. Manag.* **2008**, *49*, 3338–3342. [CrossRef]
13. Kenisarin, M.; Mahkamov, K. Solar energy storage using phase change materials. *Renew. Sustain. Energy Rev.* **2007**, *11*, 1913–1965. [CrossRef]
14. Farid, M.M.; Khudhair, A.M.; Razack, S.A.K.; Al-Hallaj, S. A review on phase change energy storage: Materials and applications. *Energy Convers. Manag.* **2004**, *45*, 1597–1615. [CrossRef]
15. Sarı, A.; Biçer, A. Thermal energy storage properties and thermal reliability of some fatty acid esters/building material composites as novel form-stable PCMs. *Sol. Energy Mater. Sol. Cells* **2012**, *101*, 114–122. [CrossRef]
16. Liu, M.; Saman, W.; Bruno, F. Review on storage materials and thermal performance enhancement techniques for high temperature phase change thermal storage systems. *Renew. Sustain. Energy Rev.* **2012**, *16*, 2118–2132. [CrossRef]
17. Liu, M.; Steven Tay, N.H.; Bell, S.; Belusko, M.; Jacob, R.; Will, G.; Saman, W.; Bruno, F. Review on concentrating solar power plants and new developments in high temperature thermal energy storage technologies. *Renew. Sustain. Energy Rev.* **2016**, *53*, 1411–1432. [CrossRef]
18. Javier Batlles, F.; Gil, B.; Ushak, S.; Kasperski, J.; Luján, M.; Maldonado, D.; Nemś, M.; Nemś, A.; Puertas, A.M.; Romero-Cano, M.S.; et al. Development and Results from Application of PCM-Based Storage Tanks in a Solar Thermal Comfort System of an Institutional Building—A Case Study. *Energies* **2020**, *13*, 3877. [CrossRef]
19. Amin, N.A.M.; Bruno, F.; Belusko, M. Effectiveness—NTU correlation for low temperature PCM encapsulated in spheres. *Appl. Energy* **2012**, *93*, 549–555. [CrossRef]
20. Yang, D.; Shi, S.L.; Xiong, L.; Guo, H.J.; Zhang, H.R.; Chen, X.F.; Wang, C.; Chen, X.D. Paraffin/Palygorskite composite phase change materials for thermal energy storage. *Sol. Energy Mater. Sol. Cells* **2016**, *144*, 228–234. [CrossRef]
21. Drissi, S.; Ling, T.C.; Mo, K.H. Thermal efficiency and durability performances of paraffinic phase change materials with enhanced thermal conductivity—A review. *Thermochim. Acta* **2019**, *673*, 198–210. [CrossRef]
22. Dogkas, G.; Koukou, M.K.; Konstantaras, J.; Pagkalos, C.; Lymperis, K.; Stathopoulos, V.; Coelho, L.; Rebola, A.; Vrachopoulos, M.G. Investigating the performance of a thermal energy storage unit with paraffin as phase change material, targeting buildings' cooling needs: An experimental approach. *Int. J. Thermofluids* **2020**, *3–4*, 100027. [CrossRef]
23. Du, K.; Calautit, J.; Wang, Z.H.; Wu, Y.P.; Liu, H. A review of the applications of phase change materials in cooling, heating and power generation in different temperature ranges. *Appl. Energy* **2018**, *220*, 242–273. [CrossRef]
24. Shamsundar, N.; Sparrow, E.M. Effect of Density Change on Multidimensional Conduction Phase Change. *J. Heat Transf.* **1976**, *98*, 550–557. [CrossRef]
25. Bareiss, M.; Beer, H. An analytical solution of the heat transfer process during melting of an unfixed solid phase change material inside a horizontal tube. *Int. J. Heat Mass Transf.* **1984**, *27*, 739–746. [CrossRef]
26. Bilir, L.; İlken, Z. Total solidification time of a liquid phase change material enclosed in cylindrical/spherical containers. *Appl. Therm. Eng.* **2005**, *25*, 1488–1502. [CrossRef]
27. Verma, P.; Varun; Singal, S.K. Review of mathematical modeling on latent heat thermal energy storage systems using phase-change material. *Renew. Sustain. Energy Rev.* **2008**, *12*, 999–1031. [CrossRef]
28. Al-Abidi, A.A.; Mat, S.; Sopian, K.; Sulaiman, M.Y.; Mohammad, A.T. Numerical study of PCM solidification in a triplex tube heat exchanger with internal and external fins. *Int. J. Heat Mass Transf.* **2013**, *61*, 684–695. [CrossRef]
29. Li, Z.X.; Sheikholeslami, M.; Shah, Z.; Shafee, A.; Al-Qawasmi, A.R.; Tlili, I. Transient process in a finned triplex tube during phase changing of aluminum oxide enhanced PCM. *Eur. Phys. J. Plus* **2019**, *134*, 173–183. [CrossRef]
30. Shah, Z.; Hajizadeh, M.R.; Ikramullah; Alreshidi, N.A.; Deebani, W.; Shutaywi, M. Entropy optimization and heat transfer modeling for Lorentz forces effect on solidification of NEPCM. *Int. Commun. Heat Mass Transf.* **2020**, *117*, 104715. [CrossRef]
31. Jourabian, M.; Farhadi, M. Melting of nanoparticles-enhanced phase change material (NEPCM) in vertical semicircle enclosure: Numerical study. *J. Mech. Sci. Technol.* **2015**, *29*, 3819–3830. [CrossRef]
32. Dubovsky, V.; Ziskind, G.; Letan, R. Analytical model of a PCM-air heat exchanger. *Appl. Therm. Eng.* **2011**, *31*, 3453–3462. [CrossRef]
33. Darzi, A.R.; Farhadi, M.; Sedighi, K. Numerical study of melting inside concentric and eccentric horizontal annulus. *Appl. Math. Model.* **2012**, *36*, 4080–4086. [CrossRef]
34. Mahdaoui, M.; Kousksou, T.; Blancher, S.; Ait Msaad, A.; El Rhafiki, T.; Mouqallid, M. A numerical analysis of solid–liquid phase change heat transfer around a horizontal cylinder. *Appl. Math. Model.* **2014**, *38*, 1101–1110. [CrossRef]
35. Regin, A.F.; Solanki, S.C.; Saini, J.S. Latent heat thermal energy storage using cylindrical capsule: Numerical and experimental investigations. *Renew. Energy* **2006**, *31*, 2025–2041. [CrossRef]
36. Czajkowski, C.; Nowak, A.I.; Pietrowicz, S. Flower Shape Oscillating Heat Pipe—A novel type of oscillating heat pipe in a rotary system of coordinates—An experimental investigation. *Appl. Therm. Eng.* **2020**, *179*, 115702. [CrossRef]
37. Pietrowicz, S.; Błasiak, P.; Czajkowski, C.; Nowak, A.I.; Rak, J.; Żuraw, A.; Barylak, S. Dispergating Mixer with Pulsating Heat Exchanger. Patent No. PL 234579, 31 March 2020.
38. [illegible], A.; Pietrowicz, S. The thermal behaviour of a special heat exchanger filled with the phase change material dedicated for low-temperature storage applications. *EPJ Web Conf.* **2019**, *201*, 01003. [CrossRef]
39. Pietrowicz, S.; Błasiak, P.; Czajkowski, C.; Nowak, A.I.; Rak, J.; Żuraw, A.; Barylak, S. Heat Storage with Phase Change Bed. Patent No. PL 234838, 30 April 2020.

40. MikroCaps d.o.o., Zasavska Cesta 95, SI-1231 Ljubljana, Slovenia. Available online: https://www.mikrocaps.com/products/microencapsulated-paraffin/ (accessed on 5 January 2018).
41. Shen, J.M.; Dong, J.G.; Li, R.Q.; Zhang, J.; Chen, X.; Qin, Y.M.; Ma, H.D. Integrated supersonic wind tunnel nozzle. *Chin. J. Aeronaut.* **2019**, *32*, 2422–2432. [CrossRef]
42. Mazzeo, D.; Oliveti, G.; Arcuri, N. A Method for Thermal Dimensioning and for Energy Behavior Evaluation of a Building Envelope PCM Layer by Using the Characteristic Days. *Energies* **2017**, *10*, 659–677. [CrossRef]
43. Zastawna-Rumin, A.; Kisilewicz, T.; Berardi, U. Novel Simulation Algorithm for Modeling the Hysteresis of Phase Change Materials. *Energies* **2020**, *13*, 1200–1214. [CrossRef]
44. Zhang, P.; Xiao, X.; Ma, Z.W. A review of the composite phase change materials: Fabrication, characterization, mathematical modeling and application to performance enhancement. *Appl. Energy* **2016**, *165*, 472–510. [CrossRef]
45. Duan, J.; Xiong, Y.L.; Yang, D. Melting Behavior of Phase Change Material in Honeycomb Structures with Different Geometrical Cores. *Energies* **2019**, *12*, 2920–2938. [CrossRef]
46. Duan, J.; Xiong, Y.L.; Yang, D. On the Melting Process of the Phase Change Material in Horizontal Rectangular Enclosures. *Energies* **2019**, *12*, 3100–3120. [CrossRef]
47. ANSYS, Inc. *Ansys® Academic Research Fluent, Release 19.0, Theory Guide*; ANSYS, Inc., Southpointe: Canonsburg, PA, USA, 2019.
48. Voller, V.R.; Swaminathan, C.R. General source-based method for solidification phase change. *Numer. Heat Transf. Part B Fundam.* **2007**, *19*, 175–189. [CrossRef]
49. Mehling, H.; Cabeza, L.F. *Heat and Cold Storage with PCM*; Springer: Berlin/Heidelberg, Germany, 2008.
50. Kean, T.H.; Che Sidik, N.A. Numerical investigation on melting of various nanoparticles enhanced phase change material inside a square enclosure. In Proceedings of the International Conference on Sustainable Energy and Green Technology, Bangkok, Thailand, 11–14 December 2019. [CrossRef]
51. Elsayed, A.O. Numerical investigation on PCM melting in triangular cylinders. *Alex. Eng. J.* **2018**, *57*, 2819–2828. [CrossRef]
52. Gasia, J.; Miró, L.; de Gracia, A.; Barreneche, C.; Cabeza, L.F. Experimental Evaluation of a Paraffin as Phase Change Material for Thermal Energy Storage in Laboratory Equipment and in a Shell-and-Tube Heat Exchanger. *Appl. Sci.* **2016**, *6*, 112–123. [CrossRef]
53. Pagkalos, C.; Dogkas, G.; Koukou, M.K.; Konstantaras, J.; Lymperis, K.; Vrachopoulos, M.G. Evaluation of water and paraffin PCM as storage media for use in thermal energy storage applications: A numerical approach. *Int. J. Thermofluids* **2020**, *1–2*, 100006. [CrossRef]
54. Bejan, A. *Convection Heat Transfer*; John Wiley & Sons, Ltd.: Hoboken, NJ, USA, 2013; Chapter 10, pp. 469–477. 9781118671627.ch10. [CrossRef]
55. Sparrow, E.M.; Broadbent, J.A. Inward Melting in a Vertical Tube Which Allows Free Expansion of the Phase-Change Medium. *J. Heat Transf.* **1982**, *104*, 309–315. [CrossRef]
56. Menon, A.S.; Weber, M.E.; Mujumdar, A.S. The dynamics of energy storage for paraffin wax in cylindrical containers. *Can. J. Chem. Eng.* **1983**, *61*, 647–653. [CrossRef]
57. Katayama, K.; Saito, A.; Utaka, Y.; Saito, A.; Matsui, H.; Maekawa, H.; Saifullah, A.Z.A. Heat transfer characteristics of the latent heat thermal energy storage capsule. *Sol. Energy* **1981**, *27*, 91–97. [CrossRef]
58. Del Rosario, M.D.L.Á.O.; Bruneau, D.; Sébastian, P.; Nadeau, J.P.; Sommier, A.; Lopez, J. Experimental study of an air-PCM heat exchanger: Melting in a cylindrical container. In Proceedings of the ESTEC Conference Proceedings 6th Engineering, Science and Technology Conference, Panama City, Panama, 11–13 October 2017; pp. 403–413. [CrossRef]
59. Cengel, Y.A. *Heat Transfer: A Practical Approach*, 2nd ed.; McGraw-Hill: New York, NY, USA, 2002.
60. Incropera, F.P.; DeWitt, D.P.; Bergman, T.L.; Lavine, A.S. *Fundamentals of Heat and Mass Transfer*; John Wiley & Sons Ltd.: Hoboken, NJ, USA, 2007; pp. 423–433.

energies

MDPI

Article

Thermodynamic Efficiency Maximum of Simple Organic Rankine Cycles

Aram Mohammed Ahmed [1,2], László Kondor [1] and Attila R. Imre [1,3,*]

1 Department of Energy Engineering, Faculty of Mechanical Engineering, Budapest University of Technology and Economics, Műegyetem rkp. 3, H-1111 Budapest, Hungary; arammohammed@energia.bme.hu (A.M.A.); kondorlaszlo94@gmail.com (L.K.)

2 Technical College of Kirkuk, Northern Technical University, Kirkuk 36001, Iraq; aram.mohammed@ntu.edu.iq

3 Centre for Energy Research, Department of Thermohydraulics, POB. 49, H-1525 Budapest, Hungary

* Correspondence: imreattila@energia.bme.hu or imre.attila@ek-cer.hu

Abstract: The increase of the maximal cycle temperature is considered as one of the best tools to increase cycle efficiency for all thermodynamic cycles, including Organic Rankine Cycles (ORC). Technically, this can be done in various ways, but probably the best solution is the use of hybrid systems, i.e., using an added high-temperature heat source to the existing low-temperature heat source. Obviously, this kind of improvement has technical difficulties and added costs; therefore, the increase of efficiency by increasing the maximal temperature sometimes has technical and/or financial limits. In this paper, we would like to show that for an ideal, simple-layout ORC system, a thermodynamic efficiency-maximum can also exist. It means that for several working fluids, the thermodynamic efficiency vs. maximal cycle temperature function has a maximum, located in the sub-critical temperature range. A proof will be given by comparing ORC efficiencies with TFC (Trilateral Flash Cycle) efficiencies; for wet working fluids, further theoretical evidence can be given. The group of working fluids with this kind of maximum will be defined. Generalization for normal (steam) Rankine cycles and CO_2 subcritical Rankine cycles will also be shown. Based on these results, one can conclude that the increase of the maximal cycle temperature is not always a useful tool for efficiency-increase; this result can be especially important for hybrid systems.

Keywords: ORC; Trilateral Flash Cycle; *T-s* diagram; adiabatic expansion; working fluid; retrofit; CO_2 power cycle

check for updates

Citation: Ahmed, A.M.; Kondor, L.; Imre, A.R. Thermodynamic Efficiency Maximum of Simple Organic Rankine Cycles. *Energies* **2021**, *14*, 307. https://doi.org/10.3390/en14020307

Received: 18 December 2020
Accepted: 5 January 2021
Published: 8 January 2021

Publisher's Note: MDPI stays neutral with regard to jurisdictional claims in published maps and institutional affiliations.

Copyright: © 2021 by the authors. Licensee MDPI, Basel, Switzerland. This article is an open access article distributed under the terms and conditions of the Creative Commons Attribution (CC BY) license (https://creativecommons.org/licenses/by/4.0/).

1. Introduction

Organic Rankine cycle-based processes are popular and effective methods to utilize heat sources with a wide range of temperature to utilize for electricity production. Most of the heat sources (like geothermal heat, industrial heat) are localized. Sometimes, there is a possibility to use an additional, non-localized heat source (like solar heat) to increase the maximal cycle temperature. For the basic thermodynamic cycle (Carnot-cycle), the first law efficiency is increasing by increasing the maximal cycle temperature while minimal cycle temperature is constant [1]. For other cycles, a similar law can be used, only in that case, maximal and minimal cycle temperatures have to be replaced by mean temperatures of heat addition and removal [2]. Using the assumption, that mean temperature of heat addition is increased when maximal cycle temperature is increased [2], a thumb-rule can be deducted, that by increasing the maximal cycle temperature, the cycle efficiency is always increased. Therefore, the increase of the temperature is always desirable in simple thermodynamic cycles, when the goal is the better performance, although other constraints (like the cost of the utilization of this added heal source) can overshadow the gain caused by the efficiency increase.

The design of an ORC-based power plant has at least four different layers [3]. The first one is about the thermodynamic cycle itself; one can design and optimize the Rankine-cycle using various working fluids and various constraints (like defining the "ideality" or "reality" of the cycle by defining the irreversibility-induced entropy-production during expansion). The second layer is the technical/engineering layer; the designed Rankine-cycle should be realized, the constraints of the previous steps have to be associated with various hardware components or process properties (like expander internal efficiency, pressure loss within the heat exchanger, etc.). An optimal second layer design might require several successive approximation steps, where the first step design has to be re-calculate due to various points (for example inaccessibility of expander with the desired internal efficiency), and the new result has to be used for the next iteration. The third layer is the economical one. One should realize that investors built power plants not to generate electricity but to generate profit. Therefore, even the most brilliant thermodynamic cycle or ORC layout might be rejected upon financial points. The fourth layer is the environmental one; it is strongly connected to the third one [4].

With a bit of oversimplification, one can say that in the first layer, the thermodynamic properties of the working fluids, especially the shape of their *T-s* diagram are the most important factor [5,6], involving even molecular properties [7,8], while in the second layer already having the working fluid, the proper choice of technological components (most often the expander) plays the leading role [9,10]. However, sometimes these two steps are very much interlocked [11]. Obviously, the final vote is always for the economic side [4].

To have successful optimization in the second layer, one should clarify the problems raised in the first layer. Although most people assume that thermodynamics already solved all related problem and no further study is necessary concerning basic cycles, it is not a valid assumption. Here, we are going to show a clear example to disprove this assumption.

ORCs are supposed to be used to utilize heat sources not utilizable with traditional steam Rankine cycle; these heat sources are most often low-enthalpy ones with relatively low temperature and sometimes with low heat flow. For this reason, ORC-based power plants have limited financial viability; investment, operational and maintenance cost should be kept as low as possible. One of the ways to do that is the use of the basic ORC layout, namely one with a heater (liquid heater plus evaporator), an expander, a condenser and a pump, without using superheater or regenerative/recuperative heat exchanges [12]. Addition of any extra component can increase investment cost and decrease the "robustness" of the design. Therefore, our goal is to solve the thermodynamic problems without the involvement of any new part, i.e., using only the basic ORC design.

In this paper, we would like to show, that using the simplest ORC (or even traditional, water/steam-based RC) layout, a thermodynamic efficiency maximum should be found in all ideal cycles using wet and in several ideal cycles using dry working fluids. To prove this statement, the efficiencies of simple ideal ORC and TFC (Trilateral Flash Cycles) processes will be studied, using working fluids from various classes. The exact location of these maxima depends on the working fluid, as well as on the minimal cycle temperature. The existence of this kind of maximum shows that the increase of the maximum cycle temperature is not always a proper tool to increase cycle efficiency; sometimes it can be contra-productive.

2. Hybrid Systems

The temperature of the heat source is an important factor for ORC applications, even though this technology can utilize sources with relatively low temperature. Geothermal energy is often considered as a low-grade energy-source; therefore, it cannot independently support high load applications. This is true even for countries with quite good geothermal potentials (like Hungary), where the well-head temperature of most of the existing geothermal wells are below 90 °C [13]. These kinds of sources are usually excluded from the pool of potential sources for electricity generation, although they can be numerous and some of them have very impressive heat-flux. In this case, one might apply hybrid systems (using secondary heat sources with a smaller heat flux but with higher temperature) to overcome the inherent weakness of the low-temperature sources.

Probably the most frequently used double-source design is the hybrid solar-geothermal installation [14–16]. For interested readers, a short overview of these kinds of systems is given in Appendix A. In hybrid solar-geothermal systems, the initial heating can be done by the low-temperature geothermal source, and then the solar heat is used to increase the maximal temperature. This can be done in two different ways. In the first solution, the solar heat can be used to "superheat" the already evaporated vapor [17]; this solution can be easily applied in retrofitted systems because only minor modifications of the existing geothermal power plant are required. In the second solution, the geothermal heat is used to preheat the compressed liquid, while the solar heat would be used to reach the maximal temperature and for evaporation. In this case, the pressure in the evaporator, as well as the input pressure of the expander, will be higher than for the same systems without solar heat; therefore, this method is not ideal for retrofitting of existing systems, but might be better for novel units [18]. The choice between the two options depends not only on the temperature values of the two sources but also on the available heat flows; in case of very small secondary heat flow, only the first case could be a plausible choice. Here, we prefer the second method, because our simple layouts would not consist of superheating units, i.e., the ORC design would remain simple.

Maximal cycle temperature, which is an important factor of the efficiency, can be increased in three ways in solar or hybrid solar systems:

- It is possible to keep the original solar collectors (i.e., total heat flux and heat source temperature remain the same), while the mass flow of the working fluid could be reduced. A small increase in the maximal cycle temperature can be obtained, but the simultaneous decrease of the pinch temperature (the minimal temperature gap in the heat exchanger) does not allow the significant increase. In this case, the investment cost of the retrofit would be minimal or zero, but the decrease of mass flow might decrease the power; therefore, relative (power-normalized) investment cost can increase.
- The number of solar collectors can be increased, increasing the collected heat but keeping the source temperature. In this way, one can keep the mass-flow of the working fluid, but pinch temperature can be slightly decreased while the maximal cycle temperature can be increased (only on a small extent). This solution, due to the extra collectors, increases the investment cost.
- The quality of solar collector system can also be improved (for example by changing the simple flat collectors to a focused system). With this solution, not only the amount of collected heat but also the temperature can be higher. In this way, because the source temperature is higher, keeping the mass-flow, the maximal cycle temperature can be remarkably increased. This solution, due to the use of better, i.e., more expensive collectors increases the investment cost.

One might expect, that just like for the ideal Carnot-cycle [1]; increasing the maximal cycle temperature for ORC or similar cycles would increase the thermal efficiency at least up to the critical point and therefore the application of an additional heat source would be limited only by technical or economic constraints [3]. Here, we are going to show that by using a basic ORC cycle, thermodynamic efficiency can have a maximum, associated with a sub-critical temperature, which depends on the material and the minimal cycle temperature.

3. Method

Concerning the basic ORC system, we are using the most straightforward and simplest layout to minimize the installation and maintenance costs for the system. The layout assuming a hypothetical dry working fluid can be seen in Figure 1a. It contains a pump, used to compress the working fluid from the low-pressure liquid state (1) to high pressure one (2). The first heat exchanger is for heating the compressed working fluid. It has two parts, the liquid heater (LH) to heat up the compressed liquid to the saturated state (2→3) and the evaporator (EV), to evaporate the saturated liquid to saturated vapor state (3→4). Cost minimization comes here first; no superheater is included, not even for wet working fluids. The fluid expands directly from this saturated vapor state (4) through the expander (E) to cold, dry vapor state (5). It will be cooled down in the condenser (C) in two steps, first by pre-cooling the vapor down to the saturated state (5→6), then by condensing it to the saturated liquid state (1). The heat extracted by the pre-cooler is lost here; for the sake of simplicity, recuperative or regenerative heat exchangers are omitted. Corresponding *T-s* diagram can be seen in Figure 1b. Using wet working fluid, the expansion line (4→5) would run in the two-phase region (wet vapor) and points 5 and 6 would be identical; therefore, pre-cooling would not be necessary.

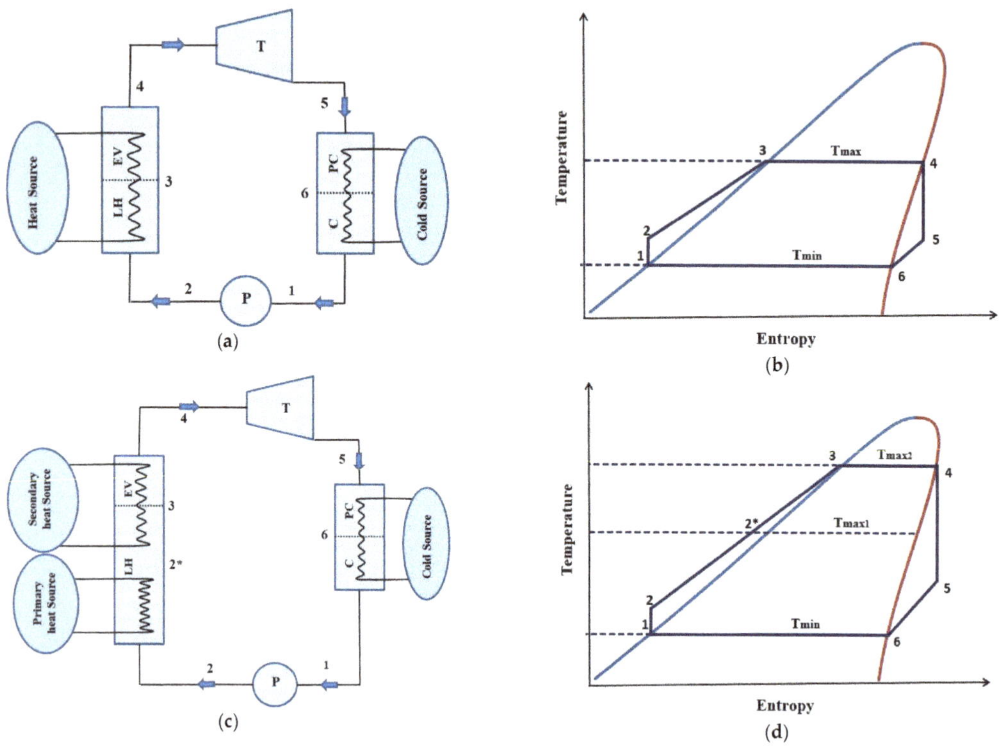

Figure 1. Schematic layouts and *T-s* diagrams for simple single-source (**a**,**b**) and hybrid (**c**,**d**) ORC systems. P: pump; LH: liquid heater; EV: evaporator; T: turbine or expander; PC: pre-cooler; C: condenser.

For the hybrid system, we have a second heat source; corresponding figures are Figure 1c,d. In this case, higher initial compression (from point 1 to 2) is used, so the pressure ratio p_2/p_1 will be higher. Pressurized working fluid (in liquid state) will be preheated by the primary heat source of the maximal temperature provided by the primary source (from point 2 to point 2*; this was the point corresponding to maximal temperature in the single source cycle). Then the secondary heat source (for example the solar one) is heating the fluid up to the maximal cycle temperature (from point 2* to point 3), and also the heat of this source is used to evaporate it (from point 3 to 4). Expansion and cooling steps (the 4-5-6-1 sequence) would be similar to the previous case, although the expansion part (from point 4 to 5) will be more significant and the cooling of the dry vapor (5–6) would start at a higher temperature.

As it has been already stated, we are discussing only the thermodynamic cycle here; therefore, all steps are considered to be ideal. Compression and expansion steps are isentropic, while for the heating and cooling steps, pressure drops are omitted, and strict isobaric heat exchange processes are assumed. Because for first-law efficiency, the knowledge of the intermediate temperature (T_{max1}) is not needed, only the effect maximal cycle temperature (T_{max2}) will be investigated.

Although wet working fluids are considered as not suitable ones for ORC applications [19], due to their unfavorable behavior (most remarkably the ones related to the appearance of droplets), in this paper we are studying the thermodynamic, rather than technical aspects of ORC. Therefore, both dry and wet working fluids were checked.

Efficiencies are going to be calculated for six different materials (see Table 1), representing a wide range of working fluids from wet ones (almost symmetrical, bell-shaped *T-s* diagrams) to very dry ones (with very skewed *T-s* diagrams), see Figure 2. Four of them might be used for hybrid geothermal/solar ORC or RC while the other two having very low critical temperature might be used in cryogenic cycles [20–22]. Therefore here—when the heat source is solar heat, presumably providing source temperatures exceeding the critical temperatures of these two materials—they are only used for demonstration purpose. Two of them are wet working fluids (using the traditional classification), or class ACZ working fluids [23]. The other four are dry working fluids. Concerning the sequence-based classification [23], dry working fluids can be divided into two classes; ACZM and AZCM types. Letters represent the special point on the *T-s* diagrams; A is the initial point of the *T-s* saturation curve, Z is the final one, C is the critical point and M is the point corresponding to maximal entropy. From these points, four-letter sequences can be built, based on the entropy values of these points. Initial point A always has the smallest entropy, while M has the highest one; therefore, only sequences starting with A and ending with M are possible. Having only two intermediate points, these are the ACZM and AMCZ. Liquid states are located in the A-C part of the curve, while vapor ones are stretching in the C-M-Z part. The crucial difference is that for ACZM, the entropy values for liquid states are always below the entropies of any of the saturated vapor states; while for AZCM, there are liquid states with entropies higher than for some vapor states. This means that for AZCM-type fluids, isentropic expansion starting from some saturated liquid states can reach the fully vaporized dry-vapor region. The ACZM-ones are closer to the wet-ones (ACZ), as it is shown in Ref. [21]. Three of the working fluids represented here are ACZM, and one for AZCM. In several fluids, a local entropy minimum (N) also can be seen; these fluids can also be classified as dry ones, or alternatively as "real isentropic ones" because, for them, it is possible to have an isentropic expansion step from saturated vapor state into another saturated vapor state [21,23,24]. In this paper, this fifth point will not be relevant.

Table 1. Working fluids used in this study with some of their relevant properties.

Name	Type	Class (Absolute/at 20 °C)	T_{cr} [K]	p_{cr} [MPa]
Water	Wet	ACZ/ACZ	647.1	22.06
Carbon dioxide	Wet	ACZ/ACZ	304.13	7.38
Hexafluoroethane (R116)	isentropic	ACNMZ/ACZ	293.03	3.05
Butane	Dry	ACNMZ/ACZM	425.12	3.8
Neopentane	Dry	AZCM/AZCM	433.74	3.2
Dodecane	Dry	AZCM/AZCM	658.09	1.82

One more piece of information is needed concerning the use of this working fluid classification [23]. The ultimate initial and final points of the saturation curve in *T-s* diagrams are related to the triple points (the last point of the liquid/vapor equilibrium; solid-phase appears at that point). The corresponding entropy values are material properties, and therefore the classification (being the location of C and M points also material-dependent) is also a material property. However, in several cases, triple points are located in technically irrelevant, low temperatures (like for butane, where it is located at 134.6 K, which is −138.5 °C). Therefore, it might be better to terminate the *T-s* curve in a technologically more relevant minimal temperature, for example, at the minimum cycle temperature. In ORC applications (with air cooling) this should be done somewhere around (or exactly at) 20 °C (293.15 K). The relevant part of the curve (relevant means that the one above 20 °C) might mimic a novel class; for example, the vicinity of the top of all the *T-s* diagrams (above point M) looks like a wet, ACZ type curve. For the sake of clarity, sometimes these "relative" classes are distinguished by marking the related points with an upper-case star (*), like A*CZ*M. Since for hybrid solar/geothermal case, usually, 20 °C can be taken as minimum temperature; therefore, both absolute and relative classes are shown in Table 1.

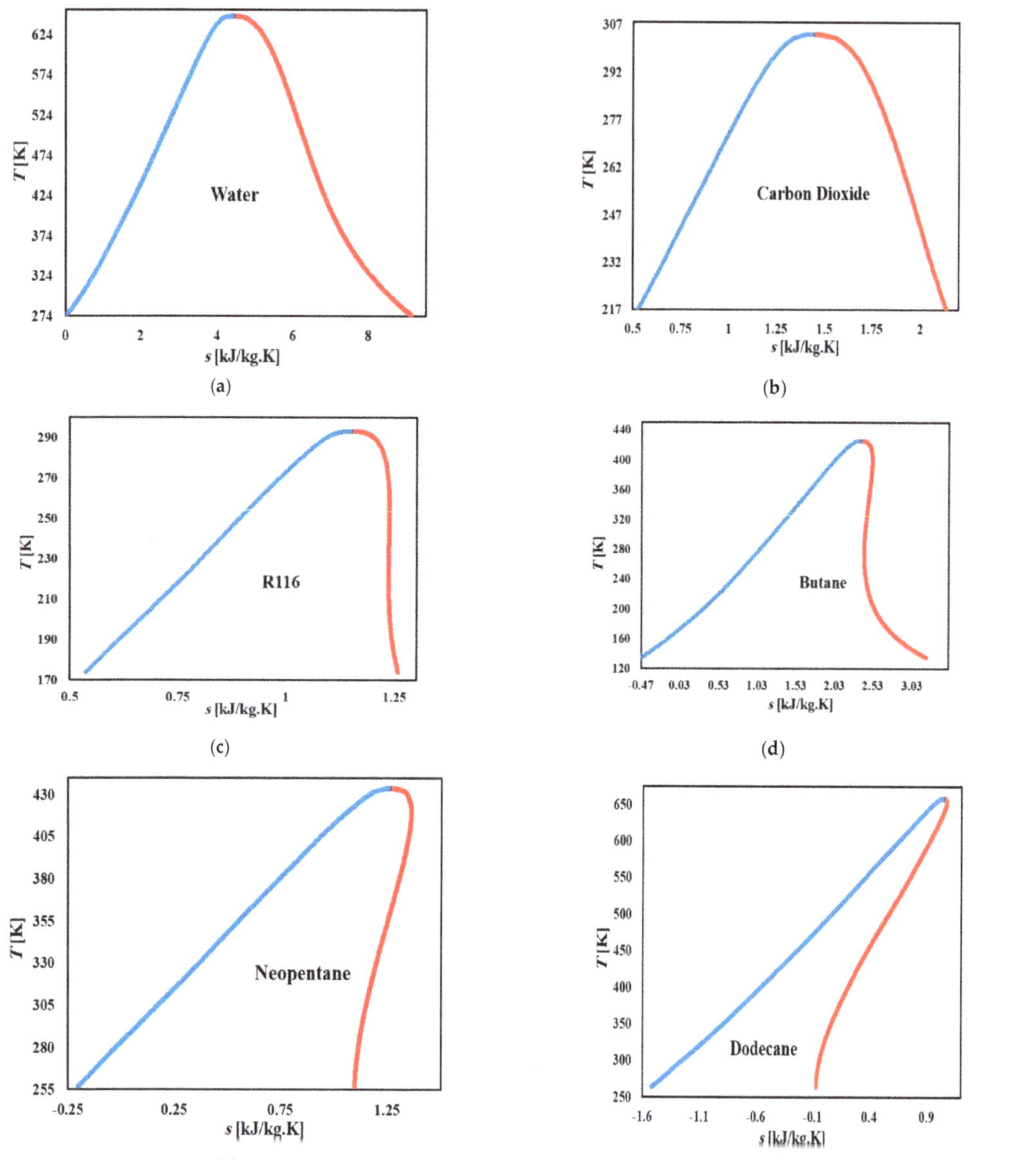

Figure 2. Temperature dependencies of ORC (or other Rankine-like) and TFC cycles for various working fluids (Water (**a**), CO_2 (**b**), R116 (**c**), Butane (**d**), Neopentane (**e**), Dodecane (**f**)); see text for further explanation.

The efficiency values for various maximal and minimal temperature pairs were calculated with MATLAB software; working fluid data were taken from the NIST Chemistry Webbook [25]. On the schematic *T-s* diagram of a wet working fluid, the blue part represents the saturated liquid state, and the red part represents the saturated vapor state.

The efficiency calculation for ORC and TFC were done in the manner shown in Figure 3. First, a minimal cycle temperature was chosen (represented by the light green line) and used as condenser temperature. For this given condenser temperature, various maximal temperature values were chosen, up to the critical temperature. For each calculation, the temperature interval between the condenser and maximal temperature was divided into 500 equal part; efficiency values were calculated for all of these 500 points, using them as maximal cycle temperatures. Some of them are shown in Figure 3a, corresponding to the evaporation plateau (ORC) and on Figure 3b, corresponding to the topmost point of the "triangle" (TFC). In this way, the efficiency of the first cycle (1-2-3-4-1) were calculated. After finishing one run, a new minimal temperature was chosen slightly above the previous one (by shifting the green base-line a bit higher), and the calculation was repeated for the cycle (1-2a-3a-4a-1) up to the last one, just one step below the critical temperature, to cycle (1-2d-3d-4d-1). In this way, separate curves representing the Tmax dependence of efficiencies were obtained for different condenser temperature.

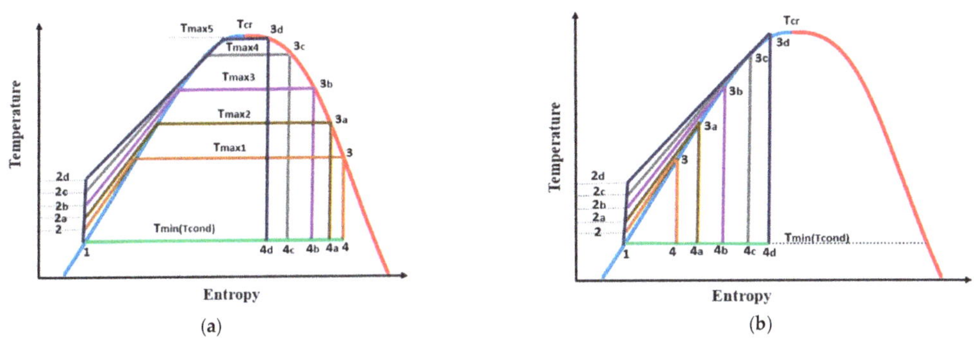

Figure 3. Steps of the efficiency-calculation for ORC (**a**) and TFC (**b**) cycles.

Concerning efficiency, the actual values were calculated by using the following equation:

$$\eta = \frac{Q_{in} - Q_{out}}{Q_{in}}, \tag{1}$$

where (concerning the isobaric heat exchanges) the heat added to the system is $Q_{in} = h_3 - h_2$ and the heat taken from the system is $Q_{out} = h_4 - h_1$; numerical enthalpy values were taken from the NIST Chemistry Webbook [25] as a function of temperature and entropy.

4. Results

On Figure 4, one can see two efficiency lines for CO_2; upper one (blue) represents the efficiency of the Rankine-like subcritical CO_2 power cycle, while the lower one (green) represents its TFC counterpart. The minimal temperature (condenser temperature) were chosen as the triple-point temperature of the CO_2 (216.6 K), and the efficiencies were calculated up to the critical point (304.13 K). On Figure 4, the whole curve can be seen, while on Figure 4b, the high-temperature region can be seen, showing the well-developed maximum on the upper curve, 4 degrees below the critical point.

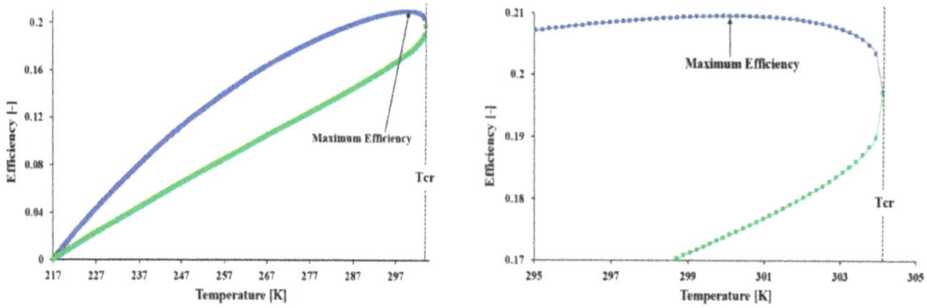

(a) (b)

Figure 4. The efficiency of a CO_2 Rankine-like Power Cycle (upper lines) and Trilateral Flash Cycle (lower lines) with fixed condenser temperature (216.6 K) and with increasing evaporator temperature. (**a**) full temperature range; (**b**) the vicinity of the critical temperature.

On Figure 5, a set of these curves (approximately 20 for each material) can be seen for all materials listed in Table 1. The first sign of this maximum was found earlier in butane [26]. For water (5a) and carbon dioxide (5b) (two typical wet working fluids), the maxima on all ORC curves are very characteristic. For CO_2, red dots are marking these maxima for better visualization; these points are omitted for the other materials to avoid them to distort the shape of the curves. The increase of the maximal cycle temperature (with a fixed minimal cycle temperature) causes an increase of the efficiency for both cycles (the Rankine and the flash-types, represented by the upper and lower curves). Reaching higher maximal temperatures, the speed of this increase will be lower and lower, and finally a maximum, then a decreasing part can be seen. It means that with a fixed condenser temperature, the increase of maximal temperature has an initially positive effect of the first-law efficiency, but when it is reaching the vicinity of the critical point, this positive effect disappears, and the further increase would be contra-productive. On the contrary, for the TFC-like cycles, the increase holds up to the critical point, i.e., having a TFC-like cycle with a wet working fluid, the increase of maximal cycle temperature always increases the efficiency. In this way, the two curves are forming an elongated rain-drop form; the pointed end is located at the condenser temperature (with $\eta = 0$). Increasing the condenser temperature, one can obtain smaller efficiencies; both the upper and lower curves are shifting down, while the location of the maximum (with smaller and smaller value) shift closer to the critical temperature (see the location of the red dots for CO_2), but always remains below that value. The rain-drop shape formed by the upper and lower curves remains, but it will be smaller and smaller. For the sake of the better visibility, the location of the maxima are marked by red dots for the CO_2 curves (Figure 5b); its movement can be clearly seen.

Hexafluoroethane (R116) is originally an almost isentropic working fluid. The *T-s* diagram has an inverse S-shaped saturated vapor part, which is very narrow, i.e., the entropy value for the vapor phase (except in the vicinity of the critical temperature) is almost constant. In the novel classification system, it is a type ACNMZ. From the mapping of various working fluid types [21], this is an almost wet isentropic fluid. The point M located at 253.5 K, i.e., choosing condenser temperature above this value, R116 acts like a wet working fluid. Point N is located at 220.9 K, i.e., choosing condenser temperature between 220.9 K and 253.5 K, this fluid can mimic a moderately dry working fluid. Efficiency-curves can be seen in Figure 5c. The curves are very similar to the ones seen for water and CO_2, the maxima are still well developed (although these points are closer to the critical point than for the two wet ones), and the joint ORC-TFC curves are also raindrop-shaped. For this material, one can also say that increasing the maximal cycle temperature has a positive effect on the first-law efficiency, but in the vicinity of the critical point, this effect disappears and changes to negative.

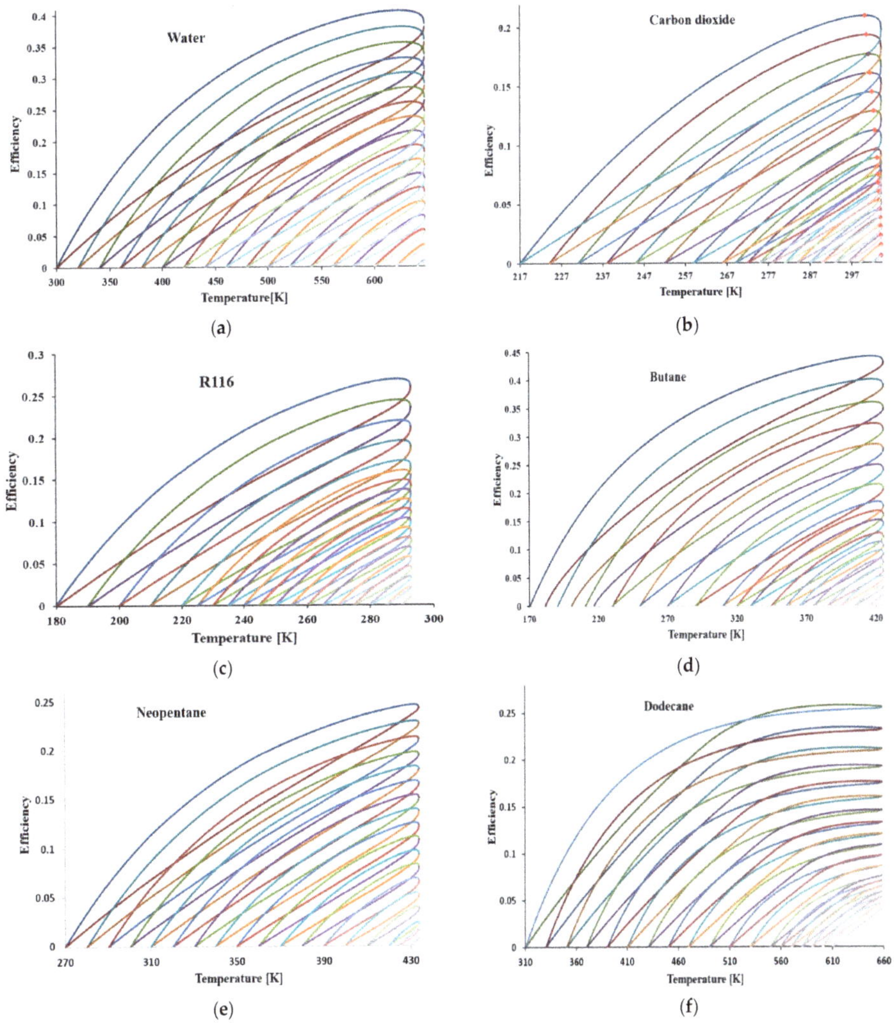

Figure 5. Temperature dependencies of ORC (or other Rankine-like cycles) and TFC cycles for various working fluids (Water (**a**), CO_2 (**b**), R116 (**c**), Butane (**d**), Neopentane (**e**), Dodecane (**f**)); see text for further explanation.

Butane is a moderately dry working fluid; it means that although the *T-s* diagram is not bell-shaped, but tilted, the titling is not very strong. ORC-TFC efficiency curves can be seen in Figure 5d. Although maxima can be still noticed, they are remarkably less developed than for the previous three materials.

For neopentane (Figure 5e), which has a more tilted *T-s* diagram that butane (i.e., it is drier), the rain-drop shape is already very distorted, and even the flat end is almost pointed (although the transition from ORC to TFC is still smooth). Very small, nearly diminished maxima can be seen, located very close to the critical temperature. In this case, one can assume, that increase of the maximal cycle temperature is always good or almost neutral for the efficiency. For the uppermost curve, the location of the maximum is only 1.97 K from the critical temperature, and even above this point, the decrease of efficiency is very small, so, one can consider it almost temperature-independent.

Finally, a very dry (on *T-s* diagram, represented by a very tilted *T-s* curve) material, dodecane (Figure 5f) were analyzed. One can see, that at low evaporation temperatures, ORC efficiency exceed TFC one, as it happened before, but increasing this temperature further, the difference will be smaller and smaller, and at a given temperature (which depend on the condenser temperature, i.e., different for each pair of curves) it disappears at the crossing of the two curves. Passing that crossing temperature, the efficiency of TFC is higher than for ORC; finally, they will join smoothly at the critical point. In this case, the efficiencies of TFC can form a shallow maximum, close to the critical curve, while the efficiency of ORC increases continuously. Therefore one can conclude, that by using very dry working fluid for an ideal (strictly isobaric heat exchanges and strictly isentropic expansion/compression) basic-layout (recuperator and superheater-free) ORC-system, the increase of maximal cycle temperature can always increase cycle efficiency, even in the vicinity of the critical temperature; additionally, small maxima might be expected for the efficiency of TFC.

3-D version of these lines for water (Figure 6a–d) and butane (Figure 6e–h) are also shown Rotating animated gif version of these figures are provided as Supplementary Material. The distortion of the initial rain-drop shape can be clearly seen.

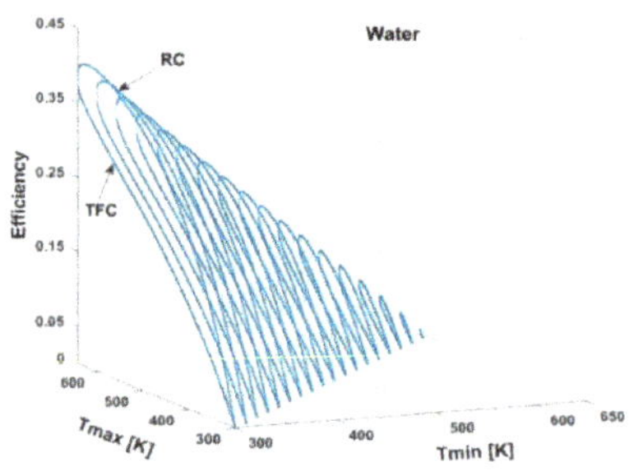

(a)

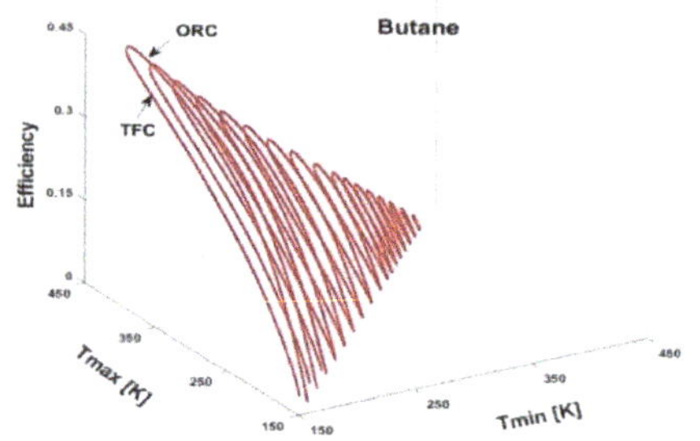

(e)

Figure 6. *Cont.*

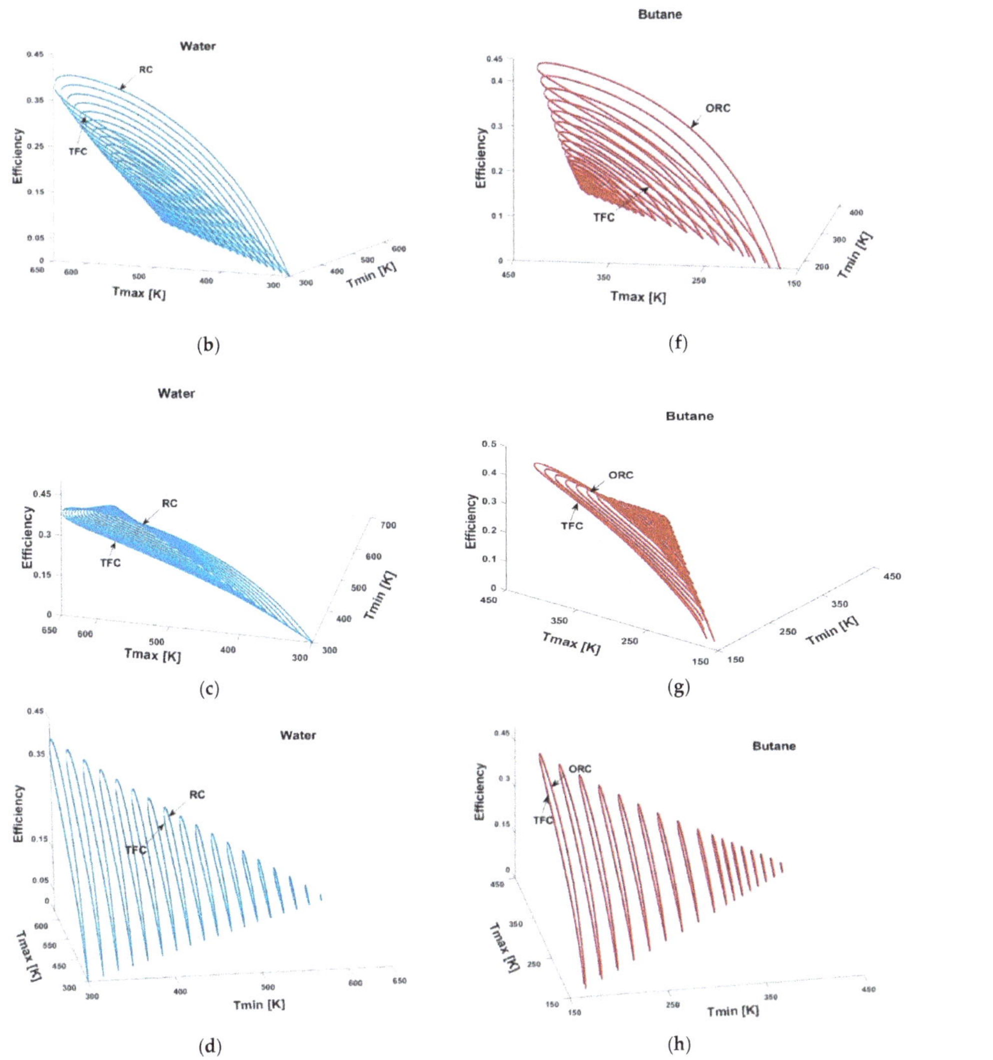

Figure 6. 3-D representation of the efficiency-curves for water (**a–d**) and butane (**e–h**). Rotating version of these figures is provided in the Supplementary Material as Figure S1a (water) and Figure S1b (butane) in animated gif-format.

5. Theoretical Reason for the Existence and Disappearance of the Maximum

A simple geometrical explanation can be given for the appearance of efficiency maximum in wet working fluids. It can be seen in Figure 7 that the ORC or any other Rankine-like cycle (steam Rankine or CO_2 power cycles) can be seen as a superposition of a TFC-like and a Carnot-like cycles, although this superposition is not simply a summing of efficiencies. The TFC-like part is the one at lower entropies, resembling a triangle, while the Carnot-part (located on higher entropies) is the rectangular one. Considering that the Carnot-cycle is the best potential cycle between the given maximal and minimal temperature, adding it to a TFC, the Carnot-like part always has a positive contribution to efficiency. In this way, for wet working fluids, η_{ORC} is always bigger, than η_{TFC}. One can see, that by increasing the maximal cycle temperature (compare Figure 7a,b), the rectangular part will be smaller and smaller, almost diminishing in the vicinity of the critical temperature (Figure 7c), and when $T_{max} = T_{cr}$, it disappears, and the TFC and ORC cycles will be indistinguishable. Therefore, at $T_{max} = T_{cr}$, $\eta_{ORC} = \eta_{TFC}$, and the transition will be smooth and continuous.

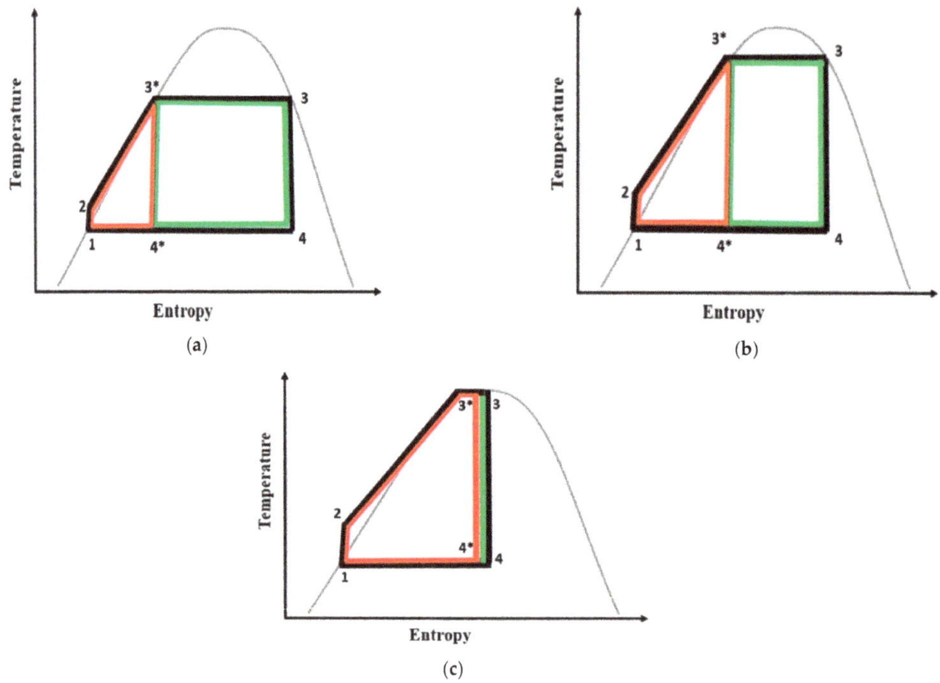

Figure 7. Rankine-cycle (black) in a wet working fluid as a superposition (but not the simple sum) of a TFC-like (red) and Carnot-like (green) parts with same condenser temperature with a medium (**a**) high (**b**) and almost critical (**c**) evaporator temperature.

This smoothness and continuity affect the shape of the efficiency-temperature curve. Previously we used T-η curves, here, for a short time, the inverse should be used, the η-T one (see Figure 8). Being $\eta_{ORC} > \eta_{TFC}$ for wet working fluids at every temperature, except T_{cr}, the low-entropy part (left side) represents the TFC, and the high-entropy part (right side) represents the ORC. In this representation, the two curves have to have end-points at the maximal temperature (T_{cr}). Continuity causes the equality of the two efficiencies at T_{cr}, i.e., the two curves have to join in this point, i.e., instead of two end-points, they will have one common end-point. Due to the smoothness, the derivative $\mathrm{d}T/\mathrm{d}\eta$ will be zero in this point; therefore, this will be a smooth maximum with positive $\mathrm{d}T/\mathrm{d}\eta$ on the left side and negative $\mathrm{d}T/\mathrm{d}\eta$ on the right one. It should be recalled, that the left side represents the efficiency of ORC; being $\mathrm{d}T/\mathrm{d}\eta$ positive on the original, T-η representation, this part would show negative slope. Therefore, in the vicinity of T_{cr}, the smooth and continuous transition between ORC and TFC cycles would cause a part, where efficiency decreases with increasing temperature. Contrary, at lower temperatures, one can see an increasing part; this can be possible only by assuming a maximum on the $\eta(T)$ curve of the ORC.

One should remember that a strong point of this reasoning was to handle ORC as a superposition of a TFC-like and a Carnot-like part. While this is true for the wet working fluid (Figure 9a), in a dry one, a third part, resembling a distorted, upside-down, inverse TFC, can also be defined, see Figure 9b–d. This part is also less efficient, that a Carnot-like one. Increasing the "dryness", i.e., having more and more tilted T-s diagrams and going from ACZM to AZCM class [23], the weight of the middle Carnot-like part will be smaller and smaller. Finally, this part can completely disappear.

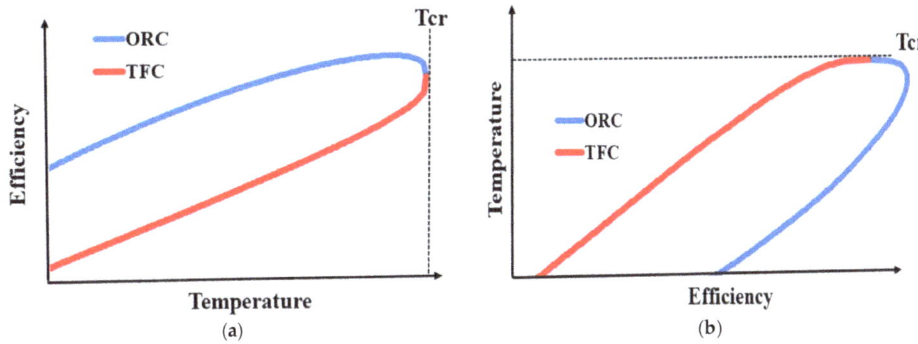

Figure 8. Schematic temperature-efficiency (**a**) and its inverse, efficiency-temperature (**b**) curves of an ORC-like and TFC-like system, explaining the necessity of the maximum. Further explanation can be found in the text.

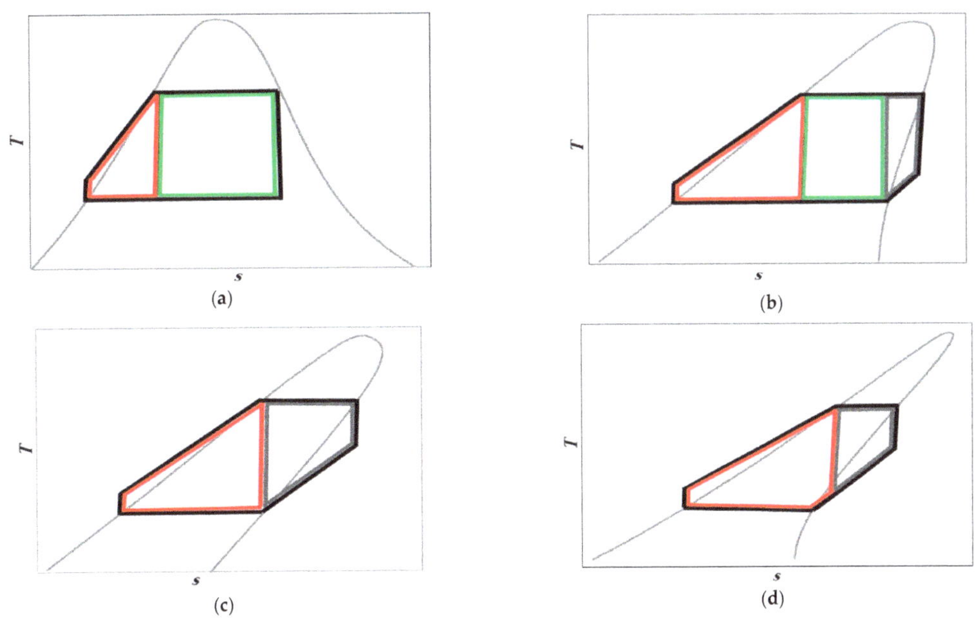

Figure 9. Different structure of an ORC cycle (black) with a wet, moderately dry, a medium dry and a very dry working fluid, divided into TFC-like (red), Carnot-like (green) and inverse TFC-like (grey) sub-cycles. The diminishment of the Carnot-like part (green) with increasing dryness can be clearly seen, as well as the loss of triangularity of TFC (red) and the appearance of the inverse TFC-like type (grey). The four figures are representing wet (**a**), moderately dry (**b**), medium dry (**c**) and very dry (**d**) working fluids.

This process can be seen in detail in Figure 9a–c. With wet working fluid (a), the ORC can be seen as a superposition of an almost perfectly triangular TFC (red) and an ideal Carnot (green) cycles, as it has been shown before (Figure 7). For moderately dry one (b), a third part (grey) appears, located at higher entropies, resembling a distorted upside-down TFC. The appearance of this part makes the Carnot-like part narrower, decreasing its superficial contribution to net first-law efficiency. For medium dry one (c) the isobaric, but not isothermal part of the heat removal is already so significant, that the Carnot-like part totally disappears and the cycle will be the superposition of an almost triangular TFC-like and an upside-down TFC-like part; because the Carnot-part disappeared, one might expect that net efficiency might be smaller than previously. Finally, with a very dry working fluid, the isobaric heat removal part preceding the condensation will be so huge, that it will reach the initial TFC-like part, cutting down the corner and terminating the "triangularity" (Figure 9d).

Because this transition happens with increasing dryness, one can expect to see the similar process by going from wet (water, CO_2) through isentropic (R116), moderately dry (butane) and dry (neopentane) working fluids to very dry one (dodecane). Upon this transition, the maximum will be less and less significant, and finally, it can disappear, as it is shown for dodecane.

6. Discussion and Conclusions

The increase of the maximal cycle temperature (which is equal to the evaporator temperature in ideal, simple, superheater-free systems) is thought to be a good way to increase the first-law efficiency in simple cycles, like steam Rankine, Organic Rankine and CO_2 power cycles, although there are signs, that problems might arise when this increase reaches the vicinity of the critical point. Some of the problems related to the properties of the materials in this region; there are significant density fluctuations, causing strong local anomalies; also, the accuracy of the Equation of States can be reduced in this region [27,28]; this can also produce problems during design and in the operational phases. Also, from the technical point of view, high evaporation temperatures, especially for dry working fluids, are also not very favorable [29,30]. For example, choosing the evaporation temperature above the temperature of local entropy maximum (the so-called point M), expansion starts in the wet region, even in dry fluids; therefore, droplets can be formed during the early stage of the expansion, causing troubles in the special "dry" expanders.

Here, we approached the problem of the increase of maximal cycle temperature from a purely thermodynamic side, by determining its effect (with fixed condenser temperature) on the first-law efficiency. Efficiencies for six different materials with increasing dryness (water, CO_2, R116, butane, neopentane and dodecane) were evaluated. Although anomalies were expected to be seen for dry working fluids [29,30], surprisingly, a very strange effect was seen for wet working fluids, namely the appearance of an efficiency maximum on the efficiency of the Rankine-like cycle (first seen for butane [26]), located close to the critical temperature. The location of the maximum depends on the condenser temperature as well as on the material. For water, taking the smallest possible condenser temperature (the triple point temperature, 273.16 K), the maximum appears 24.6 K below the critical point, while increasing the condenser temperature, it can approach the critical point. For CO_2, R116 and butane, the biggest distance between the location of the maximum from the critical point is in the order of 5–10 K (8.68 K for butane, 4.01 K for CO_2 and 4.53 K for R116); for neopentane, it is 1.97 K, and finally, for dodecane, no maximum can be seen. Concerning reduced temperatures, the biggest distances between the entropy maximum and the critical points are 0.038 (water), 0.013 (CO_2), 0.015 (R116), 0.020 (butane) and 0.005 (neopentane). Since for most real ORC and RC power plants, the operational temperature is not very near to the critical temperature; therefore, technologically, the existence of the maxima are relevant only for two cases (for water, where the maximal distance between the critical temperature end efficiency maximum is almost 25 K, and for butane, where this difference is almost 9 K). For the rest of the presented working fluids, the relevance is rather theoretical than technical. However, one might assume that the temperature differences will be sufficient to have technological importance for some of the not presented wet working fluids.

The presence of these maxima can be clearly shown for several systems by comparing the efficiency of a Rankine-like cycle with a Trilateral Flash-like Cycle (TFC). Some theoretical explanation is also given to see, that for very dry working fluids (like dodecane), the disappearance of this maximum can always be expected for simple, Rankine-like cycles, using only the basic Rankine-layout (pump, liquid heater with evaporator, expander and condenser), without recuperative unit.

The existence of the efficiency maximum for various working fluids is a disproof of the common misconception that the increase of the maximal cycle temperature always has a positive effect on first law efficiency for simple thermodynamic cycles. This disproof is interesting itself, as a strange thermodynamic fact, but also, it can give a new tool for engineers to find optimal cycle temperature for an ORC (or similar) power plant, by designating a temperature range, where a further increase of the evaporation temperature would be contra-productive.

Supplementary Materials: The following are available online at https://www.mdpi.com/1996-1073/14/2/307/s1, Figure S1a,b: Rotating 3-D representation of the efficiency-curves for water and butane, respectively (in animated gif-format).

Author Contributions: Conceptualization, A.R.I.; analysis: A.M.A., preliminary analysis: L.K., writing: A.R.I., A.M.A. and L.K. All authors have read and agreed to the published version of the manuscript.

Funding: This work was performed in the frame of the FIEK_16-1-2016-0007 project, implemented with the support provided from the National Research, Development and Innovation Fund of Hungary, financed under the FIEK_16 funding scheme. Part of the research reported in this paper and carried out at BME has been supported by the NRDI Fund (TKP2020 NC, Grant No. BME-NC) based on the charter of bolster issued by the NRDI Office under the auspices of the Ministry for Innovation and Technology.

Institutional Review Board Statement: Not applicable.

Informed Consent Statement: Not applicable.

Data Availability Statement: The data that support the findings of this study are available from the corresponding author, [A.R.I.], upon request.

Conflicts of Interest: The authors declare no conflict of interest.

Appendix A. Hybrid Solar Systems

Renewable energy is environmentally friendly; therefore, it is the most viable future option to dispose of some of the environmental problems. Solar is one of the best renewable energy sources, due to availability and high temperature, wherefore solar thermal power, or concentrated solar power (CSP), is an optimal technology to hybridize with other energy technologies for power generation. The hybrid systems using geothermal and solar energy is one of the options with good potentials. This hybrid technology has a promising future, especially in countries that have abundant solar and geothermal energy. The stand-alone geothermal energy can be classified as medium- or low-enthalpy heat source for power plants, therefore, adding solar energy is the best choice, due to increasing the temperature of a geofluid, which increases the first-law efficiency to convert heat to work (and then to electricity) [15,16,31].

In general, a hybrid system can be categorized based on the kind of used energy, such as merging the solar source with one of the renewable energy sources like geothermal, wind, and biomass (these are the so-called high-renewable hybrids); while merging it with natural gas is called medium-renewable hybrids. Finally, merging it with traditionally fueled Brayton or/and Rankine cycle, the systems are called as a low-renewable hybrid.

Reducing CO_2 emission is a crucial factor for the power plant. The lowest CO_2 emission is with high-renewable hybrids, and the rate increases with medium-renewable, and it is at its highest rate with the low-renewable hybrids; therefore, the increase of the solar part reduces the CO_2 emissions [32]. The efficiency of hybrid systems can be increased by adding thermal energy storage that uses the surplus of solar thermal energy at the night time when the temperature is low. Furthermore, the increase of the solar collector field increases the efficiency in all case with or without using thermal energy storage [33]. The second-law efficiency for the hybrid geothermal-solar system is higher than for separate geothermal and solar system at all ambient temperatures [34]. Concerning the increase of maximal temperature above the critical point, the hybrid supercritical ORC (Organic Rankine cycle) outperforms the hybrid subcritical ORC thermodynamically and economically. Therefore, the hybrid supercritical ORC has higher thermal efficiency, and the ability to produce power two to nine times more than a stand-alone subcritical geothermal plant. The geothermal-solar plant is characterized by produce relative constant power during its period operation, concerned to stand-alone solar ones [18,35].

Concerning working fluids, some preliminary result was achieved, showing that that isobutene (also called isobutylene) is a suitable working fluid in the hybrid geothermal-solar system; it has the best performance with the lower GWP, when compared to other working fluids as isobutene, *n*-butane, and 2-butene [36]. The hybrid solar-geothermal is less vulnerable to the seasonal changes in ambient temperatures than stand-alone solar one, as it was observed in an Australian geothermal-solar hybrid system case study, demonstrating that a properly designed hybrid plant can outperform a stand-alone solar thermal plant in terms of the cost of electricity production [37].

As a final issue, the thermal stability of the working fluid and the capacity of the system components might limit the solar superheating in the hybrid geothermal-solar so, to overcome this problem, water can be used as the heat transfer fluid for ORC that operates under 175 °C [38].

References

1. Struchtrup, H. *Thermodynamics and Energy Conversion*; Springer: Berlin/Heidelberg, Germany, 2016.
2. Nag, P.K. *Power Plant Engineering*, 3rd ed.; Tata McGraw-Hill: New Delhi, India, 2008.
3. Macchi, E.; Astolfi, M. *Organic Rankine Cycle (ORC) Power Systems: Technologies and Applications*; Elsevier-Woodhead Publishing: Duxford, UK, 2016.
4. Quoilin, S.; Van Den Broek, M.; Declaye, S.; Dewallef, P.; Lemort, V. Technoeconomic survey of Organic Rankine Cycle (ORC) systems. *Renew. Sustain. Energy Rev.* **2013**, *22*, 168–186. [CrossRef]
5. Morrison, G. The shape of the temperature-entropy saturation boundary. *Int. J. Refrig.* **1994**, *17*, 494. [CrossRef]
6. Chen, H.; Goswami, D.Y.; Stefanakos, E.K. A review of thermodynamic cycles and working fluids for the conversion of low-grade heat. *Renew. Sustain. Energy Rev.* **2010**, *14*, 3059–3067. [CrossRef]
7. Garrido, J.M.; Quinteros-Lama, H.; Mejía, A.; Wisniak, J.; Segura, H. A rigorous approach for predicting the slope and curvature of the temperature-entropy saturation boundary of pure fluids. *Energy* **2012**, *45*, 888–899. [CrossRef]
8. White, J.A.; Velasco, S.A. Simple Semiempirical Method for Predicting the Temperature-Entropy Saturation Curve of Pure Fluids. *Ind. Eng. Chem. Res.* **2019**, *58*, 1038–1043. [CrossRef]
9. Bao, J.; Zhao, L. A review of working fluid and expander selections for Organic Rankine Cycle. *Renew. Sustain. Energy Rev.* **2013**, *24*, 325–342. [CrossRef]
10. Kolasiński, P. The Method of the Working Fluid Selection for Organic Rankine Cycle (ORC) Systems Employing Volumetric Expanders. *Energies* **2020**, *13*, 573. [CrossRef]
11. Groniewsky, A.; Wagner, C. Investigation of the Effect of the Regenerative Heat Exchanger on the Performance of Organic Rankine Cycles Using Perturbed Chain–Statistical Associating Fluid Theory Equation of State. *Ind. Eng. Chem. Res.* **2020**, *59*, 19643–19656. [CrossRef]
12. Groniewsky, A.; Györke, G.; Imre, A.R. Description of wet-to-dry transition in model ORC working fluids. *Appl. Therm. Eng.* **2017**, *125*, 963–971. [CrossRef]
13. Tóth, A.N. *The Geothermal Atlas of Hungary*, 1st ed.; Hungarian Energy and Public Utility Regulatory Authority: Budapest, Hungary, 2016.
14. Maali, R.; Khir, T. Performance analysis of different ORC power plant configurations using solar and geothermal heat sources. *Int. J. Green Energy* **2020**, *17*, 349–362. [CrossRef]
15. Mohammadi, K.; Khanmohammadi, S.; Khorasanizadeh, H.; Powell, K. A comprehensive review of solar only and hybrid solar driven multigeneration systems: Classifications, benefits, design and prospective. *Appl. Energy* **2020**, *268*, 114940. [CrossRef]
16. Li, K.; Liu, C.; Jiang, S.; Chen, Y. Review on hybrid geothermal and solar power systems. *J. Clean. Prod.* **2020**, *250*, 119481. [CrossRef]
17. Heberle, F.; Hofer, M.; Brüggemann, D. A Retrofit for Geothermal Organic Rankine Cycles based on Concentrated Solar Thermal Systems. *Energy Procedia* **2018**, *129*, 692–699. [CrossRef]

18. Zhou, C. Hybridization of solar and geothermal energy in both subcritical and supercritical Organic Rankine Cycles. *Energy Convers. Manag.* **2014**, *81*, 72–82. [CrossRef]
19. Hung, T.-C. Waste heat recovery of organic Rankine cycle using dry fluids. *Energy Convers. Manag.* **2001**, *42*, 539–553. [CrossRef]
20. Sadaghiani, M.S.; Ahmadi, M.H.; Mehrpooya, M.; Pourfayaz, P.; Feidt, M. Process development and thermodynamic analysis of a novel power generation plant driven by geothermal energy with liquefied natural gas as its heat sink. *Appl. Therm. Eng.* **2018**, *133*, 645–658. [CrossRef]
21. Imre, A.R.; Kustán, R.; Groniewsky, A. Mapping of the Temperature–Entropy Diagrams of van der Waals Fluids. *Energies* **2020**, *13*, 1519. [CrossRef]
22. Daniarta, S.; Imre, A.R. Cold Energy Utilization in LNG Regasification System Using Organic Rankine Cycle and Trilateral Flash Cycle. *Period. Polytech. Mech. Eng.* **2020**, *64*, 342–349. [CrossRef]
23. Györke, G.; Deiters, U.K.; Groniewsky, A.; Lassu, I.; Imre, A.R. Novel Classification of Pure Working Fluids for Organic Rankine Cycle. *Energy* **2018**, *145*, 288–300. [CrossRef]
24. Imre, A.R.; Kustán, R.; Groniewsky, A. Thermodynamic Selection of the Optimal Working Fluid for Organic Rankine Cycles. *Energies* **2019**, *12*, 2028. [CrossRef]
25. NIST Chemistry WebBook. NIST Standard Reference Database Number 69. 2018. Available online: http://webbook.nist.gov/chemistry/ (accessed on 1 February 2020).
26. Kondor, L. Efficiency-increase of low-temperature. Organic Rankine Cycles with Solar Collectors—Thermodynamic and Economic Study. Master's Thesis, Budapest University of Technology and Economics, Budapest, Hungary, June 2020.
27. Kraska, T.; Deiters, U.K. An Equation of State for Pure Fluids Describing the Critical Region. *Int. J. Thermophys.* **1994**, *15*, 261–281. [CrossRef]
28. Dua, G.; Hu, J. An equation of state for accurate thermodynamic modeling of water and carbon dioxide from triple points to 647 K and 100–200 MPa. *Int. J. Greenh. Gas Control* **2016**, *49*, 94–107. [CrossRef]
29. Zhang, X.; Zhang, C.; He, M.; Wang, J. Selection and Evaluation of Dry and Isentropic Organic Working Fluids Used in Organic Rankine Cycle Based on the Turning Point on Their Saturated Vapor Curves. *J. Therm. Sci.* **2019**, *28*, 643–658. [CrossRef]
30. Zhang, X.; Zhang, Y.; Wang, J. New classification of dry and isentropic working fluids and a method used to determine their optimal or worst condensation temperature used in Organic Rankine Cycle. *Energy* **2020**, *201*, 117722. [CrossRef]
31. Powell, K.M.; Rashid, K.; Ellingwood, K.; Tuttle, J.; Iverson, B.D. Hybrid concentrated solar thermal power systems: A review. *Renew. Sustain. Energy Rev.* **2017**, *80*, 215–237. [CrossRef]
32. Pramanik, S.; Ravikrishna, R.V. A review of concentrated solar power hybrid technologies. *Appl. Therm. Eng.* **2017**, *127*, 602–637. [CrossRef]
33. Bassetti, M.C.; Consoli, D.; Manente, G.; Lazzaretto, A. Design and off-design models of a hybrid geothermal-solar power plant enhanced by a thermal storage. *Renew. Energy* **2018**, *128*, 460–472. [CrossRef]
34. Ghasemi, H.; Sheu, E.; Tizzanini, A.; Paci, M.; Mitsos, A. Hybrid solar—Geothermal power generation: Optimal retrofitting. *Appl. Energy* **2014**, *131*, 158–170. [CrossRef]
35. Bokelman, B.; Michaelides, E.; Michaelides, D. A Geothermal-Solar Hybrid Power Plant with Thermal Energy Storage. *Energies* **2020**, *13*, 1018. [CrossRef]
36. Keshvarparast, A.; Ajarostaghi, S.M.A.; Delavar, M.A. Thermodynamic analysis the performance of hybrid solar-geothermal power plant equipped with air-cooled condenser. *Appl. Therm. Eng.* **2020**, *172*, 115160. [CrossRef]
37. Zhou, C.; Doroodchi, E.; Moghtaderi, B. An in-depth assessment of hybrid solar—Geothermal power generation. *Energy Convers. Manag.* **2013**, *74*, 88–101. [CrossRef]
38. Heberle, F.; Hofer, M.; Ürlings, N.; Schr, H.; Anderlohr, T.; Brüggemann, D. Techno-economic analysis of a solar thermal retrofit for an air-cooled geothermal Organic Rankine Cycle power plant. *Renew. Energy* **2017**, *113*, 494–502. [CrossRef]

energies

MDPI

Article

Impact of Users' Behavior and Real Weather Conditions on the Energy Consumption of Tenement Houses in Wroclaw, Poland: Energy Performance Gap Simulation Based on a Model Calibrated by Field Measurements

Małgorzata Szulgowska-Zgrzywa [1,*], Ewelina Stefanowicz [1], Krzysztof Piechurski [1], Agnieszka Chmielewska [1] and Marek Kowalczyk [2]

1 Department of Air-Conditioning, Heating, Gas Engineering and Air Protection, Wroclaw University of Science and Technology, Norwida St. 4/6, 50-373 Wroclaw, Poland; ewelina.stefanowicz@pwr.edu.pl (E.S.); krzysztof.piechurski@pwr.edu.pl (K.P.); agnieszka.chmielewska@pwr.edu.pl (A.C.)

2 Department of Climatology and Atmosphere Protection, University of Wroclaw, Kosiby St. 8, 51-621 Wroclaw, Poland; marek.kowalczyk@uwr.edu.pl

* Correspondence: malgorzata.szulgowska@pwr.edu.pl

Received: 6 November 2020; Accepted: 14 December 2020; Published: 18 December 2020

Abstract: This paper presents the results of measuring the final energy consumption for heating and domestic hot water (DHW) preparation and indoor conditions in 15 apartments located in pre-war tenement houses. The measurements were compared to the computed energy consumption. The calculations ware made based on the model calibrated by field measurements. The discrepancies between measurements and calculations were assessed using the energy performance gap (EPG). Calculations were made separately for energy for heating and for DHW preparation. Additionally, the results of EPG calculations for different levels of analysis are presented aiming at assessing the impact of weather, temperature in the surrounding zones and users' behavior. Users' behaviors influencing the size of the EPG were divided into typical (energy saving or excessive energy consumption) and forced (energy poverty, response to the apartment's surroundings, technical limitations. The connection between the heating sources and the heating habits has been clearly observed in the research. The former (typical) behaviors were the origin of the energy gap in the apartments heated with natural gas and district heating. The latter (forced) were the origin of the gap in the apartments heated with mostly electricity and solid fuel (with one exception: one apartment that utilized the district heating).

Keywords: energy performance gap; user behavior; energy poverty; pre-war tenement building

---

1. Introduction

The implementation of zero-emission plans in Europe poses serious challenges for EU countries. One is that new buildings must be designed to be near zero energy [1]. However, it seems that modernizing existing facilities is even more difficult. In this case, pre-war tenement buildings with the highest energy demand appear to be the greatest challenge. In these buildings, it is necessary to change the heating systems to low-emission ones, and first to reduce their energy consumption. Each EU country faces different challenges in this area.

In Poland, an additional problem with pre-war buildings is the significant number of flats still equipped with solid-fuel heating systems. Research carried out in Wroclaw, Poland, in 2019 showed a high concentration of such heat sources in buildings built before 1945, i.e., downtown tenement houses. It turned out that as many as 64% of the apartments heated with solid fuel in the city (12,000 out of 18,700) are included in this group of buildings [2]. Carbon-based heat sources are the direct cause of emissions of substances such as particulate matter (PM2.5, PM10), benzo(a)pyrene (BaP), nitrogen oxides, sulfur dioxide, and carbon monoxide. The emission level is also directly related to the energy consumption of buildings. Estimating this energy provides a basis for building models related not only to air pollution, but also to energy modeling of municipal buildings (UBEM) [3]. Actions should be taken on many levels and interrelated goals should be defined: improving air quality in cities, reducing the energy consumption of buildings in urban agglomerations, and improving the quality of life of their inhabitants.

To conduct large-scale activities, it is necessary to properly identify the housing structure, which will allow for long-term planned research and investments. The current research directions are aimed at developing a methodology for modeling the energy demand of buildings at various scales. The micro scale includes individual premises and buildings, the meso scale determines the energy consumption of districts or building quarters, and the macro scale applies to entire cities [4]. The basic input data for modeling are meteorological data, 3D models of cities, and nongeometric properties of buildings. Such modeling also requires the prioritization of housing stock into building archetypes. Their construction is based mainly on the age, shape, method of use, and heating systems used. The process of creating archetypes is a crucial part of building models and making them reliable. The main problem indicated by the researchers is the fact that UBEM designers do not have access to the measured final energy consumption in premises or buildings. Limited access to such data and generally insufficient knowledge about the thermal properties of buildings in different age categories often make it impossible to simulate uncertainty and reduce the inaccuracies in model construction [3]. Therefore, it is necessary to provide data resulting from the calculation of energy demand, as well as on actual energy consumption in selected well-recognized buildings. There is a need to develop a way to consider user behavior that significantly affects the real energy consumption of buildings and apartments. In addition, researchers point out the need for a detailed approach to the problem of predicting energy demand profiles of building resources with detailed time resolution. These profiles should take into account geographic location, user behavior, and the effectiveness of various energy systems and technologies in cities [5].

To conclude, the literature lacks information supporting the determination of energy demand considering all of the above conditions on a micro scale, especially in older buildings. Currently, methods for energy calculation and saving mainly focus on new buildings and algorithmic determination of energy consumption, hence they often fail for existing buildings [6]. The share of historic buildings in the centers of European cities is large, which means that accurate information on their energy consumption is critical in the process of modeling and planning changes aimed at local improvement of air quality, but also to affect progressive climate change. Researchers who have access to data on the actual energy consumption of older buildings emphasize that they may use less energy than expected, pointing to the fact that the greater energy demand of these buildings leads to more conscious use of the systems [6]. Collecting as much data as possible on this topic will allow researchers to construct more accurate energy models for cities.

The actual energy consumption of a building often differs significantly from the computed value, even if it is obtained using advanced complex software for dynamic simulated energy performance. This phenomenon is commonly known and described in the literature as the energy performance gap (EPG) [7–9]. Many authors use this indicator to assess the energy situation in relation to real measurements for residential, nonresidential, and single-family buildings [10]. The energy performance gap is defined depending on the situation. For example, de Wilde [9] defines three types of gaps: between initial predictions and measurements, between machine learning and measurements, and between predictions and display certificates in legislation. The article, however, is mostly related to the design of new buildings. In the literature, this coefficient also appears in studies that describe the effects of the thermal modernization process of buildings. For example, in [11], EPG is defined as the difference between actual and design consumption as a fraction of the design consumption. Interesting research is presented in the article [8]. The size of the EPG was determined for different refurbishment solution. The scientists analyzed the impact of the occupants' interviews and surveys, and the consequential feedback to the occupants about the correct behaviors and use of the heating system. The results show that, for the buildings in which occupants learned how to optimally use the system, EPG was equal or smaller than 0% and occupant behaviors has been identified as one of the causes of the gap.

In the context of modernized buildings, it is worth noting that the literature describes a certain tendency of users to increase their needs after higher technological standards are achieved in the building, the so-called rebound effect. This effect is also known as the Jevons paradox [12]. It was found that the assumed economic use of energy in such cases is not always confirmed, and the truth is often the opposite. The rebound effect is described in two forms: indirect and direct. The direct rebound effect is described as increased efficiency and the associated reduced cost of a product or service, resulting in its increased consumption. The indirect rebound effect describes the situation where savings from reduced efficiency cost enable spending more income on other products and services [13]. A complementary concept is described in [14], in which the authors introduced the term "prebound effect". This effect measures the discrepancy between the measured and calculated energy requirements of existing buildings that have not been thermomodernized. Researchers have observed that these buildings consume less energy than calculated using the methods in energy certification. A higher prebound effect in combination with low income may be a sign of energy poverty. The authors emphasize that further research is needed to understand motivations and practices in households that show a high prebound effect.

The actual level of final energy demand of a building is influenced by users' behaviors, which is related to the tenants' individual characteristics, such as their age and habits [15–19]. For example, the heating schedule for the residences of retirees will differ significantly from those of young working people or families with children. Since the former stay in their apartments more and usually have higher temperature requirements, a correspondingly longer heating time is needed. Kashif et al. [20] stated that both time and environmental factors create a certain context, which means that residents must perform certain activities to adapt their surroundings to their needs. As inhabitants have different approaches to the balance between indoor comfort and energy consumption [21–25], they behave differently despite being in the same environment. It can be concluded that the existing studies confirm the discrepancies between expected and observed energy parameters of buildings. The energy performance gap is mainly due to differences between the assumptions for engineering calculations and reality, including, in a significant way, the behaviors of users: the savings effect or the effect of increased demand. This paper enriches this topic by providing information on the size of the difference between real and computed energy consumption in pre-war tenement houses, indicating the reasons. As it turns out, users' behaviors are very closely related to the characteristics of heating systems in buildings and reflect more problems than those mentioned in the literature so far.

2. Materials and Methods

2.1. Aim of the Research

The aim of the research was to measure the energy gap between the real and computed energy consumption of residential premises located in pre-war tenement houses and to indicate possible reasons for its occurrence. The calculations were made separately for energy for space heating and for the preparation of domestic hot water (DHW). The research was extended with a detailed analysis of the actual situation in apartments and their surroundings. Not only was energy consumption measured in the examined premises, but also the parameters of indoor and outdoor air and other factors determining the energy consumption of apartments. The collected data allowed for model development using field measurements. The model was then used to simulate energy consumption under various conditions. This in-depth analysis was aimed at assessing the impact of users' behaviors and real weather conditions on the energy consumption of tenement houses in Wroclaw.

2.2. Subject of Study

Measurements were carried out in 15 apartments (labeled A1–A15) located in pre-war tenement houses in the selected quarter of the city of Wroclaw (Poland) from 14 January to 9 March 2020. The technical standards of the facilities are varied: there are buildings that require extensive renovation (very bad condition), buildings in slightly better condition (bad condition), neglected buildings often found in Wroclaw (average condition), and those in which partial or full thermal modernization has been carried out (good condition). The facades of exemplary buildings, showing their typical technical conditions, can be seen in Figure 1. All apartments are in the same type of building (built before 1920), but they were selected due to differences on the micro scale. They differ in size, location in the building, heat source, and the tenants' situation and lifestyle. The premises were selected for research based on a pilot study conducted in 2018 [26] covering 410 residential premises. Basic data on the examined premises are presented in Table 1. Additionally, the issues important for the test results should be emphasized: there was no bathroom in apartment A9 and no DHW preparation system in apartment A7.

(a) (b) (c) (d)

Figure 1. Four specified technical conditions of tenement houses in Wroclaw (according to Table 1 and described in the text): (**a**) good; (**b**) average; (**c**) bad; (**d**) very bad.

Table 1. Basic information about the apartments.

Apartment Number	Flat Area (m^2)	Heating Source	DHW Production Source	Location/Orientation of the Window	Technical Condition of Building/U_{avg} Value [W/m^2K]	Tenants
A1	44.4	SF + EE	EE	middle floor; corner/NW	very bad/0.98	couple
A2	41.5	SF	EE	middle floor; corner/SW	very bad/1.02	retired lady
A3	40.7	SF	EE	attic; corner/SE	bad/0.97	retired lady
A4	47.4	SF	EE	middle floor; one-sided/SW	very bad/0.99	mother and daughter
A5	34.5	EE	EE	middle floor; corner/SW	very bad/0.93	elderly couple
A6	54.5	EE	EE	attic; corner/NE	very bad/0.88	mother and son
A7	62.9	EE	EE	attic; one-sided/NE	bad/1.00	single woman
A8	44.1	DH	DH	ground floor; one-sided/SW	average/0.85	couple
A9	76.1	DH	DH	middle floor; two-sided/SE	good/1.20	couple + single
A10	59.8	DH + EE	DH	ground floor; corner/NE, E, SE	good/0.68	family (4)
A11	85.5	DH	DH	mid. floor; two-sided/SW, NE	average/1.00	family (3)
A12	52.5	NG	NG	mid. floor; two-sided/SW, NE	good/0.75	young couple

Table 1. *Cont.*

Apartment Number	Flat Area (m^2)	Heating Source	DHW Production Source	Location/Orientation of the Window	Technical Condition of Building/U_{avg} Value [W/m^2K]	Tenants
A13	85.0	NG	NG	mid. floor; two-sided/SW, NE	average/1.03	multigenerational family
A14	55.9	NG	NG	middle floor; corner/SW	good/0.74	family (4)
A15	128.5	NG	NG	mid. floor; two-sided/SE, NW	good/0.67	multigenerational family

SF, solid fuel; EE, electrical energy; DH, district heating; NG, natural gas.

2.3. Measurements

Measurements were carried out using the devices listed in Table 2. Air temperature, relative humidity, and carbon dioxide concentration in the rooms were recorded. Temperature and consumption of tap water and DHW (if any) were measured in each apartment. Detailed information on electricity consumption in the apartments was also obtained. Thermal energy consumption measurements varied depending on the heat source. In the case of premises equipped with district heating, readings were obtained from individual energy meters. In the case of flats with heat sources powered by natural gas, gas flow was measured, if possible, or consumption was noted manually at daily intervals.

Table 2. Measuring Devices.

Measuring Device	Producer	Type	Measuring Accuracy
Energy data logger	Voltcraft	4000	1%
Gas volume meter	Plum	MacR6	according to meter type 1 impulse = 0.01 m^3
Two-channel temperature recorder	Testo	175-T3	±0.5 °C (−50 to +70 °C)
Clamp-on water temperature probe	Testo	K-type	class 1
One-channel temperature recorder	Testo	176-T1	±0.3 °C (−100 to +70 °C)
Two-channel temperature and humidity recorder	Testo	174H	±0.5 °C (−20 to +70 °C); ±3% RH (2 to 98% RH) at +25 °C ±0.03% RH/K
Two-channel temperature recorder with surface temperature probes	Testo	175-T2 NTC	±0.5% (−35 to +55 °C); ext. sens.: ±0.2 °C (−25 to +80 °C)
Hook scale	Steinberg	-	2–300 kg; class 3
Temperature, humidity, and CO_2 recorder	Comet System	Test-Therm	±1.8% RH; ±1.5 °C; ± (50 ppm + 2% measured value) for 25 °C and 1013 hPa
Humidity and temperature probes	Vaisala	HMP45	±0.2 °C (at +20 °C)
Pyranometer	Kipp and Zonen	CMP3	<5% (−10 to +40 °C)

RH, relative humidity.

In the case of apartments heated by electric heaters, each heat source was connected to an individual electricity meter. In flats heated with solid fuel stoves, the weight of fuel used for heating was registered and the temperature of the tiled stove surface was measured. Additionally, thermovision studies were carried out to better investigate these heat sources, and the values of ambient temperature and insolation were measured.

2.4. Computational Models

For each apartment, a calculation model of final energy demand for heating and DHW preparation was developed, based on the guidelines of the standard [27] and Polish regulations [28]. The model development was preceded by a detailed site inspection. During the field visits, the area of the premises, the level of thermal modernization of the building, the construction of building partitions, and the technical condition of windows and doors were determined. Weather conditions and conditions of use of the apartments were assumed according to the guidelines for engineering calculations [28]. Assessments made in strict accordance with these recommendations are described as Simulation 3.

To increase the accuracy of the calculations, the data of weather observed during the tests were used. It should be noted that in the analyzed period, the average ambient temperature was higher than in previous years. Figure 2 compares these values against the average values from the last 10 heating seasons in Wroclaw and the average temperature from the 30-year period, which are used in the calculations for energy certification of buildings [29].

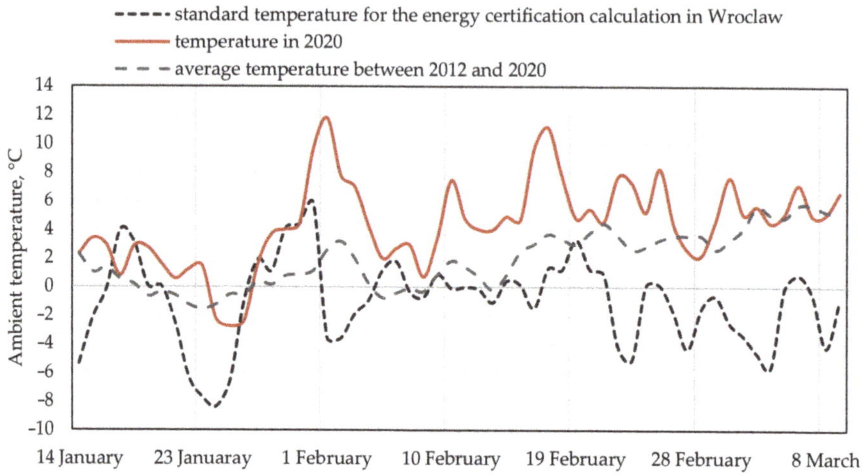

Figure 2. Average daily ambient temperatures in Wroclaw in the selected period.

The average ambient temperature during the study period was 4.4 °C, while the average for the last 10 years in the same period was 1.9 °C. An even lower value of −1.0 °C is reached with the data used in calculating audits and energy certification. Therefore, there is as much as a 5.4 °C difference in relation to the period of research. It should be emphasized that the period of the last 10 years is characterized by much milder temperatures in the heating season than in previous years. The average ambient temperature in the last 10 years in Wroclaw was 1.72 °C higher than for the same period in 2000–2009 [30] and 3.15 °C higher according to the climatic data prepared for energy audit and certification [29]. As a result, the computational energy demand for heating is overstated in relation to the real values. The scale of these discrepancies is shown in the energy gap values described in this paper.

The calibration process was based on the recognition of users' habits. In order to determine actual internal heat gains, data on the time for basic housework, home furnishings, and the use schedule of the premises were obtained. The level of ventilation was estimated based on measuring the concentration of carbon dioxide and obtaining information from residents on the organization of air exchange in the rooms, i.e., the presence of ventilation grilles and trickle vents. In the calibration process, the internal air temperature was measured in every room of the apartments, as well as in staircases, basements, and attics. As a result, the actual temperatures of the surrounding spaces were determined. For calibration of the energy consumption calculation model for DHW production, actual measurements of domestic hot water consumption were used. The most important data used in the calibration process are shown in Table 3. The calibration process was performed until the simulation result was equal to the real final energy use (q_H, q_W), also described in Table 3.

Table 3. Results of the measurements: conditions during measurements and standard of use of apartments.

Parameter	A1	A2	A3	A4	A5	A6	A7	A8	A9	A10	A11	A12	A13	A14	A15
T_{int}, °C	20.3	16.7	17.6	19.7	16.6	18.5	19.4	18.7	21.1	20.8	19.1	21.9	20.4	21.9	21.5
V_{DHW}, L/d pp	60.7	41.1	48.4	83.0	0.5	60.7	0.0	14.6	8.3	26.8	50.6	53.6	24.6	39.2	51.8
T_{DHW}, °C	49.6	46.7	44.9	39.1	39.5	50.4	-	41.8	44.3	48.0	48.8	51.9	55.8	52.0	52.0
CO_2, ppm	848	829	1860	2452	2931	1374	774	2413	1927	976	992	1025	885	855	993
q_{int}, W/m^2	3.4	2.7	2.7	3.0	5.2	2.8	1.7	4.3	3.5	4.8	2.6	3.4	3.8	5.5	2.3
q_H, kWh/m^2	62.3	35.7	26.9	20.2	10.4	15.0	17.3	19.7	14.3	25.2	22.2	27.2	30.4	25.9	24.6
q_W, kWh/m^2	6.3	2.4	2.5	5.8	0.0	2.8	0.0	2.4	1.3	7.5	7.5	5.8	2.8	6.2	6.1

Internal temperature is a very important factor that affects energy consumption for heating in residential premises. In the case of the tested apartments, a correlation between the average internal temperature and the type of heating system used was observed, as shown in Figure 3. In flats supplied from central heat sources (A8–A11) and individual gas boilers (A12–A15), this profile was correct and no significant differences in temperature were observed in different rooms. The temperature of the examined apartments oscillated between average and maximum values. The inhabitants were able to maintain thermal comfort in the rooms without much effort. Indoor temperature is not only based on thermal comfort needs, but is also an indicator of problems related to energy poverty or the inability to provide adequate temperature [31]. In some of the apartments, the average interior temperature during the research period was about 16 °C, and the lowest recorded in the bathroom was 13 °C. This is mainly due to the heating characteristics of solid fuel stoves (A1–A4) and the most expensive electric-powered systems in Poland (A5–A7). In apartments heated with systems powered by solid fuel or electricity, the interior temperature profile is not correct. Significant differences in the measured air temperature in different rooms in the apartment were observed, and temperature changes oscillated between average and minimum observed values.

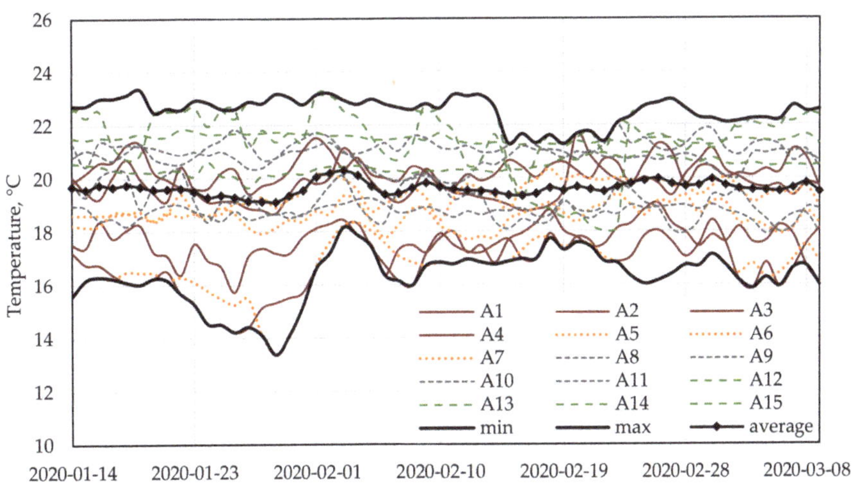

Figure 3. Average indoor air temperatures in apartments during the tests: 14 January to 9 March 2020.

2.5. Simulations

The final energy consumption simulation for space heating and DHW preparation in apartments was carried out for various variants of the use of premises (simulations 0–4), as follows:

- Simulation 0: The results show the actual final energy consumption for space heating and DHW preparation during the research period (14 January to 9 March 2020). The results of this simulation correspond to the apartment use conditions (see Table 3) and weather conditions observed during the tests.

- Simulation 1: The results show the final energy consumption for space heating and DHW preparation in the research period for the weather conditions observed during the tests, but for the proper use of the apartment. The proper use is defined by interior temperature, ventilation level and DHW consumption as described below the interior temperature in the apartments was changed to comply with the requirements of Polish regulations [32], i.e., 20 °C in residential rooms and 24 °C in bathrooms. The ventilation level was assumed to meet the hygienic minimum [33], which should allow the CO_2 concentration to be maintained at the appropriate level. DHW consumption was assumed at the level of 40 L/day per person (which corresponds to the average DHW consumption in all analyzed premises).
- Simulation 2: The results show the final energy consumption for space heating and DHW preparation in the research period for proper use of the apartment, as described in simulation 1, but for weather conditions corresponding to the data used in engineering calculations [28].
- Simulation 3: The results show the final energy consumption for space heating and DHW preparation in the research period for the weather conditions corresponding to the data used in engineering calculations [28] and for the use of the premises described as standard in engineering calculations of energy demand [32].
- Simulation 4: The results show the final energy consumption for space heating and DHW preparation in the research period for the weather conditions observed during the tests and for the use of the premises described as standard in engineering calculations of energy demand [32].

2.6. Methods of Energy Performance Gap Calculation

The energy performance gap is determined by the difference between the measured and calculated heating energy consumption in a given period. In this paper, the *EPG* indicator is defined at several levels.

$EPG_{h+w(0\to3)}$ in Equation (1) describes the gap between the measurement results (simulation 0) and engineering calculations (simulation 3) of total final energy consumption for space heating and DHW preparation ($h + w$). This indicator includes all factors that affect the size of the energy gap.

$$EPG_{h+w(0\to3)} = \frac{q_{h,0} + q_{w,0} - \left(q_{h,3} + q_{w,3}\right)}{q_{h,3} + q_{w,3}} \tag{1}$$

$EPG_{h(0\to3)}$ in Equation (2) describes the gap between the measurement results (simulation 0) and engineering calculations (simulation 3) of final energy consumption for space heating (h).

$$EPG_{h(0\to3)} = \frac{\left(q_{h,0} - q_{h,3}\right)}{q_{h,3}} \tag{2}$$

$EPG_{w(0\to3)}$ in Equation (3) describes the gap between the measurement results (simulation 0) and engineering calculations (simulation 3) of final energy consumption for DHW preparation (w).

$$EPG_{w(0\to3)} = \frac{\left(q_{w,0} - q_{w,3}\right)}{q_{w,3}} \tag{3}$$

$EPG_{h(1\to3)}$ in Equation (4) describes the gap between the results of simulation 1 and engineering calculations (simulation 3) of final energy consumption for space heating (h). This indicator determines the energy gap between engineering calculations and the simulation of the proper use of the apartment in the observed meteorological conditions.

$$EPG_{h(1\to3)} = \frac{\left(q_{h,1} - q_{h,3}\right)}{q_{h,3}} \tag{4}$$

$EPG_{w(1\to3)}$ in Equation (5) describes the gap between the results of simulation 1 and engineering calculations (simulation 3) of final energy consumption for DHW preparation (w). This indicator determines the energy gap between engineering calculations and the simulation of the proper domestic hot water consumption.

$$EPG_{w(1\to3)} = \frac{\left(q_{w,1} - q_{w,3}\right)}{q_{w,3}} \tag{5}$$

$EPG_{h(2\to3)}$ in Equation (6) describes the gap between the results of simulation 2 and engineering calculations (simulation 3) of final energy consumption for space heating (h). This indicator determines the energy gap between engineering calculations and the simulation of the proper use of the apartment for the meteorological conditions corresponding to the data used in engineering calculations.

$$EPG_{h(2\to3)} = \frac{\left(q_{h,2} - q_{h,3}\right)}{q_{h,3}} \tag{6}$$

$EPG_{h(4\to3)}$ in Equation (7) describes the gap between the results of simulation 4 and engineering calculations (simulation 3). This indicator can be understood as the energy gap resulting from the difference in the meteorological conditions used for the calculations.

$$EPG_{h(4\to3)} = \frac{\left(q_{h,4} - q_{h,3}\right)}{q_{h,3}} \tag{7}$$

$EPG_{w(2\to3)}$ and $EPG_{w(4\to3)}$ have not been defined, because, in these cases, the use of the DHW system (the consumption of domestic hot water) in apartments does not change, and the influence of the ambient temperature on energy consumption for DHW preparation is not included in the calculations.

3. Results and Discussion

3.1. Final Energy Use: Measurement vs. Calculation

The actual unit consumption of final energy for space heating and DHW preparation and the demand calculated according to the assumptions of simulation 3 (engineering calculations) are shown in Figure 4. The results for individual apartments are very different. These differences were expected due to the location of the apartment in the building and its insulation. For example, apartments A10, A12, A14, A15 are in a good technical condition and have lower than other average heat transfer coefficient (Table 1), so their unit energy demand should be the lowest. Despite this, the lowest values are observed in A5–A9, and the most significant differences between energy consumption compared to expectations are in A1–A10. The reasons for such unlike energy consumption compared to expectations are very low temperature of indoor air in the premises, insufficient ventilation, and/or the influence of the environment, i.e., the flow of thermal energy from zones adjacent to the apartment. Little insulation of interior walls causes intensive heat exchange between zones in the building. In the case of A9 these zones are exceptionally warm (for example the temperature in the staircase adjacent to apartment A9 exceeded 22 °C), so its unit energy demand is low. The opposite effect was observed in apartment A1, where the adjacent spaces are extremely underheated. This shows how important it is to analyze in detail the temperature of adjacent spaces, such as staircases, basements, and attics. These are important elements that influence the energy gap, which are discussed in detail later in the paper.

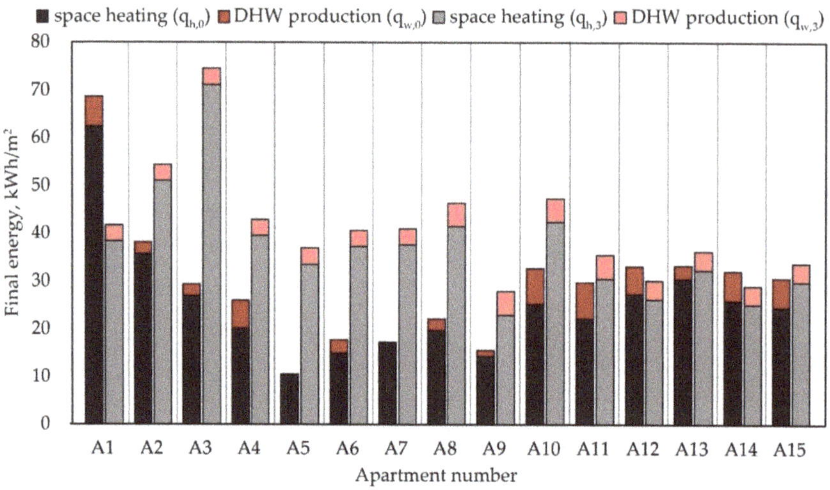

Figure 4. Final energy for space heating and domestic hot water (DHW) production: measurement vs. calculation.

3.2. Energy Performance Gap

The results of energy performance gap (EPG) calculations for different levels of analysis are presented in Table 4.

Table 4. Energy performance gap for different levels of analysis.

Parameter	A1	A2	A3	A4	A5	A6	A7	A8	A9	A10	A11	A12	A13	A14	A15
$EPG_{h+w(0\to3)}$	0.30	−0.44	−0.69	−0.52	−0.78	−0.65	−0.67	−0.62	−0.56	−0.45	−0.34	−0.13	−0.27	−0.13	−0.28
$EPG_{h(0\to3)}$	0.29	−0.44	−0.70	−0.60	−0.75	−0.68	−0.64	−0.62	−0.51	−0.53	−0.42	−0.18	−0.26	−0.19	−0.35
$EPG_{h(1\to3)}$	0.53	0.32	−0.18	−0.18	0.19	−0.29	−0.35	−0.17	−0.39	−0.52	−0.10	−0.16	−0.29	−0.28	−0.44
$EPG_{h(2\to3)}$	1.12	0.83	0.15	0.21	0.69	0.02	−0.04	0.20	−0.05	−0.28	0.29	0.21	0.06	0.12	−0.17
$EPG_{h(4\to3)}$	−0.34	−0.33	−0.31	−0.36	−0.33	−0.32	−0.35	−0.33	−0.39	−0.33	−0.35	−0.35	−0.36	−0.38	−0.34
$EPG_{w(0\to3)}$	0.47	−0.44	−0.42	0.36	−0.99	−0.34	−1.00	−0.62	−0.80	0.19	0.21	0.19	−0.42	0.27	0.24
$EPG_{w(1\to3)}$	0.25	−0.33	−0.32	0.17	0.61	−0.49	−0.56	0.26	0.10	0.86	−0.03	0.06	−0.02	0.99	0.08

The energy gap for space heating and DHW preparation between measurement and engineering calculation results ($EPG_{h+w(0\to3)}$) is significant and is shown in Figure 5. Apartments A1–A4 are heated with solid fuel. In this case, the energy gap is −0.55 on average (excluding A1, for which the EPG value is positive at 0.3). A5–A7 are heated with electricity and are characterized by an energy gap of −0.7 on average. A8–A11 are heated with district heat and are characterized by a slightly smaller energy gap of −0.49 on average. A11–A15, for which the energy gap is the smallest at −0.2 on average, are heated with natural gas. There are different reasons for the discrepancy between the energy gap calculated for space heating and DHW preparation. A detailed analysis of this problem was performed and described below.

In Table 4 an expected energy gap between measurement and engineering calculations, due to significant differences in ambient temperature (see Figure 2), is presented ($EPG_{h(4\to3)}$). These values range from −0.31 to −0.39. Comparing the results of $EPG_{h(4\to3)}$ and $EPG_{h(0\to3)}$, Table 4, by extending the analysis with the actual parameters of the use of the premises, most of the apartments (A2–A11) have a greater negative EPG, which means that they actually consume less energy than expected. Several apartments (A1, A12–A14) have higher values, i.e., they consume more energy than calculated.

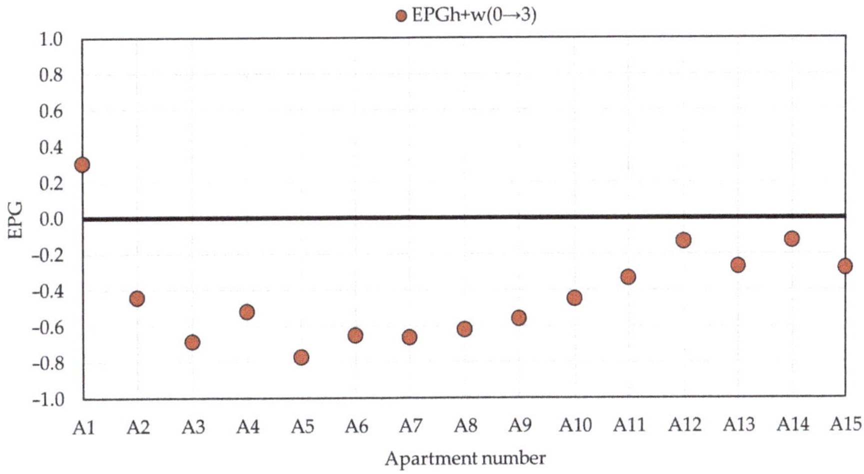

Figure 5. Energy performance gap for heating and DHW preparation.

To understand the causes of this condition, it is necessary to examine residents' behavior. The indoor temperature and the ventilation level in many apartments differ significantly from the norm. If they were correct, the energy performance gap would be smaller. This situation is marked and calculated as ($EPG_{h(1\rightarrow3)}$) and presented in Figure 6. It is worth noting that the difference in EPG values calculated for the real ($EPG_{h(0\rightarrow3)}$) and proper ($EPG_{h(1\rightarrow3)}$) operation is the highest for apartments A2–A8 and A11. These are heated with solid fuel and electricity, and two with district heat. The differences are significant and mostly cannot be classified as positively understood energy saving. In many apartments, the effect of energy poverty is noticeable, as manifested by extreme underheating and/or insufficient ventilation. In some premises heated with solid fuel (A2, A4), energy saving may be additionally forced by difficulty with operating the system, and not by energy poverty. In only one case, A11, the energy saving effect was not forced by the financial situation and technical problems.

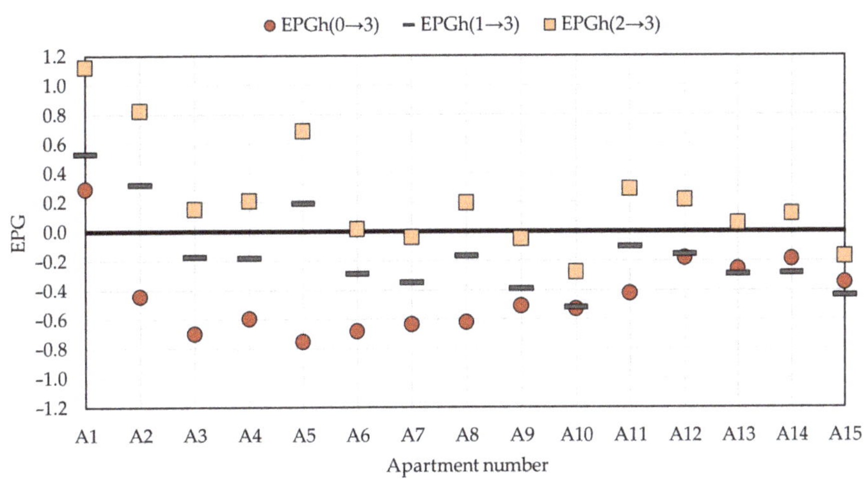

Figure 6. Energy performance gap for space heating.

Figure 6 also shows the value of the energy performance gap between the results of simulations 2 and 3 ($EPG_{h(2\rightarrow3)}$). Despite the fact that both simulations were performed with the same meteorological conditions, the EPG value in many cases does not approach 0. The obtained values are both strongly positive (1.12 for apartment A1) and negative (−0.28 for A10). These values illustrate the effects of imprecise assumptions regarding ventilation of rooms, internal heat gains, and temperatures in internal zones surrounding the apartment on the calculations. The latter especially contributes to significant differences between actual consumption, both positive and negative, and expected values, which is most noticeable in A1, A2, and A5, which are in buildings in very serious technical condition. Extremely underheated spaces and empty premises surrounding these flats increase their actual energy consumption.

The results of EPG calculations related to DHW preparation are shown in Figure 7. The discrepancies between measurement and calculation are significant (see $EPG_{w(0\rightarrow3)}$). As with space heating, in some cases they are not dictated by typical user behavior, but by energy poverty and technical limitations (e.g., lack of bathrooms). This is the case of apartments A5, A7, A8, and A9. The EPG values calculated for the situation where all inhabitants consume similar amounts of DHW (the average of the measurements) are also significant ($EPG_{w(1\rightarrow3)}$). This means that methods of estimating DHW consumption, at least in the case of Poland, require many studies, such as those shown in [34]. It is necessary to clarify the calculation methods because the current ones do not meet expectations. In pre-war tenement houses, due to their significant energy consumption related to space heating, this problem may not be very important, but in new buildings, the EPG related to DHW preparation systems may have a significant impact on the assessment of the entire facility.

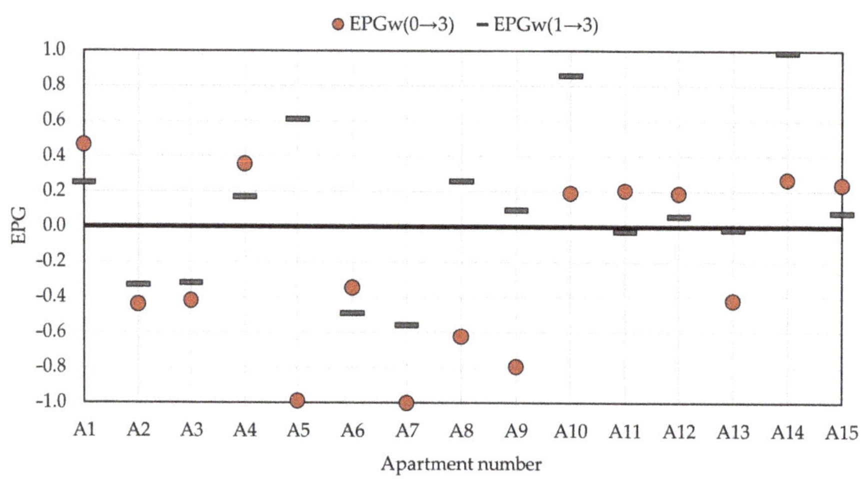

Figure 7. Energy performance gap for DHW production.

4. Conclusions

The conducted in situ research made it possible to identify user behavior and helped to calibrate energy models of apartments in pre-war tenement houses, that allowed to perform simulations helping to analyze the energy performance gap on several levels connected to the apartment's utilization and ambient temperature. On this basis, the energy performance gap for heating and DHW production has been analyzed. The calculated values turned out to be significant and negative, reaching even −0.78. Only in one considered case, the value obtained was positive and equaled to 0.3. The detailed analysis allowed us to refine crucial issues. The size of the energy gap is influenced not only by the ambient and internal temperature, but also by the temperature of surrounding spaces (for example: empty premises or stairways), which is different from the assumptions for the calculations. User behaviors that influence the size of the EPG can be divided into typical (energy saving or excessive energy consumption) and forced (energy poverty, response to the environment around the apartment, technical limitations).

The former (typical) behaviors were the origin of the energy gap in the apartments heated with natural gas and district heating (A9–A15). The latter (forced) were the origin of the gap in the apartments heated with mostly electricity and solid fuel (A1–A8). The only exception was the apartment A8, that utilized the district heating. The most important forced aspects observed during the research were as follows: not heating some rooms, providing additional heating with gas cookers, covering the ventilation grids, limiting ventilation of the rooms, and consuming very little domestic hot water. Correctly determining and classifying the energy performance gap between expected and actual energy consumption can be important for the proper planning and implementation of thermal modernization in pre-war buildings. It may allow for better determination of energy and cost savings by considering the specific nature of the use of such buildings. It is important, because the high energy consumption of premises in tenement houses and the related high costs are the common problems faced by the residents. The connection between the heating resources cost and the heating habits has been clearly observed in the research. Solid fuel is a cheap energy resource in Poland, and it is currently used by the poorest part of the society. Electricity is expensive and requires significant savings. Heating with district heat usually allows residents to maintain thermal comfort, while natural gas, the cheapest source of thermal energy, combined with the good condition of tenement houses allows residents to freely manage their resources. The research results show that a good energy standard of a building together with a modern and cost-acceptable heating source help to eliminate the forced energy gap. Further research on the impact of resident's behavior on the energy consumption and the size of the energy gap may be crucial for micro-scale modeling. Studying these behaviors requires an interdisciplinary approach, in particular a combination of engineering and social sciences. Energy behavior is not static; it changes with accumulated experience and financial and technical possibilities and is often inconsistent. These factors increase research difficulties and should be considered in any energy behavior analysis to minimize inaccuracies in assessing the situation.

Author Contributions: Conceptualization, M.S.-Z., E.S. and K.P; Methodology and Validation, M.S.-Z., E.S., K.P., A.C.; Formal Analysis and Investigation, M.S.-Z., E.S. and K.P; Resources and Data Curation, M.S.-Z., E.S., K.P., A.C. and M.K.; Writing—Original Draft Preparation, E.S., M.S.-Z.; Writing—Review & Editing, E.S., M.S.-Z.; Visualization, K.P.; Supervision, M.S.-Z.; All authors have read and agreed to the published version of the manuscript.

Funding: The project was implemented thanks to the "Divercity 4—Polish-Norwegian cooperation in the field of creating modern development solutions in cities" project with an EUR 300,000 grant from Iceland, Lichtenstein, and Norway through the EEA and Norway Grants. The initiative aims to identify the most effective solutions and create a cooperation network as well as exchange experiences between Polish self-governments and the Donor States, with particular emphasis on the subject of social participation in urban renewal and improvement of the urban environment. The funders had no role in the design of the study; in the collection, analyses, or interpretation of data; in the writing of the manuscript, or in the decision to publish the results.

Conflicts of Interest: The authors declare no conflict of interest.

Nomenclature

DHW	Domestic hot water
$A1-A15$	Apartment numbers
U_{avg}	Average heat transfer coefficient, W/m^{2}K
T_{int}	Internal temperature, °C
V_{DHW}	Domestic hot water consumption, L/d per person
T_{DHW}	Domestic hot water temperature, °C
q_{int}	Internal gains, W/m^2
EPG_h	Energy performance gap for space heating
EPG_w	Energy performance gap for domestic hot water preparation
EPG_{h+w}	Energy performance gap for space heating and domestic hot water production
q_h	Final energy for space heating, kWh
q_w	Final energy for domestic hot water preparation, kWh
Subscripts :	
$1-4$	Simulation numbers

References

1. Directive (EU) 2018/2002 of the European Parliament and of the Council of 11 December 2018 Amending Directive 2012/27/EU on Energy Efficiency. 2018. Available online: http://data.europa.eu/eli/dir/2018/2002/oj (accessed on 18 December 2020).
2. Baborska-Narożny, M.; Szulgowska-Zgrzywa, M.; Piechurski, K.; Stefanowicz, E.; Fidorów-Kaprawy, N.; Laska, M.; Machyńska, A.; Chmielewska, A.; Wójcik, F. Źródła Ciepła na Paliwo Stałe w Budynkach Mieszkalnych we Wrocławiu. 2019. Available online: https://bip.um.wroc.pl/attachments/download/80120. (accessed on 17 December 2020).
3. Reinhart, C.F.; Davila, C.C. Urban building energy modeling—A review of a nascent field. *Build. Environ.* **2016**, *97*, 196–202. [CrossRef]
4. Frayssinet, L.; Merlier, L.; Kuznik, F.; Hubert, J.-L.; Milliez, M.; Roux, J.-J. Modeling the heating and cooling energy demand of urban buildings at city scale. *Renew. Sustain. Energy Rev.* **2018**, *81*, 2318–2327. [CrossRef]
5. Kazas, G.; Fabrizio, E.; Perino, M. Energy demand profile generation with detailed time resolution at an urban district scale: A reference building approach and case study. *Appl. Energy* **2017**, *193*, 243–262. [CrossRef]
6. Aksoezen, M.; Daniel, M.; Hassler, U.; Kohler, N. Building age as an indicator for energy consumption. *Energy Build.* **2015**, *87*, 74–86. [CrossRef]
7. Zou, P.X.W.; Xu, X.; Sanjayan, J.; Wang, J. Review of 10 years research on building energy performance gap: Life-cycle and stakeholder perspectives. *Energy Build.* **2018**, *178*, 165–181. [CrossRef]
8. Calì, D.; Osterhage, T.; Streblow, R.; Müller, D. Energy performance gap in refurbished German dwellings: Lesson learned from a field test. *Energy Build.* **2016**, *127*, 1146–1158. [CrossRef]
9. De Wilde, P. The gap between predicted and measured energy performance of buildings: A framework for investigation. *Autom. Constr.* **2014**, *41*, 40–49. [CrossRef]
10. Kowalski, P.; Szałański, P. Computational and the real energy performance of a single-family residential building in Poland—An attempt to compare: A case study. *E3S Web Conf.* **2017**, *17*, 00045. [CrossRef]
11. Galvin, R. Making the "rebound effect" more useful for performance evaluation of thermal retrofits of existing homes: Defining the "energy savings deficit" and the "energy performance gap". *Energy Build.* **2014**, *69*, 515–524. [CrossRef]
12. Jevons, W.S. *The Coal Question: An Inquiry Concerning the Progress of the Nation, and the Probable Exhaustion of Our Coal-Mines*; Macmillan and Co.: London, UK; Cambridge, UK, 1865.
13. Maxwell, D.; Owen, P.; McAndrew, L.; Muehmel, K.; Neubauer, A. *Addressing the Rebound Effect, A Report for the European Commission DG Environment*; European Commission DG ENV: Ivry-sur-Seine, France, 2011.
14. Galvin, R.; Sunikka-Blank, M. Quantification of (p)rebound effects in retrofit policies—Why does it matter? *Energy* **2016**, *95*, 415–424. [CrossRef]
15. Pisello, A.L.; Piselli, C.; Cotana, F. Influence of human behavior on cool roof effect for summer cooling. *Build. Environ.* **2015**, *88*, 116–128. [CrossRef]
16. Al-Mumin, A.; Khattab, O.; Sridhar, G. Occupants' behavior and activity patterns influencing the energy consumption in the Kuwaiti residences. *Energy Build.* **2003**, *35*, 549–559. [CrossRef]
17. De Meester, T.; Marique, A.-F.; De Herde, A.; Reiter, S. Impacts of occupant behaviours on residential heating consumption for detached houses in a temperate climate in the northern part of Europe. *Energy Build.* **2013**, *57*, 313–323. [CrossRef]
18. Yu, Z.; Fung, B.C.M.; Haghighat, F.; Yoshino, H.; Morofsky, E. A systematic procedure to study the influence of occupant behavior on building energy consumption. *Energy Build.* **2011**, *43*, 1409–1417. [CrossRef]
19. Guerra Santin, O.; Itard, L.; Visscher, H. The effect of occupancy and building characteristics on energy use for space and water heating in Dutch residential stock. *Energy Build.* **2009**, *41*, 1223–1232. [CrossRef]
20. Kashif, A.; Ploix, S.; Dugdale, J.; Le, X.H.B. Simulating the dynamics of occupant behaviour for power management in residential buildings. *Energy Build.* **2013**, *56*, 85–93. [CrossRef]
21. De Groot, E.; Spiekman, M.; Opstelten, I. Dutch research into user behavior in relation to energy use of residences. In Proceedings of the PLEA 2008—25th Conference on Passive and Low Energy Architecture, Dublin, Ireland, 22–24 October 2008.
22. Menezes, A.C.; Cripps, A.; Bouchlaghem, D.; Buswell, R. Predicted vs. actual energy performance of non-domestic buildings: Using post-occupancy evaluation data to reduce the performance gap. *Appl. Energy* **2012**, *97*, 355–364. [CrossRef]

23. Dall'O', G.; Sarto, L.; Galante, A.; Pasetti, G. Comparison between predicted and actual energy performance for winter heating in high-performance residential buildings in the Lombardy region (Italy). *Energy Build.* **2012**, *47*, 247–253. [CrossRef]
24. Lopes, M.A.R.; Antunes, C.H.; Martins, N. Energy behaviours as promoters of energy efficiency: A 21st century review. *Renew. Sustain. Energy Rev.* **2012**, *16*, 4095–4104. [CrossRef]
25. Paone, A.; Bacher, J.-P. The Impact of Building Occupant Behavior on Energy Efficiency and Methods to Influence It: A Review of the State of the Art. *Energies* **2018**, *11*, 953. [CrossRef]
26. Baborska-Narożny, M.; Szulgowska-Zgrzywa, M.; Fidorów-Kaprawy, N.; Chmielewska, A.; Laska, M.; Stefanowicz, E.; Piechurski, K. Ciepło w Domu Gdy Zimno na Dworze. 2018. Available online: h8p://w-r.com.pl/projekty/cieplozimno/ (accessed on 20 October 2020).
27. International Organization for Standardization. *ISO 52016-1:2017, Energy Performance of Buildings—Energy Needs for Heating and Cooling, Internal Temperatures and Sensible and Latent Heat Loads—Part 1: Calculation Procedures*; ISO: Geneva, Switzerland, 2017. (In Polish)
28. Polish Minister of Infrastructure and Economic Development. Regulation of the Polish Minister of the Infrastructure and Economic Development of 21 February 2015 on the methodology for determining of building or part of the building and energy performance certificates. *J. Laws* **2015**, *376*. (In Polish)
29. Ministry of Investment and Development. Data for Buildings Energy Calculations. Available online: https://dane.gov.pl/pl/dataset/797,typowe-lata-meteorologiczne-i-statystyczne-dane-klimatyczne-dla-obszaru-polski-do-obliczen-energetycznych-budynkow (accessed on 22 October 2020).
30. Remund, J.; Müller, S.; Schmutz, M.; Gra, P. Meteonorm: Global Meteorogical Database for Solar Energy and Applied, 7-11/IX/2020, Version 8. In Proceedings of the 35th European PV Solar Energy Conference and Exhibition (EUPVSEC 2018), Brussels, Belgium, 24–28 September 2018.
31. Baborska-Narożny, M.; Szulgowska-Zgrzywa, M.; Mokrzecka, M.; Chmielewska, A.; Fidorow-Kaprawy, N.; Stefanowicz, E.; Piechurski, K.; Laska, M. Climate justice: Air quality and transitions from solid fuel heating. *Build. Cities* **2020**, *1*, 120–140. [CrossRef]
32. Polish Minister of Infrastructure and Economic Development. Regulation of the Polish Minister of the Infrastructure and Economic Development of 12 April 2002 on the Technical Conditions, Which Are to Be Met by Buildings and Their Location. *J. Laws* **2002**. (In Polish)
33. Polish Normalization Committee. PN-83/B-03430/Az3:2000 Ventilation in Residential Buildings of Collective Residence and Public Utility Buildings—Requirements. 2000; (In Polish). Available online: http://www.iko.pwr.wroc.pl/PracowDrct/Agnieszka_Zajac/_PN-83_B-03430_doc.pdf (accessed on 18 December 2020).
34. Chmielewska, A.; Szulgowska-Zgrzywa, M.; Danielewicz, J. Domestic hot water consumption in multiapartment buildings. *E3S Web Conf.* **2017**, *17*, 00014. [CrossRef]

Publisher's Note: MDPI stays neutral with regard to jurisdictional claims in published maps and institutional affiliations.

© 2020 by the authors. Licensee MDPI, Basel, Switzerland. This article is an open access article distributed under the terms and conditions of the Creative Commons Attribution (CC BY) license (http://creativecommons.org/licenses/by/4.0/).

energies

MDPI

Article

Are BBQs Significantly Polluting Air in Poland? A Simple Comparison of Barbecues vs. Domestic Stoves and Boilers Emissions

Artur Badyda [1,*], Piotr Krawczyk [2], Jan Stefan Bihałowicz [3], Karolina Bralewska [3], Wioletta Rogula-Kozłowska [3], Grzegorz Majewski [4], Przemysław Oberbek [5], Andrzej Marciniak [3] and Mariusz Rogulski [1]

1 Faculty of Building Services, Hydro- and Environmental Engineering, Warsaw University of Technology, 00-653 Warszawa, Poland; mariusz.rogulski@pw.edu.pl

2 Faculty of Power and Aeronautical Engineering, Warsaw University of Technology, 00-665 Warszawa, Poland; piotr.krawczyk@pw.edu.pl

3 Institute of Safety Engineering, The Main School of Fire Service, 01-629 Warszawa, Poland; jbihalowicz@sgsp.edu.pl (J.S.B.); kbralewska@sgsp.edu.pl (K.B.); wrogula@sgsp.edu.pl (W.R.-K.); amarciniak@sgsp.edu.pl (A.M.)

4 Institute of Environmental Engineering, Warsaw University of Life Sciences, 02-787 Warszawa, Poland; grzegorz_majewski@sggw.pl

5 Central Institute for Labour Protection–National Research Institute, 00-701 Warszawa, Poland; ober@ciop.pl

* Correspondence: artur.badyda@pw.edu.pl; Tel.: +48-22-234-59-50

Received: 23 October 2020; Accepted: 19 November 2020; Published: 26 November 2020

Abstract: The study attempts to compare the total annual emissions of selected air pollutants emitted during occasional grilling and the emission of the same pollutants from small domestic heating installations. For this purpose, in the absence of any data on the emission of pollutants during grilling processes, tests were carried out consisting of measuring the concentration of air pollutants in exhaust streams from two types of grills (solid fuel grill powered by charcoal briquette and gas grill powered by liquid propane), using popularly prepared dishes (previously marinated meat and raw, seasoned mixed vegetables). The concentrations of $PM_{2.5}$, CH_4, CO, CO_2, H_2O, NH_3, N_2O, NO, NO_2, SO_2 were measured in the exhaust stream from both grills using a particulate matter (PM) measuring device and a portable spectrometer, separately while grilling the same portions of meat and vegetables. Then, considering the available data on Poles' barbecue habits, the emissions that are released into the air during occasional grilling were estimated. The calculated emissions were compared with the data on emissions from domestic heating installations used in Poland. It has been shown that during grilling, as much as 2.30, 92.07, 4.11, 3.83, 2.96, and 9.81 Gg of $PM_{2.5}$, CO, NO_x, SO_2, NH_3, and CH_4 may be released into the atmosphere in Poland, respectively. In the case of PM, the amount of the pollutant emitted to the air is over 100 times lower than the emissions caused by the operation of small heating installations. In the case of other pollutants, the differences are smaller. Nevertheless, emissions from grills should not be underestimated as, in certain periods of the year, these sources may be responsible for not meeting the air quality standards in selected areas of the country, and thus the excessive exposure of people to pollutants resulting in negative health consequences. Therefore, attention was paid to the legitimacy of abandoning the use of charcoal and charcoal briquette grills and replacing them with gas-powered grills or electric ones, not only due to the health benefits of food and lower human exposure, but also by the reason of ecological values.

Keywords: grill; barbecue; stove; boiler; charcoal briquettes; liquid propane; particulate matter; gaseous air pollutants

1. Introduction

According to the data of The National Centre for Emissions Management (KOBiZE), the main source of air pollution in Poland is the so-called low-stack emission, i.e., the emission of harmful gases and particles from emitters not exceeding 40 m above the ground, resulting mainly from combustion processes in local stoves/boilers rooms and domestic furnaces and fireplaces, i.e., the municipal and households sector [1,2]. Due to lower price and widespread access, coal, charcoal, and wood are mostly used in domestic furnaces, and in mining regions also brown coal, silts, and fleets [3]. In Poland, the problem of low-stack emissions is aggravated by a large number of coal stoves, burning of low-quality coal, wood, and sometimes even municipal waste, the use of inefficient and old-type boilers with manual stoking and natural draft, and a long heating season (from October to March) [3,4]. The phenomenon of emissions from the municipal and households sector occurs mainly in the areas of single-family housing and applies to the whole year. The most unfavorable situation, however, appears in winter, when there is the highest heat demand. The scale of the problem is proved by seasonal changes of particulate matter, polycyclic aromatic hydrocarbons (PAH), and gaseous pollutants (NO_x, SO_2) levels [5–9].

Low-stack emission is also a significant source of air pollution in many regions worldwide, including Canada, the United States, Australia, Ireland, Austria, Italy, where it is mainly associated with wood burning in fireplaces and wood burning stoves [10–17]. Fireplaces are inefficient combustion devices with high uncontrolled excess air rates and without any sort of secondary combustion, resulting in a significant quantity of unburnt combustibles. Several studies indicate that burning in fireplaces is accompanied by the emission of a wide spectrum pollutants, i.e., fine and coarse particles (particulate matter—PM), carbon monoxide (CO), carbon dioxide (CO_2), nitrogen dioxide (NO_2), polycyclic aromatic hydrocarbon (PAHs), and volatile organic compounds (VOCs) [13,15–20]. Research by Guo et al. [13] conducted in 27 randomly selected Irish houses, found that concentrations of CO and particulate matter inside homes with fireplaces were several times higher than the average concentrations in a normal house with central heating. Alves et al. [15] who compared fine particle emission from cast iron woodstove operated manually in batch mode, with handheld control of combustion air and from a traditional Portuguese brick open fireplace operated manually in batch mode, with no control of combustion air, showed that $PM_{2.5}$ emission factors ($g \times kg^{-1}$ fuel burned, dry basis) were higher for the fireplace and the woodstove.

Another, increasingly important source of atmospheric air pollutants including PM, carbon monoxide (CO), nitrogen oxides (NO_x), volatile organic compounds (VOCs), carbonyls, trace metals, and polycyclic aromatic hydrocarbons (PAHs) are barbecue grills powered by lump charcoal and charcoal briquettes [21–27]. According to the reports of Hearth, Patio and Barbecue Association almost 64% of U.S. adults own a grill, while in Canada—about 72%. More than half of them use a charcoal grill [28]. About 75% of owners grill during the whole year, at least 1–2 times per week, especially during birthday parties, camping trips, holidays, or sports events. In Poland, according to the research conducted by the TNS Polska (Public Opinion Research Centre) in 2015, the barbecue season lasts from late spring to early autumn. Then, 30.7% of Poles grill at least once a week, and another 43.6% 2–3 times a month. Herein, the vast majority also use traditional charcoal or charcoal briquette fired grills for this purpose. In Poland, the problem of air pollution due to grilling occurs especially in resorts, recreation places, i.e., clearings outside the city, or on beaches with designated places for this purpose. Air pollutant emissions from grilling may be a significant source of health exposure, and the mechanism of the impact of pollutants on human health may be similar to emissions from the combustion of solid fuels in household furnaces (coal boilers/stoves or fireplaces), except that the emissions from the grill takes place at the level of the human respiratory tract, and the emission from the boiler 10–40 m above the ground. Preliminary measurements results of air pollutants emitted from selected grill furnaces [24] showed that this method of food preparation favors the emission of significant amounts of solid particles, which, especially when using traditional fuels (charcoal and briquette), contain groups of polycyclic aromatic hydrocarbons (PAHs) in high concentrations. Research on pollutant emissions related to grilling mainly concerns indoor PM and PAHs concentrations during grilling in residences and restaurants [23,29–32], as well as dietary exposure, and health risk related to polycyclic aromatic hydrocarbons in grilled meals [24,33–37]. There is much less research to investigate the influence of outdoor barbecue smoke on atmospheric air quality [26,27,38].

Investigations of Rahman and Kim [39] indicate that charcoal-fueled grills are also a source of offensive odorants, including aldehydes, VOCs, carbonyls, reduced sulfur compounds, and ammonia. The emission of odorants also accompanies the burning of wood in domestic stoves [40]. While emissions from a stack appear on the Environmental Protection Act (EPA) [41] list of factors that can cause odors, grilling is not included in this list. A relatively small number of publications on the subject of odors accompanying burning in grills and stoves allows to conclude that this problem is not fully understood, which is the result of the high level of complexity of research in this subject, significant costs of their conduct, and the lack of uniform legal regulation for odor standards. The EU suggests introducing local regulations in this matter, therefore the permissible concentrations of odorous substances in most EU countries result from internal, national legal acts regulating the permissible concentrations of air pollutants. In Germany, France, the Netherlands, Great Britain, Japan, the United States, and Canada, legal solutions are available covering both regulations on emission standards, limit levels, as well as research methods and modelling of odor spread, as well as nuisance assessment using sociological research. However, they mainly focus on the emission of odorous substances in industry and agriculture, as well as in other branches of the economy. The problem of air pollution and odors accompanying grilling and burning in domestic stoves is becoming more and more noticeable. The nuisance of air pollution and odors accompanying the use of grills, wood-fired stoves is evidenced by solutions introduced in some districts, housing communities, public places prohibiting grilling or allowing the use of gas or electric grills. In some Polish provinces, a periodic ban on burning in fireplaces is also introduced, especially in the winter season when the risk of smog episodes appearance increases.

The aim of this study was to present, analyze, and discuss the results of experimental measurements, in which the scope of the first studies [24] was extended to include the analysis of the concentrations of particulate matter and selected gaseous pollutants emitted from grill furnaces powered by charcoal and gas. For comparative purposes, the results of measurements of pollutants in emission streams generated during the combustion of solid fuels in selected domestic heating devices are also presented. The experiment was conducted in Warsaw because the emission structure in this city is the closest to the emission structure in other European cities.

2. Materials and Methods

2.1. Instruments and Measurements

The study covering emissions of selected combustion products of grilling ($PM_{2.5}$, CH_4, CO, CO_2, H_2O, NH_3, N_2O, NO, NO_2, SO_2) was conducted in the summer period, in typical conditions for a barbecue party. The measurements were completed in allotment gardens located in one of the districts of Warsaw, capital of Poland, at a considerable distance (about 1500 m) from the nearest air pollutants emission sources (road) that may potentially interfere with the measurement results. The investigation was made on 6 September 2018 (average daily temperature: 18.5 °C, no precipitation, air pressure at station level 1001 hPa, relative humidity 75%, average daily wind velocity 2.0 m/s) and 11 September 2018 (average daily temperature: 17.9 °C, no precipitation, air pressure at station level 1008 hPa, relative humidity 68%, average daily wind velocity 1.9 m/s). We used two types of grills: charcoal briquette-fueled (B) and liquid propane-fueled (G). Measurements were done each day on the different type of grill in order to minimize interference of different exhausts in the measurement device (Figure 1).

Figure 1. Measuring points and devices.

We analyzed emission from empty grills (i.e., without any food) and during the preparation of meat and vegetables separately. The type and weight of dishes were identical during both days of grilling—they were the barbeque sets from a popular grocery supermarket. In the exhaust stream we put particulate matter probe, $PM_{2.5}$ aspirator, and spectrometer which was continuously measuring the concentration of CH_4, CO, CO_2, H_2O, NH_3, N_2O, NO, NO_2, SO_2.

The SidePak AM510 Personal Aerosol Monitor from TSI Inc. (Shoreview, MN, USA) was selected for continuous measurements of $PM_{2.5}$ concentrations over the grill hearth. The device uses 90° light scattering technology with a 670 nm laser diode to determine the mass concentration of the selected particles fraction (from 0.1 to 2.5 μm in this case). By means of an automatic pump installed in the device, a continuous stream of air along with the aerosol is introduced into the measuring chamber. The measuring range is from 0 to 20,000 μg/m^3 and the resolution 1 μg/m^3. The air flow rate through the apparatus was set at the level of 1.7 dm^3/min and data were collected with a 1-min time resolution. Zero stability was checked at the beginning of each measurement procedure.

The measurement of the concentration of gaseous substances was done using gas analyzer Gasmet DX 4000 (Vantaa, Finland). It is a portable FTIR (Fourier transformed infrared spectroscopy) gas analyzer. This system utilizes hot-and-wet measurement principle (no drying or dilution). The measuring range for individual gases is: water vapor from 0 to 10%, carbon dioxide from 0 to 25%, carbon monoxide from 0 to 400 $mg/m^3{}_n$, nitrogen monoxide from 0 to 250 $mg/m^3{}_n$, nitrogen dioxide from 0 to 200 $mg/m^3{}_n$, sulfur dioxide from 0 to 14,000 $mg/m^3{}_n$, ammonia from 0 to 200 $mg/m^3{}_n$ and methane from 0 to 7000 $mg/m^3{}_n$. Other characteristic parameters of the analyzer are: scanning frequency 10 spectra/s, wave spectrum range 900–4200 cm^{-1}, and gas flow 140 dm^3/h. Measured gas pressure is the atmospheric pressure.

2.2. Data Analysis

For all the collected data we calculated basic descriptive measures: minimum, quartiles, median, maximum, and mean values. During the experiment 40 data series were collected. The lengths of the series were different. Using the Shapiro–Wilk test we checked each set whether the collected data met the assumption of a normal distribution [42]. If the concentrations of given substances for different grilling types were characterized by a normal distribution, we used the Welch t-test [43] to compare whether the distributions were similar or not. If for a given substance any of four sets did not met the hypothesis of distribution normality, we chose Mann–Whitney U test to evaluate the similarity of sets [44]. The choice of the U test is a result of different sizes of samples. We performed these calculations using Python 3.7 [45] with packages pandas [46], SciPy [47], and Matplotlib [48].

We also estimated the total emission of substances during grilling different kinds of meals (meat and vegetables). The concentrations were observed as time series, so we were able to find total emission during the whole grilling process. The emission EM of a given substance in the specific process of grilling i is defined by Riemann sum [49] over partition defined by the sampling rate of concentration.

$$EM_i = \sum_{k=1}^{n_i} C_i(t_k) q_i(t_k) \Delta t_{ik}, \tag{1}$$

where k indexes time series, n_i is a length of the i-th time series, C_i is a concentration of a given substance, q_i is the flow of gas above the grill, and Δt_{ik} is a time step in i-th time series. If we assume constant flow of air in all-time series since the temperature of grills was kept the same in all measurements, the formula simplifies, and we can calculate the sum. It does not allow us to give an exact numerical value of emitted kg during the grilling, however, we can compare the emissions between grilling of meat and vegetables on briquette or gas-fueled grill.

2.3. Methodology of Emission Estimation

We estimated the annual emission from the stoves and compared it with the emissions from briquette-fueled and gas-fueled grills.

For stoves, we used the concentration of total suspended particles (TSP), carbon monoxide (CO), nitrogen oxides (NO_x), and sulfur dioxide (SO_2) in the exhaust gases according to [50]. Since we had only data about the TSP from the stoves, we used values provided in [51] in order to calculate $PM_{2.5}$ concentration based on the TSP concentration. According to data, the $PM_{2.5}$ to TSP ratio is 10% for biomass stove, 30% for old type stove, and 60% for modern (eco) stove. We calculated the flow of the exhaust emitted per fuel mass and hence we were able to calculate the emission factors of these substances. The data on the total annual consumption of wood and coal in domestic stoves/boilers in Poland were also used [52]. According to the report of Institute of Environmental Economics about the energy efficiency in Poland in 2017 [53], we assumed the division of coal-fueled stoves into old type (81.5%) and the new type (18.5%). The same study provides information about the share of the given quality of coal (86.1% good quality, 8.3% poor quality, 5.6% unknown). For the bio-mass fueled stoves we assumed that 30% of fuel mass is wet and 70% is dry. All these data and assumptions allowed us to calculate the annual emission of selected pollutants in Poland for a given type of stove. We compared the emission to the data published by Statistics Poland [54] about the total household emissions of the same pollutants.

In the case of grill analysis, we used the average concentrations of $PM_{2.5}$, CO, SO_2, NH_3, CH_4 calculated from our experimental data. The concentration of NOx we calculated according to Formula (2). Again, we calculated the exhaust flow and emission factors of these pollutants per kg of mass of fuel.

$$C_{NO_x} = \frac{46}{30} C_{NO} + C_{NO_2} \quad (2)$$

We estimated the amount of the briquette used for grills according to the national survey "How do Poles grill?" conducted in April 2015 by the TNS Polska (Public Opinion Research Centre, Warsaw, Poland) on a sample of 1000 Poles over 15 years old [55]. We assumed that grilling season lasts 20 weeks, from the so-called long weekend related to national holidays at the beginning of May and ends just after the end of summer break in schools, in the first week of September. We estimated the annual number of grilling events in Poland taking into consideration the frequency of grilling activities according to the responses given in the above-mentioned TNS survey. The data on a normalized share of meat (66.7%) and vegetables (13.5%) used for grilling purposes were based also on the results of the same study [55]. We also considered results of another survey, "Barbecue habits of Poles," completed in March 2017 by IQS commissioned by Coca-Cola Poland Services (Warsaw, Poland) on a representative nationwide sample of 500 Poles aged 16–49 [56] as the input data for calculation of the duration of grilling processes. The briquette consumption for a single grilling process was estimated on the results of our experiment. Consumption was assumed at the level of 1.3 kg/h. All the aforementioned data were used to estimate the total annual emissions from all grilling activities in Poland.

The data about the gaseous grills are very limited since gas grills are not popular in Poland. The data provided by [57] indicate that 90% of grills are powered by charcoal or charcoal briquettes, and 2% are powered by liquid gas (usually liquid propane). We decided therefore that the annual duration of all gas grilling processes in Poland is the total duration calculated for the charcoal and charcoal briquettes grilling divided by 45 (ratio reflecting the ownership shares). The propane consumption of grills we estimated at level 0.365 kg/h based on technical details of selected types of gaseous grills available on the Polish market [58]. These data were used for the calculation of the total consumption of liquid propane for barbecue purposes and of annual emission from gas-fueled grills.

3. Results and Discussion

We analyzed the concentration data collected during the experiment. Figure 2 presents the matrix of 10 plots, one for each analyzed substance. For each substance, four data sets are presented—the concentrations above the grill hearth during the preparation of meat (M) or vegetables (V) on charcoal briquette-fueled grill (B) or gas-fueled grill (G). The data sets were typically 170 samples long. The concentrations of pollutants and green-house gases exhibited high variability depending on the type of fuel burned and the kind of prepared meal (meat vs. vegetables). Concentrations of almost all of the analyzed substances, except for NO_2, were much higher during the preparation of meals with the use of charcoal briquette than with the liquid propane. The most pronounced differences concern CO concentrations, which for meat and vegetables, respectively, were 155 and 189 times higher during briquette combustion than during gas combustion, then NH_3 (36-times for meat, six-times for vegetables), CH_4 (15-times for meat, five-times for vegetables), and $PM_{2.5}$ (four-times meat, 13-times vegetables). Although the results present different variability, we observed two main patterns in concentrations: BM > BV > GM > GV or BM > BV > GV > GM (for example for CO) and BM > GM > BV > GV (for example for $PM_{2.5}$ and in the inverse order for NO_2). The first pattern suggests that the concentration is mainly influenced by the type of fuel, and the second suggests the type of grilled material as the main cause of emission. We found that the first pattern is typical for the concentrations of CH_4, CO, NH_3, SO_2, while the second one for $PM_{2.5}$ and NO_2 (reversed order).

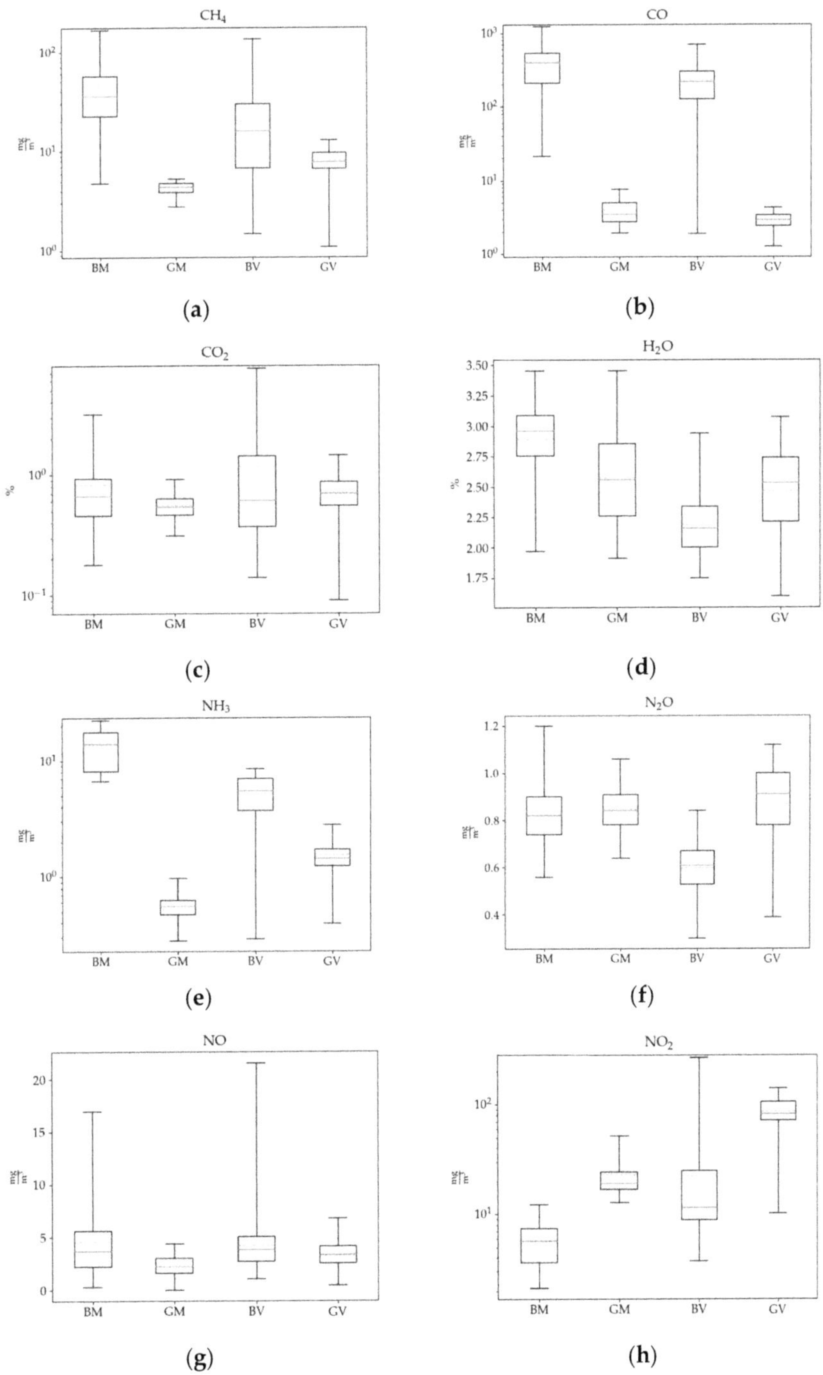

Figure 2. *Cont.*

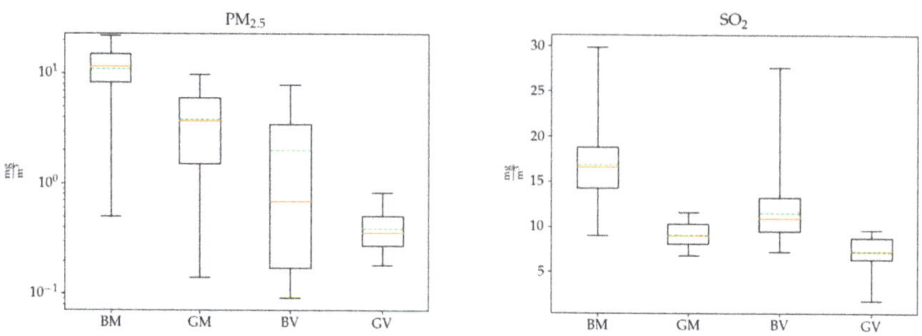

(i) (j)

Figure 2. Concentration of selected pollutants and green-house gases directly above the grill during preparation of BM—meat on coal briquette, GM—meat on gas-fueled grill, BW—vegetables on coal briquette, GW—vegetables on gas-fueled grill: (**a**) CH_4 in mg/m^3, (**b**) CO in mg/m^3, (**c**) CO_2 in %, (**d**) H_2O in %, (**e**) NH_3 in mg/m^3, (**f**) N_2O in mg/m^3, (**g**) NO in mg/m^3, (**h**) NO_2 in mg/m^3, (**i**) $PM_{2.5}$ in mg/m^3, (**j**) SO_2 in mg/m^3. Whiskers are from the minimum to maximum values, the box is from the first quartile to the third quartile, the orange line represents the median, and the green, dashed line represents the average value.

We performed the Shapiro–Wilk test to check which of the 40 data sets are taken from the normal distribution. The result showed that only five sets represent normal distribution for $\alpha = 0.1$: $PM_{2.5}$ BM, CO GV, NO GV, NO_2 BM, and NH_3 GM (for details see Appendix A). We decided to use only the Mann–Whitney U test for a comparison of four data sets in each of the 10 groups. The test showed that only eight pairs of sets reveal similarities at $\alpha = 0.01$ (detailed results of tests are presented in Appendix A). These are for H_2O%: GM-GV, for CO_2% BM-BV, BM-GV, GM-BV, BV-GV, for N_2O BM-GM, for NO BM-BV, BM-GV.

We calculated the total emissions according to Formula (1). The emissions per unit flow are presented in Figure 3. During the combustion of briquettes, the highest emissions of CO (1299–1578 g × s/m^3) and CH_4 per unit flow (112–173 g × s/m^3) were noted. High emissions per unit flow, which were observed for carbon monoxide, may indicate an incorrect combustion process. CO is a product of incomplete combustion, generally associated with the smoldering phase. Burning briquettes in inefficient devices such as grills generate high levels of CO due to relatively low temperatures and a reduced or almost no visible flame above the layer [59]. Methane is an important component released during biomass combustion [60]. In the case of gas-fueled grills, the highest emission per unit flow was recorded for NO_2 (60–210 g × s/m^3). NO_2 emissions from gas combustion are mostly associated with thermal and prompt formation, whereas the dominant mechanism of NO_2 formation during the combustion of briquettes is from the fuel nitrogen content since the temperatures achieved in grills are usually too low for prompt and thermal formation [61,62]. Summarizing, we see similar tendencies as in the concentration box plots (Figure 2). The order of the emissions of CO, $PM_{2.5}$, SO_2, and N_2O is as follows: BM > BV > GM > GV. It means that the total emission is firstly dependent on fuel type and secondarily on the type of grilled products. In case of NO, NH_3, and CH_4 the situation is different and fuel type has a crucial influence on emission (i.e., B > G) while it was not possible to determine the impact of the kind of grilled dishes. The total emission of NO_2 during the grilling shows that there is no clear relation of emission to dish or fuel.

The total emissions of pollutants and gases observed over the charcoal briquette-fueled grill hearth are comparable to that reported by Vicente et al. and Yu et al., who tested the emission of the gaseous pollutants and particulate matter ($PM_{2.5}$) from charcoal combustion in a typical brick barbecue grill [27,62]. There are few data on pollutant emissions depending on the type of meal being prepared. The results in this regard were presented in [63], where it was shown that marinated meat yielded an increased total VOC and total PM emissions compared to non-marinated meat and there were no significant differences in emission rates between chicken and beef.

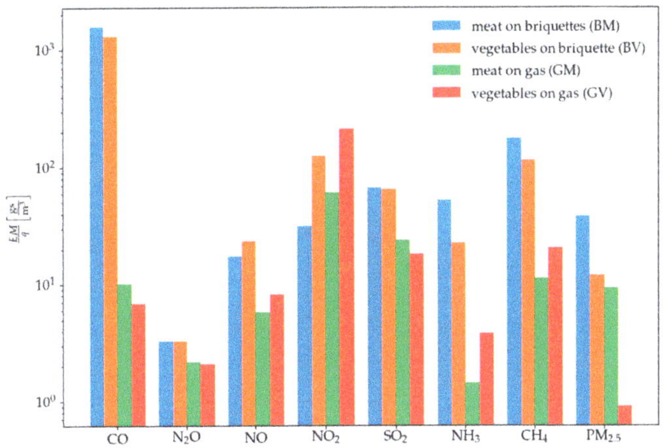

Figure 3. The total emissions of substances per unit flow over the grill. The figure allows us to compare the relative total emissions of substances in different grilling conditions.

Table 1 summarizes the annual emissions of individual pollutants released during meat and vegetables grilling on two types of grills (charcoal briquette and gas-fueled) and the annual emissions related to the combustion of coal and wood in three types of stoves/boilers (bio-mass, old type, Ecoproject stove) taking into account the quality of the fuel (high vs. poor). Table 2 compares the emissions from briquette and gas-fueled grills with the total emissions from three types of stoves and with the data from the Statistics Poland [54] on emissions of the analyzed pollutants from households. In addition, Table 2 shows the percentage of individual pollutants from grills in the annual emission resulting from combustion in the above-mentioned stoves/boilers. According to the data presented in the Tables 1 and 2, CO has the highest mass share in the total mass of pollutants emitted during the year due to grilling, while $PM_{2.5}$ the lowest. In case of wood and coal combustion in stoves, the largest mass share also belongs to CO, and the lowest to NO_x (in case of old type stoves) and $PM_{2.5}$ (in case of Ecoproject stoves), but the authors did not reach the data on NH_3 and CH_4 emissions related to combustion in different kind of stoves. Emissions of pollutants released by grilling account for a small percentage (2–8%) of the annual emissions of pollutants resulting from combustion of coal and wood in stoves. However, it should not be interpreted that the emissions of pollutants from the grilling processes are negligible. It results from the intensive use of domestic stoves and boilers herein and the associated high PM and gaseous emissions [1]. The comparison of the calculation data on emissions from grills and boilers with the statistical data of Statistics Poland and the fact that they are of the same order of magnitude confirms the justifiability of the assumptions made in the calculations. Although the total emissions from grills is definitely lower compared to the annual emissions from stoves/boilers, it can be treated as a significant cause of seasonal changes in air quality, i.e., in the grill season, which in Poland lasts from spring to early autumn. Frequent grilling can deteriorate the air quality, especially at campsites, single-family houses, holiday cottages estates, suburban clearings, arcades with restaurants, or beaches with a designated barbecue area, where dozens of outdoor barbecue events take place almost every day. It can also be an important cause of health exposure and the resulting adverse health effects, which in the case of air quality are of great importance in Poland [64,65]. Barbecue cooking can have a significant influence on ambient air quality due to the uncontrolled emissions and subsequent diffusion and transport mechanism with the wind [62,66]. It should be noted that the study did not take into account the concentrations of other pollutants released during grilling, e.g., polycyclic aromatic hydrocarbons, or trace metals—the concentration of which during grilling were higher compared to the ambient atmosphere according to research [23,27]. In the case of emissions from home furnaces and stoves, the combustion process is much more refined and controlled. Air quality management is not accidental as it is when burning solid fuel on a grill. In modern stoves/boilers (meeting the requirements of the Ecoproject Directive), not only the value of the excess air coefficient is controlled, but also its distribution in the furnace chamber. As a result, $PM_{2.5}$, CO, NO_x, SO_2 emissions are significantly lower than in the case of burning fuel on the grill (Tables 1 and 2). The PM2.5 concentration was calculated according to [51].

Table 1. The annual emissions of pollutants released during grilling processes taking into account the percentage of the type of prepared meals (meat vs. vegetables) and annual emissions of the same pollutants released from three types of stoves (bio-mass, old type, and Ecoproject) taking into account the quality of the fuel used (high vs. poor).

	Briquette-Fueled Grills		Gas-Fueled Grills		Bio-Mass Stoves		Old Type Stoves		Ecoproject Stoves	
Share	Meat 67% [1]	Vegetables 14% [1]	Meat 67% [1]	Vegetables 14% [1]	Dry Wood 70%	Wet Wood 30%	High Quality Coal 86%	Poor Quality Coal 8%	High Quality Coal 86%	Poor Quality Coal 8%
	Gg/Year									
$PM_{2.5}$	2.21	0.08	0.006	0.000	9.78 [2]	6.77 [2]	39.61 [2]	5.36 [2]	0.86 [2]	0.08 [2]
CO	82.89	9.17	0.006	0.001	608.58	423.83	1320.35	178.80	53.52	5.16
NO_x	2.91	1.14	0.039	0.024	6.52	2.79	26.72	2.58	10.70	1.03
SO_2	3.36	0.45	0.013	0.002	2.54	1.86	125.75	26.67	0.00	0.00
NH_3	2.75	0.21	0.001	0.000	NA	NA	NA	NA	NA	NA
CH_4	9.00	0.80	0.006	0.002	NA	NA	NA	NA	NA	NA

[1] Share of meat and vegetables in grilling; [2] $PM_{2.5}$ emission calculated as fraction of TSP emission according to [51].

Table 2. The comparison of summed up annual emissions from grills (briquette and gas-fueled) with the total emissions from three types of stoves and the percentage of individual pollutants from grills in the annual emissions of pollutants resulting from combustion in stoves.

	Briquette-Fueled Grills [Gg/Year]	Gas-Fueled Grills [Gg/Year]	Stoves [Gg/Year]	Statistics Poland Data Stoves [Gg/Year]	Ratio of Briquette-Fueled Grills Emission/Stoves Emission [%]
$PM_{2.5}$	2.29	0.006	62.46	121.95	4
CO	92.06	0.006	2590.24	1365.95	4
NO_x	4.11	0.063	50.35	61.62	8
SO_2	3.82	0.015	156.81	132.31	2
NH_3	2.95	0.001	NA	NA	NA
CH_4	9.80	0.009	NA	NA	NA

Obtained results indicate that further studies on the concentrations of various pollutants released during grilling and the assessment of health exposure should be investigated. The calculations described above are based on emission data only for Poland, where the measurements were carried out. In future research, based on the data on low-stack emissions in other countries, e.g., UK, Ireland, Germany [67,68], and based on barbecue habits of their citizens, a similar analysis could be carried out using the methodology proposed in the manuscript.

4. Conclusions

The tests carried out in the described investigation consisted of measuring selected gaseous pollutants and particulates from two types of grills during the preparation of typical types of food used while grilling. Considering the measurement time and assuming evenly distributed in time emissions during grills usage, PM and considered gaseous pollutants emission factors were calculated for each type of grill and fuel. They were compared with the emission factors from small domestic furnaces powered by hard coal and biomass (wood).

Research results demonstrated that the pollutants present in the emission streams over the grill hearths powered by traditional fuels (charcoal briquettes) are present in lower concentrations than in the case of domestic coal or wood stoves/boilers. The emissions from grills, however, occur at the breathing height in the direct presence of the grill operator. The results of the analyses show that the concentrations of some pollutants from grills show a greater dependence on the type of fuel used (in case of CH_4, CO, NH_3, SO_2), and others on the type of grilled food (in case of $PM_{2.5}$ and NO_2). However, regardless the pattern, research results indicate that in the spring and summer period, emissions from grills may locally be a serious air pollutant, and therefore may also be an important health risk factor resulting from inhalation of pollutants from grilling processes.

Author Contributions: Conceptualization: A.B., G.M., W.R.-K., P.K.; methodology: A.B., G.M., P.K., W.R.-K.; software: J.S.B.; investigation, measurement: A.B., G.M., W.R.-K., P.K., P.O., M.R., J.S.B., K.B., A.M.; validation: A.B., W.R.-K., J.S.B., K.B.; data curation: J.S.B., P.K.; funding acquisition: A.B., G.M., W.R.-K.; writing—original draft preparation: A.B., W.R.-K., J.S.B., K.B.; writing—review and editing: A.B., G.M.; visualization: J.S.B.; supervision: A.B. All authors have read and agreed to the published version of the manuscript.

Funding: This research received no external funding. Research was financed from the private resources of the authors. Costs of the manuscript preparation and publishing were partially covered by The Main School of Fire Service and Warsaw University of Technology.

Conflicts of Interest: The authors declare no conflict of interest.

Appendix A

The Appendix A contains the detailed results of statistical analyses.

Table A1. The *p*-value of the Shapiro–Wilk test for samples. Significant values at $\alpha = 0.1$ are marked with green.

	BM	BV	GM	GV
CH_4	0.000	0.000	0.001	0.001
CO	0.000	0.000	0.000	0.179
CO_2%	0.000	0.000	0.065	0.007
H_2O%	0.000	0.000	0.003	0.000
N_2O	0.002	0.001	0.036	0.000
NH_3	0.000	0.000	0.227	0.000
NO	0.000	0.000	0.057	0.108
NO_2	0.239	0.000	0.000	0.003
$PM_{2.5}$	0.168	0.000	0.033	0.025
SO_2	0.000	0.000	0.001	0.000

Table A2. The *p*-value of the Mann–Whitney U test for $PM_{2.5}$, values significant at $\alpha = 0.1$ are marked with green, at $\alpha = 0.05$ yellow, and at $\alpha = 0.01$ orange.

	BM	BV	GM	GV
BM	1.0	1.6×10^{-10}	5.8×10^{-19}	1.2×10^{-15}
BV		1.0	0.00018	8.0×10^{-8}
GM			1.0	0.0096
GV				1.0

Table A3. The *p*-value of the Mann–Whitney U test for $H_2O\%$, values significant at $\alpha = 0.1$ are marked with green, at $\alpha = 0.05$ yellow, and at $\alpha = 0.01$ orange.

	BM	BV	GM	GV
BM	1.0	1.4×10^{-12}	3.1×10^{-55}	9.4×10^{-20}
BV		1.0	4.3×10^{-21}	0.079
GM			1.0	9.7×10^{-14}
GV				1.0

Table A4. The *p*-value of the Mann–Whitney U test for CO_2, values significant at $\alpha = 0.1$ are marked with green, at $\alpha = 0.05$ yellow, and at $\alpha = 0.01$ orange.

	BM	BV	GM	GV
BM	1.0	1.5×10^{-5}	0.47	0.47
BV		1.0	0.022	1.88×10^{-8}
GM			1.0	0.33
GV				1.0

Table A5. The *p*-value of the Mann–Whitney U test for CO, values significant at $\alpha = 0.1$ are marked with green, at $\alpha = 0.05$ yellow, and at $\alpha = 0.1$ orange.

	BM	BV	GM	GV
BM	1.0	3.6×10^{-48}	1.3×10^{-14}	3.4×10^{-43}
BV		1.0	2.7×10^{-52}	6.2×10^{-8}
GM			1.0	2.4×10^{-47}
GV				1.0

Table A6. The *p*-value of the Mann–Whitney U test for N_2O, values significant at $\alpha = 0.1$ are marked with green, at $\alpha = 0.05$ yellow, and at $\alpha = 0.01$ orange.

	BM	BV	GM	GV
BM	1.0	0.016	4.7×10^{-47}	0.00018
BV		1.0	3.5×10^{-50}	0.0010
GM			1.0	1.3×10^{-29}
GV				1.0

Table A7. The *p*-value of the Mann–Whitney U test for NO, values significant at $\alpha = 0.1$ are marked with green, at $\alpha = 0.05$ yellow, and at $\alpha = 0.01$ orange.

	BM	BV	GM	GV
BM	1.0	8.6×10^{-11}	0.18	0.028
BV		1.0	3.9×10^{-26}	4.9×10^{-12}
GM			1.0	0.00013
GV				1.0

Table A8. The *p*-value of the Mann–Whitney U test for NO_2, values significant at $\alpha = 0.1$ are marked with green, at $\alpha = 0.05$ yellow, and at $\alpha = 0.01$ orange.

	BM	BV	GM	GV
BM	1.0	3.3×10^{-17}	2.8×10^{-13}	6.1×10^{-17}
BV		1.0	8.0×10^{-12}	8.0×10^{-38}
GM			1.0	1.1×10^{-44}
GV				1.0

Table A9. The p-value of the Mann–Whitney U test for SO_2, values significant at $\alpha = 0.1$ are marked with green, at $\alpha = 0.05$ yellow, and at $\alpha = 0.01$ orange.

	BM	BV	GM	GV
BM	1.0	3.0×10^{-45}	3.2×10^{-42}	1.1×10^{-46}
BV		1.0	6.8×10^{-20}	8.9×10^{-18}
GM			1.0	5.3×10^{-45}
GV				1.0

Table A10. The p-value of the Mann–Whitney U test for NH_3, values significant at $\alpha = 0.1$ are marked with green, at $\alpha = 0.05$ yellow, and at $\alpha = 0.01$ orange.

	BM	BV	GM	GV
BM	1.0	3.6×10^{-48}	4.4×10^{-52}	2.3×10^{-46}
BV		1.0	5.6×10^{-47}	3.2×10^{-38}
GM			1.0	2.2×10^{-39}
GV				1.0

Table A11. The p-value of the Mann–Whitney U test for CH_4, values significant at $\alpha = 0.1$ are marked with green, at $\alpha = 0.05$ yellow, and at $\alpha = 0.01$ orange.

	BM	BV	GM	GV
BM		1.5×10^{-47}	2.4×10^{-20}	3.4×10^{-36}
BV		1.0	1.5×10^{-31}	4.7×10^{-36}
GM			1.0	8.9×10^{-12}
GV				1.0

References

1. Ministry of the Environment. *National Air Protection Program until 2020 (with a Prospect until 2030)*; Air Protection, Ministry of the Environment: Warsaw, Poland, 2015.
2. Zieliński, E.; Wielgus, A.; Dreliszak, J.; Zukow, W. Air pollution-selected health effects in Poland. *J. Educ. Health Sport* **2018**, *8*, 641–648.
3. Proszak-Miąsik, D.; Rabczak, S. Methods for reducing low emissions from heating devices in single- family housing. *E3S Web Conf.* **2018**, *45*, 00069. [CrossRef]
4. Adamczyk, J.; Piwowar, A.; Dzikuć, M. Air protection programmes in Poland in the context of the low emission. *Environ. Sci. Pollut. Res.* **2017**, *24*, 16316–16327. [CrossRef] [PubMed]
5. Rogula-Kozłowska, W.; Klejnowski, K.; Rogula-Kopiec, P.; Ośródka, L.; Krajny, E.; Błaszczak, B.; Mathews, B. Spatial and seasonal variability of the mass concentration and chemical composition of PM2.5 in Poland. *Air Qual. Atmos. Health* **2014**, *7*, 41–58. [CrossRef] [PubMed]
6. Jędruszkiewicz, J.; Czernecki, B.; Marosz, M. The variability of PM10 and PM2.5 concentrations in selected Polish agglomerations: The role of meteorological conditions, 2006–2016. *Int. J. Environ. Health* **2017**, *27*, 441–462. [CrossRef]
7. Majewski, G.; Rogula-Kozłowska, W.; Rozbicka, K.; Rogula-Kopiec, P.; Mathews, B.; Brandyk, A. Concentration, chemical composition and origin of PM1: Results from the first long-term measurement campaign in Warsaw (Poland). *Aerosol Air Qual. Res.* **2018**, *18*, 636–654. [CrossRef]
8. Wiśniewska, K.; Lewandowska, A.U.; Staniszewska, M. Air quality at two stations (Gdynia and Rumia) located in the region of Gulf of Gdansk during periods of intensive smog in Poland. *Air Qual. Atmos. Health* **2019**, *12*, 879–890. [CrossRef]
9. Chambers, S.D.; Podstawczyńska, A. Improved method for characterising temporal variability in urban air quality part II: Particulate matter and precursors in central Poland. *Atmos. Environ.* **2019**, *219*, 117040. [CrossRef]
10. Dasch, J.M. Particulate and gaseous emissions from wood-burning fireplaces. *Environ. Sci. Technol.* **1982**, *16*, 643–667. [CrossRef]

11. Gullett, B.K.; Touati, A.; Hays, M.D. PCDD/F, PCB, HxCBz, PAH, and PM Emission factors for fireplace and woodstove combustion in the San Francisco Bay Region. *Environ. Sci. Technol.* **2003**, *37*, 1758–1765. [CrossRef]
12. Zuk, M.; Rojas, L.; Blanco, S.; Serrano, P.; Cruz, J.; Angeles, F.; Tzintzun, G.; Armendariz, C.; Edwards, R.D.; Johnson, M.; et al. The impact of improved wood-burning stoves on fine particulate matter concentrationsin rural Mexican homes. *J. Expo. Sci. Environ. Epidemiol.* **2007**, *17*, 224–232. [CrossRef]
13. Guo, L.; Lewis, J.O.; McLaughlin, J.P. Emissions from Irish domestic fireplaces and their impact on indoor air quality when used as supplementary heating source. *Glob. NEST J.* **2008**, *10*, 209–216.
14. Caseiro, A.; Bauer, H.; Schmidl, C.; Pio, C.A.; Puxbaum, H. Wood burning impact on PM10 in three Austrian regions. *Atmos. Environ.* **2009**, *43*, 2186–2195. [CrossRef]
15. Alves, C.; Gonçalves, C.; Fernandes, A.P.; Tarelho, L.; Pio, C. Fireplace and woodstove fine particle emissions from combustion of western Mediterranean wood types. *Atmos. Res.* **2011**, *101*, 692–700. [CrossRef]
16. Evtyugina, M.; Alves, C.; Calvo, A.; Nunes, T.; Tarelho, L.; Duarte, M.; Prozil, S.O.; Evtuguin, D.V.; Pio, C. VOC emissions from residential combustion of Southern and mid-European woods. *Atmos. Environ.* **2014**, *83*, 90–98. [CrossRef]
17. Gennaro, G.; Dambruoso, P.R.; Gilio, A.; Palma, V.; Marzocca, A.; Tutino, M. Discontinuous and continuous indoor air quality monitoring in homes with fireplaces or wood stoves as heating system. *Int. J. Environ. Res. Public Health* **2016**, *13*, 78. [CrossRef]
18. Purvis, C.R.; McCrillis, R.C.; Kariher, P.H. Fine particulate matter (PM) and organic speciation of fireplace emissions. *Environ. Sci. Technol.* **2000**, *34*, 1653–1658. [CrossRef]
19. Fine, P.M.; Cass, G.R.; Simoneit, B.R.T. Chemical characterization of fine particle emissions from the fireplace combustion of wood types grown in the Midwestern and Western United States. *Environ. Eng. Sci.* **2004**, *21*, 387–409. [CrossRef]
20. Ozil, F.; Tschamber, V.; Haas, F.; Trouve, G. Efficiency of catalytic processes for the reduction of CO and VOC emissions from wood combustion in domestic fireplaces. *Fuel Process. Technol.* **2009**, *90*, 1053–1061. [CrossRef]
21. Kabir, E.; Kim, K.H.; Ahn, J.W.; Hong, O.F.; Sohn, J.R. Barbecue charcoal combustion as a potential source of aromatic volatile organic compounds and carbonyls. *J. Hazard. Mater.* **2010**, *174*, 492–499. [CrossRef]
22. Chen, J.W.; Wang, S.L.; Hsientang Hsieh, D.P.; Yang, H.H.; Lee, H.L. Carcinogenic potencies of polycyclic aromatic hydrocarbons for back-door neighbors of restaurants with cooking emissions. *Sci. Total Environ.* **2012**, *417*, 68–75. [CrossRef] [PubMed]
23. Huang, H.L.; Lee, W.M.G.; Wu, F.S. Emissions of air pollutants from indoor charcoal barbecue. *J. Hazard. Mater.* **2016**, *302*, 198–207. [CrossRef] [PubMed]
24. Badyda, A.J.; Widziewcz, K.; Rogula-Kozłowska, W.; Majewski, G.; Jureczko, I. Inhalation exposure to PM-bound polycyclic aromatic hydrocarbons released from barbecue grills powered by gas, lump charcoal, and charcoal briquettes. *Adv. Exp. Med. Biol.* **2017**, *1023*, 11–27.
25. Lango-Reynoso, V.; López-Spiegel, J.; Lango-Reynoso, F.; Castañeda-Chávez, M.D.R.; Montoya-Mendoza, J. Estimation of CO_2 Emissions Produced by Commercial Grills in Veracruz, Mexico. *Sustainability* **2018**, *10*, 464. [CrossRef]
26. Song, Y.; Sun, L.; Wang, X.; Zhang, Y.; Wang, H.; Li, R.; Xue, L.; Chen, J.; Wang, W. Pollution characteristics of particulate matters emitted from outdoor barbecue cooking in urban Jinan in eastern China. *Front. Environ. Sci. Eng.* **2018**, *12*, 14. [CrossRef]
27. Yu, K.P.; Chen, Y.C.; Miao, Y.J.; Siregar, S.; Tsai, Y.W.; Lee, W.M.G. Effects of oil drops and the charcoal's proximate composition on the air pollution emitted from charcoal barbecues. *Aerosol Air Qual. Res.* **2020**, *20*, 1480–1494. [CrossRef]
28. Hearth, Patio and Barbecue Association, 2020—State of the Barbecue Industry. Available online: https://www.hpba.org/Resources/PressRoom/ID/1911/2020-State-of-the-Barbecue-Industry (accessed on 2 October 2020).
29. Lee, S.C.; Li, W.M.; Yin Chan, L. Indoor air quality at restaurants with different styles of cooking in metropolitan Hong Kong. *Sci. Total Environ.* **2001**, *279*, 181–193. [CrossRef]
30. Ho, S.S.H.; Yu, J.Z.; Chu, K.W.; Yeung, L.L. Carbonyl emissions from commercial cooking sources in Hong Kong. *J. Air Waste Manag.* **2006**, *56*, 1091–1098. [CrossRef] [PubMed]

31. Taner, S.; Pekey, B.; Pekey, H. Fine particulate matter in the indoor air of barbeque restaurants: Elemental compositions, sources and health risks. *Sci. Total Environ.* **2013**, *454*, 79–87. [CrossRef] [PubMed]
32. Alves, C.A.; Duarte, M.; Nunes, T.; Moreira, R.; Rocha, S. Carbonaceous particles emitted from cooking activities in Portugal. *Glob. NEST J.* **2014**, *16*, 412–420.
33. Aaslyng, M.D.; Duedahl-Olesen, L.; Jensen, K.; Meinert, L. Content of heterocyclic amines and polycyclic aromatic hydrocarbons in pork, beef and chicken barbecued at home by Danish consumers. *Meat Sci.* **2013**, *93*, 85–91. [CrossRef] [PubMed]
34. Torkmahalleh, M.A.; Ospanova, S.; Baibatyrova, A.; Nurbay, S.; Zhanakhmet, G.; Shah, D. Contributions of burner, pan, meat and salt to PM emission during grilling. *Environ. Res.* **2018**, *164*, 11–17. [CrossRef] [PubMed]
35. Jiang, D.; Wang, G.; Li, L.; Wang, X.; Li, W.; Li, X.; Shao, L.; Li, F. Occurrence, dietary exposure, and health risk estimation of polycyclic aromatic hydrocarbons in grilled and fried meats in Shandong of China. *Food Sci. Nutr.* **2018**, *6*, 2431–2439. [CrossRef] [PubMed]
36. Cheng, J.; Zhang, X.; Ma, Y.; Zhao, J.; Tang, Z. Concentrations and distributions of polycyclic aromatic hydrocarbon in vegetables and animal-based foods before and after grilling: Implication for human exposure. *Sci. Total Environ.* **2019**, *690*, 965–972. [CrossRef]
37. Iko Afé, O.H.; Saegerman, C.; Kpoclou, Y.E.; Anihouvi, V.B.; Douny, C.; Igout, A.; Mahillon, J.; Hounhouigan, D.J.; Scippo, M.L. Polycyclic aromatic hydrocarbons contamination of traditionally grilled pork marketed in South Benin and health risk assessment for the Beninese consumer. *Food Addit. Contam. Part A* **2020**, *37*, 742–752. [CrossRef]
38. Wu, C.; Bao, L.J.; Guo, Y.; Li, S.M.; Zeng, E.Y. Barbecue fumes: An overlooked source of health hazards in outdoor settings? *Environ. Sci. Technol.* **2015**, *49*, 10607–10615. [CrossRef]
39. Rahman, M.; Kim, K.H. Release of offensive odorants from the combustion of barbecue charcoals. *J. Hazard. Mater.* **2012**, *215*, 233–242. [CrossRef]
40. Kistler, M.; Schmidl, C.; Padouvas, E.; Giebl, H.; Lohninger, J.; Ellinger, R.; Bauer, H.; Puxbaum, H. Odor, gaseous and PM10 emissions from small scale combustion of wood types indigenous to Central Europe. *Atmos. Environ.* **2012**, *51*, 86–93. [CrossRef]
41. Parliament of the United Kingdom. *Environmental Protection Act 1990*; The Stationery Office: London, UK, 1990.
42. Shapiro, S.S.; Wilk, M.B. An analysis of variance test for normality (complete samples). *Biometrika* **1965**, *52*, 591–611. [CrossRef]
43. Welch, B.L. The generalization of 'student's' problem when several different population variances are involved. *Biometrika* **1947**, *34*, 28–35. [CrossRef]
44. Man, H.B.; Whitney, D.R. On a test of whether one of two random variables is stochastically larger than the other. *Ann. Math. Stat.* **1947**, *18*, 50–60. [CrossRef]
45. Python Software Fundation. *Python Language Reference, Version 3.7*; Python Software Foundation: Wilmington, NC, USA, 2020.
46. McKinney, W. Data structures for statistical computing in Python. In Proceedings of the 9th Python in Science Conference, Austin, TX, USA, 28 June–3 July 2010.
47. Virtanen, P.; Gommers, R.; Oliphant, T.E.; Haberland, M.; Reddy, T.; Cournapeau, D.; Burovski, E.; Peterson, P.; Weckesser, W.; Bright, J.; et al. SciPy 1.0: Fundamental algorithms for scientific computing in Python. *Nat. Methods* **2020**, *17*, 261–272. [CrossRef] [PubMed]
48. Hunter, J.D. Matplotlib: A 2D graphics environment. *Comput. Sci. Eng.* **2007**, *9*, 90–95. [CrossRef]
49. Maurin, K. *Analysis Part One: Elements*, 1st ed.; Springer Netherlands: Heidelberg, Germany, 1976.
50. Kubica, K. *Proposal of Energy and Emission Criteria for Central Heating Boilers Fired with Solid Fuels Promoted under the Low-Stack Emission Reduction Program (PONE)*; Polish Chamber of Ecology: Katowice, Poland, 2013; (Elaboration in Polish).
51. Stanek, W.; Kubica, R.; Plis, M.; Bogacz, W.; Falecki, A. *Wielowariantowa Analiza Eliminowania Przestarzałych, Niskoefektywnych Energetycznie i Wysokoemisyjnych Źródeł Wytwarzania Energii Użytkowej ze Spalania Węgla w Indywidualnych Gospodarstwach Domowych, STOPSMOG Elaboration*; Institute of Environmental Protection-National Reseach Institute: Warsaw, Poland, 2017; (Elaboration in Polish).

52. Stala-Szlugaj, K. Analysis of the municipal and housing hard coal consumers sector. *Energy Policy J.* **2017**, *20*, 117–133.
53. Dreger, M.; Lewandowski, P.; Mazur, J.; Pytliński, Ł.; Sałach, K.; Skoczkowski, T.; Sokulska, A.; Śniegocki, A.; Toborek-Mazur, J.; Walczak, E. *et al. Energy efficiency in Poland. 2017 Review*; Institute of Environmental Economics: Cracow, Poland, 2018; (Elaboration in Polish).
54. Statistics Poland. *Environment 2019*; Zakład Wydawnictw Statystycznych: Warsaw, Poland, 2019; (Elaboration in Polish).
55. TNS Poland. Available online: www.wirtualnemedia.pl/artykul/3-4-polakow-regularnie-grilluje-kielbasa-najpopularniejsza (accessed on 20 October 2020).
56. Barbecue Habits of Poles. Available online: https://www.wiadomoscihandlowe.pl/artykul/grillowe-zwyczaje-polakow-raport-na-majowke (accessed on 18 October 2020).
57. Barbecue Trends in the UK & Europe. Available online: https://www.slideshare.net/KatyPhillips6/barbecue-trends-in-the-uk-europe (accessed on 18 October 2020).
58. How much gas does a grill consume and what does a gas grill cost? Available online: https://www.broilking.pl/blog/ile-gazu-zuzywa-grill-i-jaki-jest-koszt-grillowania-na-gazie (accessed on 18 October 2020).
59. MacCarty, N.; Still, D.; Ogle, D. Fuel use and emissions performance of fifty cooking stoves in the laboratory and related benchmarks of performance. *Energy Sustain. Dev.* **2010**, *14*, 161–171. [CrossRef]
60. Bhattacharya, S.C.; Albina, D.O.; Abdul Salam, P. Emission factors of wood and charcoal-fired cookstoves. *Biomass Bioenergy* **2002**, *23*, 453–469. [CrossRef]
61. Brunner, T.; Obernberger, I.; Scharler, R. Primary measures for low-emission residential wood combustion—Comparison of old with optimised modern systems. In Proceedings of the 17th European Biomass Conference & Exhibition, Hamburg, Germany, 29 June–3 July 2009.
62. Vicente, E.D.; Vicente, A.; Evtyugina, M.; Carvalho, R.; Tarelho, L.A.C.; Oduber, F.I.; Alves, C. Particulate and gaseous emissions from charcoal combustion in barbecue grills. *Fuel Process. Technol.* **2018**, *176*, 269–306. [CrossRef]
63. United States Environmental Protection Agency. *Emissions from street vendor cooking devices (charcoal. Grilling)*; Office of Research and Development: Washington, DC, USA, 1999.
64. Adamkiewicz, Ł.; Badyda, A.J.; Gayer, A.; Mucha, D. Disability-adjusted life years in the assessment of health effects of traffic-related air pollution. *Adv. Exp. Med. Biol.* **2015**, *834*, 15–20.
65. Badyda, A.J.; Grellier, J.; Dąbrowiecki, P. Ambient PM2.5 exposure and mortality due to lung cancer and cardiopulmonary diseases in polish cities. *Adv. Exp. Med. Biol.* **2017**, *944*, 9–17.
66. Loomis, D.; Grosse, Y.; Lauby-Secretan, B.; El Ghissassi, F.; Bouvard, V.; Benbrahim-Tallaa, L.; Guha, N.; Baan, R.; Mattock, H.; Straif, K. The carcinogenicity of outdoor air pollution. *Lancet Oncol.* **2013**, *14*, 1262–1263. [CrossRef]
67. Mitchell, E.J.S.; Lea-Langton, A.R.; Jonesa, L.M.; Williams, A.; Layden, P.; Johnson, R. The impact of fuel properties on the emissions from the combustion of biomass and other solid fuels in a fixed bed domestic stove. *Fuel Process. Technol.* **2016**, *142*, 115–123. [CrossRef]
68. Viana, M.; Alastuey, A.; Querol, X.; Guerreiro, C.; Vogt, M.; Colette, A.; Collet, S.; Albinet, A.; Fraboulet, I.; Lacome, J.; et al. *Contribution of Residential Combustion to Ambient Air Pollution and Greenhouse Gas Emissions*; Technical Report 2015/1; European Topic Centre on Air Pollution and Climate Change Mitigation: Bilthoven, The Netherlands, 2016.

Publisher's Note: MDPI stays neutral with regard to jurisdictional claims in published maps and institutional affiliations.

© 2020 by the authors. Licensee MDPI, Basel, Switzerland. This article is an open access article distributed under the terms and conditions of the Creative Commons Attribution (CC BY) license (http://creativecommons.org/licenses/by/4.0/).

energies

MDPI

Article

The Analysis of the Effectiveness of Implementing Emission Reduction Measures in Improving Air Quality and Health of the Residents of a Selected Area of the Lower Silesian Voivodship

Izabela Sówka [1,*], Małgorzata Paciorek [2], Krzysztof Skotak [3], Dominik Kobus [4], Maciej Zathey [5,6] and Krzysztof Klejnowski [7]

1 Department of Environment Protection Engineering, Faculty of Environmental Engineering, Wroclaw University of Science and Technology, Pl. Grunwaldzki 9, 50-377 Wroclaw, Poland

2 BSiPP EKOMETRIA Sp. z o.o., ul. Orfeusza 2, 80-299 Gdańsk, Poland; malgorzata.paciorek@ekometria.com.pl

3 Integrated Environmental Monitoring Department, Institute of Environmental Protection—National Research Institute, ul. Krucza 5/11d, 00-548 Warsaw, Poland; krzysztof.skotak@ios.edu.pl

4 inFAIR, ul. Puszczyka 10/55, 02-785 Warsaw, Poland; dominik.kobus@infair.eu

5 Department of Spatial Planning and Settlements Processes, Faculty of Architecture, Wroclaw University of Science and Technology, B. Prusa 53/55, 50-317 Wroclaw, Poland; maciej.zathey@irt.wroc.pl

6 Institute for Territorial Development, J. Wl. Dawida 1a, 50-527 Wroclaw, Poland

7 Institute of Environmental Engineering, Polish Academy of Sciences in Zabrze, M. Skłodowskiej-Curie 34 Str., 41-819 Zabrze, Poland; krzysztof.klejnowski@ipis.zabrze.pl

* Correspondence: izabela.sowka@pwr.edu.pl; Tel.: +48-71-320-25-60

Received: 7 July 2020; Accepted: 27 July 2020; Published: 3 August 2020

Abstract: The case study selected in order to analyze and evaluate the effectiveness of implemented solutions for improving air quality with the WRF-CALMET/CALPUFF modeling system as an element of decision support was the subject of this paper. Its character can be considered unique due to its geographical location, topography and the functioning PGE GiEK S.A. Turów Power Complex (ELT), and, in particular, the PGE GiEK S.A. Turów Coal Mine (KWBT). The conducted analyses have defined three scenarios of emission changes: (1) scenario related to the activities of the energy complex resulting from the minimizing measures indicated in the report on the environmental impact of the mine, (2) scenario resulting from the so-called "anti-smog" regional resolution and (3) scenario compiling the abovementioned scenarios. According to the analyses, the lowest values of the annual mean $PM_{2.5}$ concentration were noted in the eastern part of the studied area and did not exceed 14 μg/m^3 (56% of the permissible level). The best results in improving air quality were proven for scenario 3, i.e., a 10% reduction in concentration values over the entire analyzed area of the commune. In the case of this scenario, as the most effective and health-promoting solution, only in 25% of the area was the improvement in the residents' health below 5%, while the reduction in the estimated number of premature deaths by over 15% was observed in nearly one third of the studied area.

Keywords: lignite mine; $PM_{2.5}$; PM_{10}; WRF-CALMET/CALPUFF; health risk

1. Introduction

Ambient air pollution is one of the main environmental problems in Europe, and some pollutants are the issue primarily in Poland. The European Union has implemented legislation that sets standards and goals for many air pollutants in the form of Directives 2008/50/EC and 2004/107/EC [1,2]. These provisions have been transposed into national law, primarily as the provisions of the Act on

Environmental Protection Law and the Regulation of the Minister of the Environment on the levels of certain substances in the air [3,4]. In addition to standards for the concentrations of selected pollutants, the provisions also regulate the methods of air quality assessment and management, e.g., through the development and implementation of recovery programs and air quality plans.

International organizations dealing with air quality issues and assessing their impact on society, health and the environment, such as the World Health Organization (WHO) and the European Environment Agency (EEA), study and analyze the impact of air pollution on health. According to EEA estimates, there were around 412,000 premature deaths attributable to ambient $PM_{2.5}$ pollution in 41 European countries in 2016, of which 43,100 deaths occurred in Poland (premature deaths are deaths that occur before reaching the average age defined by life expectancy. Premature deaths are considered avoidable if their causes were eliminated) [5]. Particulate matter, both less than 2.5 μm and less than 10 μm in diameter, is the source of pollution that causes the most problems and exceeds permissible air quality standards.

Adverse health effects of air pollution, particularly regarding particulate matter, are observed as a result of both long- and short-term exposure. For this reason, WHO recommendations, as well as air quality standards introduced by the European Union legislation and transposed into the laws of individual Member States, define annual mean and daily limit values. According to Polish regulations, pursuant to Directive 2008/50/EC, the annual mean limit value concentration for $PM_{2.5}$ is 25 $\mu g/m^3$ (since 1 January 2020, the so-called second phase of the standard is 20 $\mu g/m^3$). For PM_{10}, the annual mean is 40 $\mu g/m^3$ and the 24-h mean is 50 $\mu g/m^3$ (allowing 35 days with exceedances of the limit value in a calendar year). In addition, selected European countries have introduced information and alarm thresholds for PM_{10} concentration. As of 2019, these thresholds in Poland are respectively 100 $\mu g/m^3$ and 150 $\mu g/m^3$ for the daily mean concentration [6].

The results of the air quality assessments carried out annually in Poland by the Chief Inspectorate for Environmental Protection indicate a bigger problem with meeting the PM_{10} daily limit value. In 2018, the norm was exceeded in 39 of the 46 assessed zones in the country. Annual means were exceeded in nine zones. The limit value was exceeded during this period at 160 and 25 measuring sites in the country out of 227 included in the assessment, for both standardized averaging times, respectively [7]. Therefore, actions have been taken to improve air quality in Poland [8]. One example of these activities is restrictions and bans on the operation of fuel-burning installations, effective also in the Lower Silesian Voivodship. This is due to three so-called "anti-smog" resolutions adopted by the Lower Silesian Regional Assembly on 30 November 2017 [9–11]. One of the abovementioned resolutions covers the territory of the entire voivodship, excluding the city of Wrocław and 11 health resorts of the Lower Silesian Voivodship. The provisions introduced by this resolution are binding in Bogatynia commune and introduce restrictions and bans on the operation of fuel-burning installations. As of 1 July 2018, in Bogatynia commune, it is forbidden to use the following:

- lignite and solid fuels manufactured with lignite
- coal sludges, coal flotoconcentrates and mixtures manufactured with them
- hard coal in fine form (fine coal) with grain size less than 3 mm
- solid biomass (including wood) with moisture content above 20%.

The resolution introduces a gradual withdrawal of out-of-class installations, and as of 1 July 2018, it is allowed to install only such new boilers and local air heaters (fireplaces) that meet the ecodesign requirements regarding particulate matter emissions [12]. As of 1 July 2024, the resolution introduces a ban on the use of solid fuel installations that do not meet a minimum of third class requirement according to PN-EN 303-5:2012. The deadline for the implementation of the resolution and the resulting bans is 1 July 2028. From then on, the use of solid fuel installations that do not meet the minimum emission standards of class 5 in terms of particulate matter emission limits according to PN-EN 303-5:2012 will be prohibited. Introducing these restrictions for the combustion of solid fuels and the use of installations is expected to bring a significant improvement in air quality and thus reduce the likelihood of developing air pollution-related diseases.

In Poland, according to official data, the majority of PM_{10} emissions to the atmosphere (approximately 47%) are generated by non-industrial combustion processes, including in the municipal and residential sector related to solid fuel biomass combustion for heating and hot water preparation [13]. The next three groups with the largest share in PM_{10} emissions are combustion processes in industry (approximately 14%), road transport (approximately 8%) and manufacturing (approximately 7%). These data are from 2017 and concern the entire country [14]. The situation may be different in individual regions or may be considered locally, where the impact of individual sectors may vary. This may be connected with the presence of a specific source or group of industrial sources, or the density of the high-traffic road network, which occurs primarily in the central areas of the agglomeration and large cities. The impact of specific groups of sources is also time-varying and depends, e.g., on the season of the year (heating and non-heating period) or day (variability of household heating systems activity or the volume of traffic) [15].

The analysis of the reasons for high levels of particulate matter concentration, including the exceedance of daily limit values, must also consider the episode period, location of the area of exceedance in relation to emission sources and local topographic conditions affecting ventilation possibilities, as well as meteorological conditions conducive to the accumulation of pollution, such as low wind speed or temperature inversion phenomena. In the case of large urban centers, peripheral areas are generally more exposed to the influence of heating sources, while car transport may be of more importance in the city centers. In special meteorological situations and certain areas, the movement of pollutants (inflows) and their accumulations affect the range of impact of individual emission sources [16].

One of the elements of ambient air quality management in a given area is its assessment and diagnosis of the conditions, taking into account possible exceedances of the limit values. In accordance with the current laws, as part of the State Environmental Monitoring coordinated by the Chief Inspectorate of Environmental Protection (CIEP), this assessment is carried out using three basic groups of methods [17,18]:

1. measuring (monitoring) concentrations of selected pollutants,
2. mathematical modeling of transformation in the atmosphere and transport of pollution,
3. objective estimation, based on various methods and information, including the spatial distribution and activity of pollution emission sources.

Actions connected with the following aspects are examples of the air quality management process:

- air quality plans and programs for corrective measures aimed at achieving and/or maintaining appropriate level of air quality
- analysis of the effects of implementing specific solutions (scenario analysis)
- analysis of the impact of air pollution on specific social groups and other environment elements, such as vegetation or materials (e.g., buildings or technical facilities)
- air quality forecasting, considering emission sources and meteorological conditions
- informing the public and policymakers about historical, current and predicted air quality.

All of the above elements occur at various levels: international (e.g., European), national, regional (e.g., voivodship) and local (e.g., urban or a specific industrial plant or a single installation). All of them also use various air quality modeling techniques, such as chemical modeling of transport and transformation of pollutants, receptor or statistical modeling, also using, e.g., artificial neural networks [16,19,20].

Table 1 summarizes examples of the use of modeling techniques at various levels and for the purpose of achieving various main objectives of air quality management. It presents general examples and, in selected cases, references to specific projects or applied solutions. Considering the comfort of life and health of people, especially in the areas close to objects that can significantly affect the environment, activities aimed at minimizing the onerous impact of those objects are particularly important. Therefore, in order to analyze and evaluate the effectiveness of the implemented solutions in the context of improving air quality using modeling, the area selected as a case study is the commune in Poland of a nature unique to Poland and Europe, both due to its geographical location bordering with two countries—Czechia and Germany—and its characteristically diverse terrain. The selected area is an interesting case also because it contains one of the largest energy complexes in Poland, which includes an open-pit lignite mine with a large lignite-fired fuel combustion facility. In addition, ambient air quality in the analyzed area is influenced by combustion processes in household boilers. Therefore, the analysis of the distribution of pollutant concentrations and the assessment of the effectiveness of measures for reducing air pollutant emissions, including health risk assessment, are extremely important.

Table 1. Examples of the use of modeling techniques at various levels for the purposes of achieving the main objectives of air quality management with particular emphasis on the area of Poland.

Purpose of Activities	International Level	National Level	Regional Level	Local Level
Air quality assessment (diagnosis)	Analyses regarding air quality in Europe and its impact on health and the mortality rate of society conducted by the European Environment Agency and the cooperating European Topic Centre on Air Pollution, Transport, Noise and Industrial pollution https://www.eea.europa.eu/themes/air Air quality analyses on a continental and regional scale, developed and published as part of the European CAMS (Copernicus Atmosphere Monitoring Service) project https://atmosphere.copernicus.eu/ Air quality analyses in Europe carried out under the EMEP Programme implemented under the Convention on Long-Range Transboundary Air Pollution (LRTAP) https://emep.int/mscw/mscw_moddata.html	Assessment of air quality carried out as part of the SEM * by CIEP **, including identification and analysis of situations and areas of exceedance in individual zones of the country, e.g., with the use of modeling provided by IEP-NRI *** http://powietrze.gios.gov.pl/pjp/publications/card/19100 Identification of episodes of high concentrations of particulate matter in Poland and neighboring countries, considering the aspect of transboundary pollution transfer http://powietrze.gios.gov.pl/pjp/publications/card/2052 Analyses of air quality in Poland performed by West and East Meteorological Synthesizing Centers operating under the Convention on Long-Range Transboundary Air Pollution (LRTAP) https://www.emep.int/mscw/mscw_publications.html http://en.msceast.org/index.php/publications/reports?id=204	Additional diagnostic analyses performed for the purposes of developing Air Protection Programmes for voivodships (e.g., identification of the share of emission sources) http://bip.umwd.dolnyslask.pl/dokument,iddok,51944,idmp,293,r,r	Air quality analyses at city level, with impact assessment of specific emission sources. https://journals.pan.pl/dlibra/show-content?id=102486 & https://laqm.defra.gov.uk/ Impact analysis for specific groups of emission sources, industrial facilities or installations, with health impact assessment http://life-apis.meteo.uni.wroc.pl/
Programming and designing of activities	Sectoral analyses at European Union level, e.g., emissions control strategy analyses using the GAINS **** model http://gains.iiasa.ac.at/gains/emissions.EU/index.menu?page=303 Analyses related to the development and implementation of legislative solutions (e.g., EU directives), e.g., Air Quality–Fitness Check of the AAQ Directives ***** conducted by the European Commission https://ec.europa.eu/environment/air/quality/aqd_fitness_check_en.htm	Sectoral analyses on a national scale https://bip.mos.gov.pl/strategie-plany-programy/krajowy-program-ograniczania-zanieczyszczenia-powietrza/ https://nfosigw.gov.pl/download/gfx/nfosigw/pl/nfoekspertyzy/858/92/1/2011-160.pdf https://www.gov.pl/web/aktywa-panstwowe/zaktualizowany-projekt-polityki-energetycznej-polski-do-2040-r Scenarios and long-term forecasts related to adopting legal and organizational solutions at national level, e.g., implemented at the request of CIEP http://powietrze.gios.gov.pl/pjp/content/air_quality_forecast_long_term	Development of air protection programs, including scenario analysis and assessment of the effectiveness of corrective actions at the voivodship level. https://www.mazovia.pl/ekologia-i-srodowisko/ochrona-powietrza/ Analyses related to the development of regional anti-smog resolutions https://irt.wroc.pl/pliki/ekspertyza_wskazujaca_efekt_ekologiczny_w_dolnoslaskich_uzdrowiskach/index.html	Analyses of local corrective action scenarios https://bip.um.wroc.pl/artykul/643/25539/ograniczenia-niskiej-emisji-z-indywidualnego-ogrzewania-weglowego-na-terenie-wroclawia-w-latach-2016-2020 Estimating the impact of implementing local policies and programs on air quality (e.g., in commune, city or health resort) https://bip.um.wroc.pl/artykul/643/25539/ograniczenia-niskiej-emisji-z-indywidualnego-ogrzewania-weglowego-na-terenie-wroclawia-w-latach-2016-2020 Environmental impact assessment of projects, including the potential impact of planned installations on air quality https://bip.gmstrzelin.finn.pl/res/serwisy/pliki/17237287?version=1.0 Analysis of the effectiveness of designed mechanisms and techniques for reducing air emissions http://raportyspoleczne.pl/wp-content/uploads/raports/efebccc9f3e11a50f68503b53485a5c9.pdf Estimating the effects of implementing organizational and technological solutions at the city level (e.g., changing the communication system, introducing Intelligent Transport Systems, etc.) https://its.tychy.pl

Table 1. *Cont.*

Purpose of Activities	International Level	National Level	Regional Level	Local Level
Forecasting and information	Regional air quality forecasts published as part of the CAMS project https://atmosphere.copernicus.eu/ Information on air quality in Europe (European Air Quality Index) based on measurement and modeling results (forecasts) implemented by the European Environment Agency https://airindex.eea.europa.eu/ Current and forecasted air quality information obtained using modeling, made available on an open and commercial basis on the BreezoMeter website https://breezometer.com/	Short-term air quality forecasts on a national scale performed by the IOŚ-PIB and published by CIEP http://powietrze.gios.gov.pl/pjp/airPollution.	Regional air quality forecasts and information services http://powietrze.podkarpackie.pl/ https://powietrze.malopolska.pl/jakosc-powietrza/ http://powietrze.gios.gov.pl/pjp/airPollution?woj=mazowieckie&rwms=true	Local small-scale air quality forecasts, considering specific conditions of topography, development and arrangement of emission sources https://air.wroclaw.pios.gov.pl/prognozyhttp://life-apis.meteo.uni.wroc.pl/ http://powietrze.pwr.edu.pl/ https://www.londonair.org.uk

* SEM—State Environmental Monitoring; ** CIEP—Chief Inspectorate of Environmental Protection; *** IEP-NRI—Institute of Environmental Protection—National Research Institute; **** GAINS model —Greenhouse gas—Air pollution INteractions and Synergies model; ***** AAQ Directives - EU Ambient Air Quality Directives

2. Characteristics of the Research and Analysis Area

The area of Bogatynia commune (Figure 1) located within three basic geographical regions: the mesoregion of the Zittau Basin, the Izera Foothills and Izera Mountains and the Nysa Łużycka Valley were selected for the case study. The mesoregion of the Zittau Basin has a lower terrain in relation to the Izera Foothills and the Izera Mountains surrounding it from the northeast and the east. On the west side, the area is limited by the Nysa Łużycka Valley, behind which the Zittau Valley extends into the Lusatian Foothills. Such location and character of the area mean that its diverse terrain can significantly shape air mass flows within it.

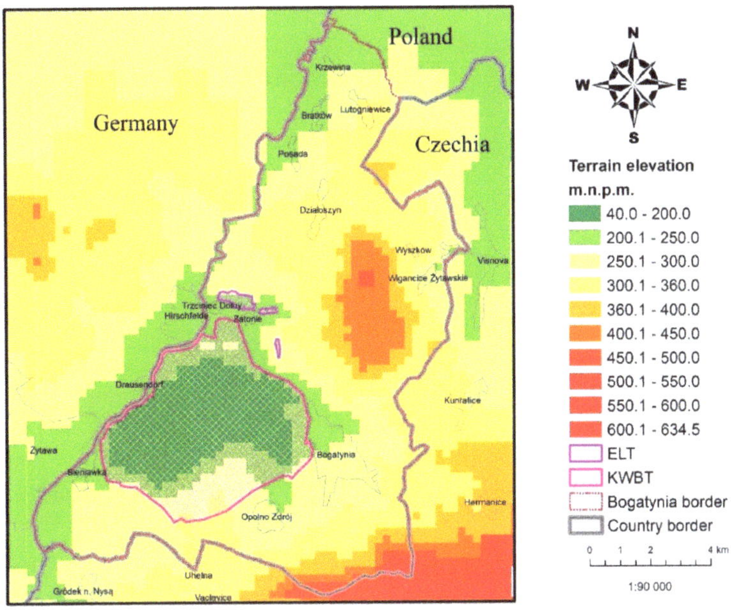

Figure 1. Natural topography in Bogatynia area (source: own study).

The largest cities of the studied region are Polish Bogatynia in the center and German Zittau (Żytawa) and Czech Hrádek nad Nisou (Gródek na Nysą) (Figure 2) in the southwest, already outside Poland. The other places are of a rural character.

Three low-volume voivodship roads, with a volume below 10,000 vehicles a day, run through the studied area. The DW352 road connects Zgorzelec with the state border at the Kunratice/Bogatynia crossing. The DW354 road runs from the Bogatynia–Zatonie district to the west, and further south to the town of Sieniawka, along the border with Germany, where it crosses the border. Road DW332 is a short section connecting route 178 (on the German side) and a border crossing with Germany in the town of Sieniawka, as well as a border crossing with Czechia in the direction of Hrádek nad Nisou, where on the Czech side it turns into route 35. Another regional road is route 99 running on the German side of the border along the Nysa Łużycka River. Other roads are local and have basically no influence on local air quality.

The analyzed area holds one of the largest energy production complexes in Poland, which supplies around 8% of the energy production to the national energy system. It consists of a conventional block heat and power plant (ELT) located in the north, with interstage steam superheating and a closed cooling water system, whose basic fuel is lignite. Currently, the installed capacity of the PGE GiEK S.A. Turów Power Plant Complex (ELT) is 1498.8 MW in six power units of 235 MW and 260 MW capacity. Coal is supplied directly by belt conveyors from the PGE GiEK S.A. Turów Coal Mine (KWBT) located in the south. The surface of the open-pit excavation area with an internal backfill area currently covers around 26 km^2. The open pit is directly adjacent to the town of Bogatynia (in the east) and the state border (in the west). Two districts of Bogatynia—Trzciniec Dolny and Zatonie, located between the power plant and the mine, are significantly exposed to the impact of both facilities. Currently, the bottom of the open-pit mine is around 10 m.a.s.l., and the elevations around it are at 225–300 m.a.s.l. The surrounding mountains and foothills lie at heights above 500 m.a.s.l. in the southern part and in the range 400–500 m.a.s.l. in the northern part.

By 2028, the ambient air quality in the analyzed area will have been shaped by a number of changes significantly affecting the size of emission balances in Bogatynia commune. These activities are discussed in detail in Section 3.1.3.

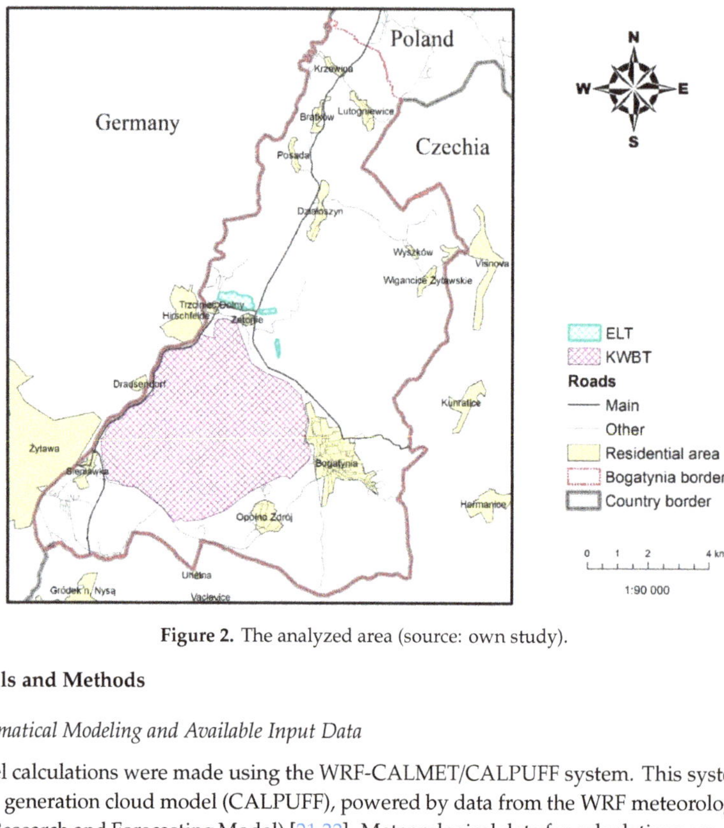

Figure 2. The analyzed area (source: own study).

3. Materials and Methods

3.1. Mathematical Modeling and Available Input Data

Model calculations were made using the WRF-CALMET/CALPUFF system. This system is based on the 2nd generation cloud model (CALPUFF), powered by data from the WRF meteorological model (Weather Research and Forecasting Model) [21,22]. Meteorological data for calculations are prepared by the CALMET preprocessor, which determines the time and space variables of meteorological parameters with the grid resolution specified by the user. The conducted analyses defined three scenarios of emission changes: (1) scenario (scenario 1) related to changes in emissions in the studied mine resulting from the minimizing measures indicated in the report on the mine's environmental impact [23], (2) scenario resulting from the anti-smog resolution in force in the Lower Silesian Voivodship [11] (scenario 2) and (3) scenario compiling the abovementioned scenarios (scenario 3).

For calculations on a local scale, smaller mesh sizes with detailed information about the terrain and land use are applied, as these parameters can significantly affect the shape of the pollution field, especially in mountainous areas. For the purposes of this study, a system of nested grids with resolution sizes of 0.25–1 km was used. The domain in which the meteorological parameter fields were calculated covered the area within 10 km of the border of Bogatynia commune. Meteorological data from 2018 were used, which were adapted to two nested grids with a resolution of 0.5 km in Bogatynia commune and 1 km in the rest of the computational domain (Figure 3a). Information about the terrain (as the average in a grid) and land use (as the prevailing value) was implemented into the model with the same resolution [24,25].

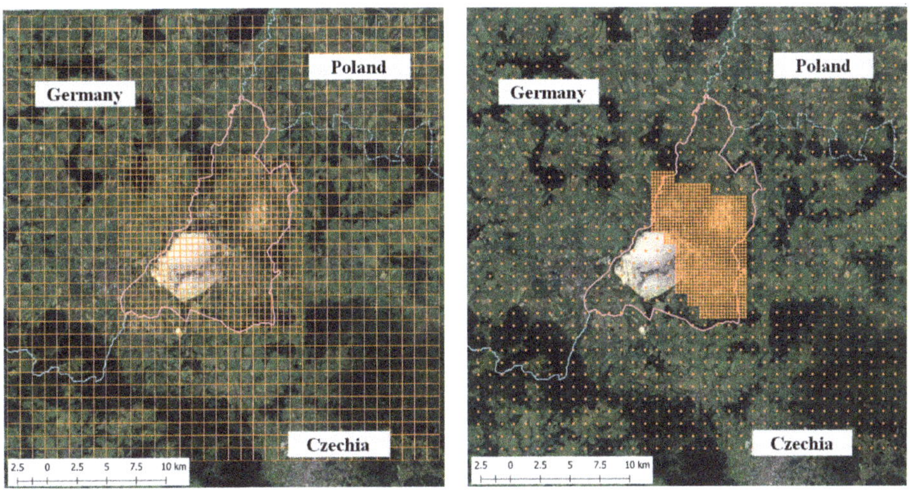

(a) (b)

Figure 3. Parametrization of grids and receptors in the model: (**a**) meteorological grid; (**b**) discrete receptors (source: own study; map background: https://www.google.pl/maps).

Calculations of pollution concentrations were carried out based on grids in two resolutions: 0.25 km in the area of Bogatynia commune and 1 km in the rest of the computational domain (Figure 3b). The obtained results were visualized using inverse distance weighting (IDW). According to this method, the value for each interpolated cell is calculated based on the values of neighboring points weighed by the inverse of their distance. For such a dense receptor network, this is the most optimum interpolation method [26].

In the air quality modeling, the influx of pollutants from outside the examined area was also taken into account as boundary conditions varying in time and space derived from chemistry transport model (CTM) calculations. Data from the Copernicus project were used [27].

3.1.1. Emission Data

Due to the location of the studied area (direct neighborhood of Poland, Czechia and Germany), and the resulting problems with the unification of emission databases, data from three sources described below were used for more comprehensive analyses.

In the area of Poland, data from a detailed emission inventory were used for annual air quality assessments performed up to 2017 by the Voivodship Inspectorate for Environmental Protection in Wrocław (currently the Regional Department of Environmental Monitoring of the Chief Inspectorate for Environmental Protection) [28,29]. This database contains data on all types of emissions from the voivodship, including emissions from residential heating (SNAP0202), transport (SNAP07), agriculture (SNAP10) and industry (SNAP0201, SNAP01, SNAP03, SNAP04). Inventories took into account the activity of sources. Due to the fact that the above database has been created continuously since 2012, data contained therein can be considered of a good quality. The database was additionally validated with annual model calculations [30].

In the calculations, an indispensable element is capturing the temporal and spatial variability of the emission field, because such changes together with the variability of meteorological conditions practically determine the final distribution of pollution concentrations in the studied area. Considering the temporal variability of emissions can also capture the cases of high concentration episodes and link them to the types of emissions responsible for poor air quality. Therefore, the concentration calculations were implemented in an emission model, which includes primarily variations dependent on changes in meteorological conditions (e.g., temperature for emissions from residential heating and precipitation for emissions from transport, agriculture) or the mode of source operation. This methodology was used, among others, in modeling air quality for the needs of annual air quality assessments in the Lower Silesian Voivodship [29]. The results of the calculations obtained in the context of the balance of specific sources in $PM_{2.5}$ and PM_{10} emissions in Bogatynia commune are presented in Figure 4.

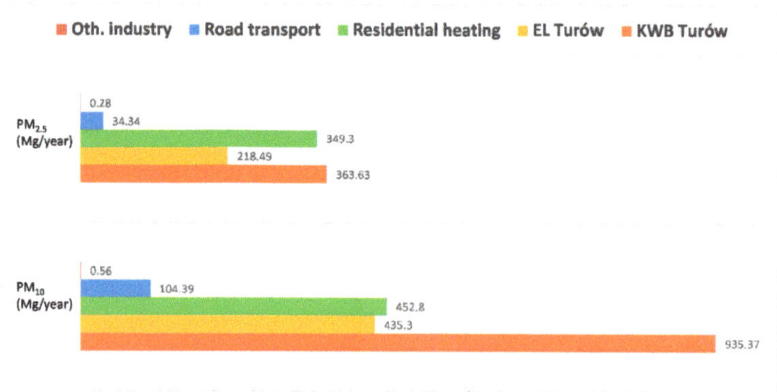

Figure 4. Particulate matter emissions for different emission sources inventoried in Bogatynia commune (source: own study).

The performed analyses show that Turów power complex, particularly the mine, has the largest share in the total PM_{10} emission in Bogatynia commune. In the case of $PM_{2.5}$ emissions, the communal and household sector is clearly the main source of emissions in the studied area. While emissions from household sources, or from transport, can be considered a seasonal variable (heating and post-heating season, daily and weekly transport cycle) and of a quasi-uniform nature, the emission associated with industrial sources may be characterized by short-term changes and be dependent on a number of factors resulting from, e.g., technology, specificity of the sources or a specific mode of operation of an industrial installation. Therefore, considering the impact of industrial sources on air quality, it was necessary to build an emission model dedicated to the examined facility (in this case, ELT), which was presented in the report on the environmental impact of the KWBT [23].

In the case of the ELT power plant, organized emissions resulting mainly from the boiler operating conditions were taken into account. In addition, in the case of ELT, fugitive emission sources were considered, i.e., combustion waste buffer area, where ash from boilers is temporarily stored before being mixed with overburden in an open pit. This happens when it cannot be collected by the mine (KWBT). Emissions for this source were determined on the basis of United States Environmental Protection Agency (US-EPA) emission factors adapted to the characteristics of the source in terms of quantity and quality of the stored material, including machine operation [23]. Examples of emission values included in the calculation scenarios are presented in Table 2.

Table 2. PM_{10} and $PM_{2.5}$ emission from ELT in 2018.

Emission Sources	PM_{10} (Mg/Year)	$PM_{2.5}$ (Mg/Year)
Point source emitters	375.6	194.7
Fugitive emission	60.3	23.8

In the case of KWBT, four basic groups of emission sources were identified: emissions related to the mining process, storage and reloading of coal, transport and entrainment of material from exposed surfaces. The emissions related to the mining process covered the exploitation area (northern part of the open pit) and the backfilling area (southern part of the open pit). It was assumed that the volume of emissions from the exploitation area compared to the backfilling area would be significantly lower due to the quarried material, which is very moist and heavy. In the backfilling area, the quarried material is much more volatile because it consists mainly of overburden loam and ash from power plant boilers. Emissions related to storage and reloading concerned a coal bin located in the north of the facility. In the area of the bin, there are two trenches in which higher quality coal is deposited and two coal sales points—wholesale and retail. In addition to emissions from coal loading and unloading, emissions associated with wind entrainment are an important factor in this case. The last group is formed by emissions related to transport, both through belt conveyors and through vehicles and railways (wholesale and retail). The emission model for the mine was based on indicators determined with reverse modeling, US-EPA methodologies, considering the specificity of meteorological conditions and mineral material mined and backfilled in the open pit, and the time of operation of the machines [23,30]. Examples of emission values included in the calculation scenarios are presented in Table 3.

Table 3. PM_{10} and $PM_{2.5}$ emission from ELT in 2018 by sector (on the basis of [23,30]).

Emission Sources	PM_{10} (Mg/Year)	$PM_{2.5}$ (Mg/Year)
Mining	47.7	12.4
Backfilling	860.90	344.36
Coal bin with a coal transfer point	22.93	5.94
Coal sorting area	0.11	0.03
Retail point of sales area	0.0007	0.0002
Wheeled transport	3.73	0.9
Total	935.37	363.63

The last group of data is information on emissions from Czechia and Germany from the emission inventory posted on the public websites of the Czech Hydrometeorological Institute and the German Ministry of the Environment [31,32]. In Germany, the inventory included sources related to residential heating (SNAP 0202) and SNAP 07 road transport (also SNAP 08 other transport), and the total emission of PM_{10} and $PM_{2.5}$ was 44.6 Mg. For Czechia, the data concerned industry (SNAP01, SNAP03, SNAP04), residential heating (SNAP0202) and road transport (SNAP07). Total emissions in the studied area are around 182 Mg for PM_{10} and 169 Mg for $PM_{2.5}$.

3.1.2. Validation of Modeling Results in the Base Year

The obtained calculation results were validated using the available measurement data from measuring stations located in Poland, Czechia and Germany (Figure 3, Table 2). The stations were selected based on the availability of results in publicly available databases (e.g., AIRBASE, EEA European Air Quality Portal, JPOAT via the Air Quality Portal of the Chief Inspectorate for Environmental Protection). It was also assumed that the completeness of the measurement series must meet the requirements of the CAFE directive [1]. Ultimately, the analysis included measurements from six stations: four in Poland (Wyszków, Jasna Góra, Bogatynia, Działoszyn), one in Czechia (Frýdlant) and one in Germany (Zittau), the location of which is shown in Figure 5. Three of the selected stations located in Poland belong to the measurement network of ELT, and the station in Działoszyn is included in the network of the State Environmental Monitoring (SEM) run by the Chief Inspectorate of Environmental Protection under the national code DSDzialoszyn. The stations in Zittau and Frýdlant are national stations of the German and Czech networks.

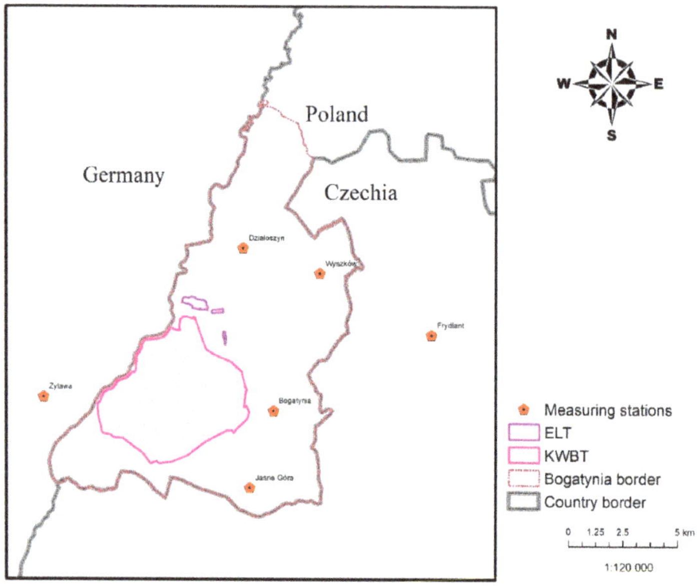

Figure 5. Location of measuring stations (source: own study).

The relative error rate of the modeling result in relation to the measurement was the basic measure of the correctness of the results of model calculations. According to the CAFE directive, this parameter for particulate matter (including PM_{10} and $PM_{2.5}$) should not exceed 50% for annual mean values [1]. Negative values indicate an underestimation of concentrations.

The values obtained as a result of the comparison allowed us to determine the uncertainty of the model (relative errors) and are listed in Table 4. In most cases, the uncertainty of the model did not exceed 13%. The highest compliance was obtained at the station in Bogatynia (for both PM_{10} and $PM_{2.5}$) and at stations in Zittau and Frýdlant, where the relative errors did not exceed 5%. The highest 32% underestimation of results was obtained at the station in Działoszyn. At the same time, it can be stated that a match of results was much better for $PM_{2.5}$ concentrations than for PM_{10}.

Table 4. Annual mean concentrations of PM_{10} and $PM_{2.5}$ modeling results and measurement.

Monitoring Station Name	Parameter	Measurement (μg/m^3)	Model (μg/m^3)	Relative Error (%)
Wyszków (ELT station)	PM_{10} year	20.84	18.70	−10
	$PM_{2.5}$ year	15.52	14.77	−5
Jasna Góra, ul. Sportowa (ELT Station)	PM_{10} 24 h	36.54	31.89	−13
	PM_{10} year	22.58	19.83	−12
Bogatynia, ul. Chopina (ELT station)	PM_{10} year	30.20	30.15	0
	$PM_{2.5}$ year	20.75	20.61	−1
Działoszyn/DSDzialoszyn (SEM station)	PM_{10} year	28.93	19.81	−32
Zittau (DE)	PM_{10} year	21.70	22.65	4
Frýdlant (CZ)	PM_{10} year	18.0	17.4	−3

3.1.3. Emission Change Scenarios

The assessment of the effectiveness of planned long-term actions or ad hoc measures limiting the emission of pollutants is carried out, among others, through changes in the emission introduced into the model, activity of sources and through a possible modification of the original assumptions regarding, e.g., operating mode or frequency of preventive measures used. These changes may concern both the volume of emission loads and the modification of time variations. Then, based on the new emission values, re-calculations were made, assuming no changes for the remaining emission data and/or meteorological parameters. This study discusses three scenarios for emission changes. The first scenario is related to changes in the volume of emissions from the mine, indicated in the report on the mine's Environmental Impact Assessment [23]. The second scenario evaluates the effectiveness of implementing the anti-smog resolution in force in the Lower Silesian Voivodship, which will result in a deep modification in the profile of residential heat sources [11]. The last scenario is an assessment of the implementation of both previously mentioned scenarios. The data included in the calculations for the three assumed scenarios are summarized in Table 5.

Table 5. Emission for individual sources included in the calculations for individual scenarios.

Emission Source	PM_{10} (Mg/Year)	$PM_{2.5}$ (Mg/Year)	PM_{10} (Mg/Year)	$PM_{2.5}$ (Mg/Year)	PM_{10} (Mg/Year)	$PM_{2.5}$ (Mg/Year)
	Scenario 1		Scenario 2		Scenario 3	
KWBT	521.51	207.88	935.37	363.63	521.51	207.88
ELT	435.3	218.49	435.3	218.49	435.3	218.49
Household heating	452.8	349.3	19	18.1	19	18.1
Road transport	104.39	34.34	104.39	34.34	104.39	34.34
Oth. industry	0.56	0.28	0.56	0.28	0.56	0.28
Czechia and Germany	226.8	213.65	226.8	213.65	226.8	213.65
Total	1741.36	1023.94	1721.42	848.49	1307.56	692.74

As a result of the analyses presented in the report on the environmental impact of the mine [23], in order to avoid a significant impact of the object on the neighboring areas, it was necessary to indicate possible additional mitigation measures (Table 6).

In addition, in 2019, a new ash conveyor was created directly from the power plant to the open pit, which will eliminate the impact of the storage area. It was estimated that, as a result of the implemented measures, the total emissions from the mine will be reduced from 935.4 Mg by approximately 44% for PM_{10} and from 363.6 Mg by approximately 43% for $PM_{2.5}$ and will be 521.5 Mg and 207.9 Mg, respectively.

Another important change in the emission characteristics of the commune is the implementation of the anti-smog resolution, according to which, by 2028, boilers of a class lower than 5 will not be operating in the entire Lower Silesian Voivodship. Within the voivodship, it is allowed to burn solid fuels in devices from which "particulate matter emissions do not exceed the emission threshold values set out in Commission Regulation (EU) 2015/1189 of 28 April 2015 implementing Directive 2009/125/EC of the European Parliament and of the Council with regard to ecodesign requirements for solid fuel boilers" [12]. The implementation of the resolution is estimated to result in around a 95% reduction in dust pollution. Total PM_{10} emissions are to be reduced from 452.8 to around 19 Mg and $PM_{2.5}$ from 394.3 to 18.1 Mg. The values of emission totals and the scale of emission reduction were estimated on the basis of an emission database from a detailed inventory of emissions prepared for annual air quality assessments carried out until 2017 by the Voivodship Inspectorate for Environmental Protection in Wrocław (currently the Regional Department of Environmental Monitoring of CIEP). This database contains, among others, information on surfaces heated using old-type boilers fed with coal and wood.

Table 6. Measures to minimize the impact of KWBT on air quality in Bogatynia commune (on the basis of [23]).

Action	Location of Action	Effectiveness of Action
Spraying of working levels (water cannons)	Open-pit area (backfilling and operation area)	Up to 60%
Securing the top of the backfill area, which will be backfilled again or redeposited	The open-pit region	Depending on the degree of land cover (reduction of emissions to 100%)
Organization of work control system depending on weather conditions	Open-pit area and coal bin	20–100%
Housing of selected sections of conveyor belts with particular emphasis on ash lines	Open-pit area and coal bin	Up to 70–85%
Water mist installations in transfer nodes	Open-pit area and coal bin	Up to 90%
Limiting the height of free falling of dusting material	Open-pit area and coal bin	Depending on the degree of reduction, e.g., reduction of height by 50%; reduction of emissions by around 60%
Building on selected transfer node	Coal bin	100%
Windscreen assembly	Coal bin	70–80%
Sprinkling of technical roads	Coal bin/transport	Up to 60%
Regular road washing	Coal bin/transport	50%
Reduction of car traffic on the "coal road" by 50%	Coal bin/transport	40–50%
Reducing brown coal sales including liquidation of sales to domestic retail customers	Coal bin/retail sale	Up to 90%

3.2. Health Risk Assessment

The health risk assessment associated with long-term (annual) $PM_{2.5}$ exposure for the analyzed scenarios was performed using the dose–response function and relative risk index (RR (95% CI) = 1.062 (1.040–1.083) for every 10 μg/m^3) [33,34].

The crucial element of health risk assessment for the analyzed area was accurate mapping of the exposure of the population at risk. For this reason, the health assessment was performed for each of the separated areas adopted as the air quality model grid, based on obtained results of the level of annual mean $PM_{2.5}$ concentration and demographic indicators assigned to the model grid (Figure 6). The number of exposed people and the number of deaths classified by causes were determined on the basis of data from official statistics with a division of the population into age groups [35]. For more detailed analysis, data from spatial distribution were used, considering the current administrative division of the country according to Nomenclature of Territorial Units for Statistics (NUTS) level 5 and the state register of administrative borders [36–39]. For this purpose, among others, the dasymetric map of population density developed by the European Environment Agency—raster layer "Population density grid of EU-27+, version 4 and 5"—was used. The dasymetric map and population density were corrected in line with the current land use information. To verify the population in the model grid, the National Official Register of the Territorial Division of the Country was used with the distinction of towns with district rights, urban, rural and urban-rural communes as well as towns and rural areas in urban-rural communes.

The analyses were performed for the baseline condition without considering emission reduction scenarios and for each scenario separately. The results are presented both in the form of the estimated number of premature deaths related to the exposure of the general population to $PM_{2.5}$ and relative changes in the impact assessment for each scenario.

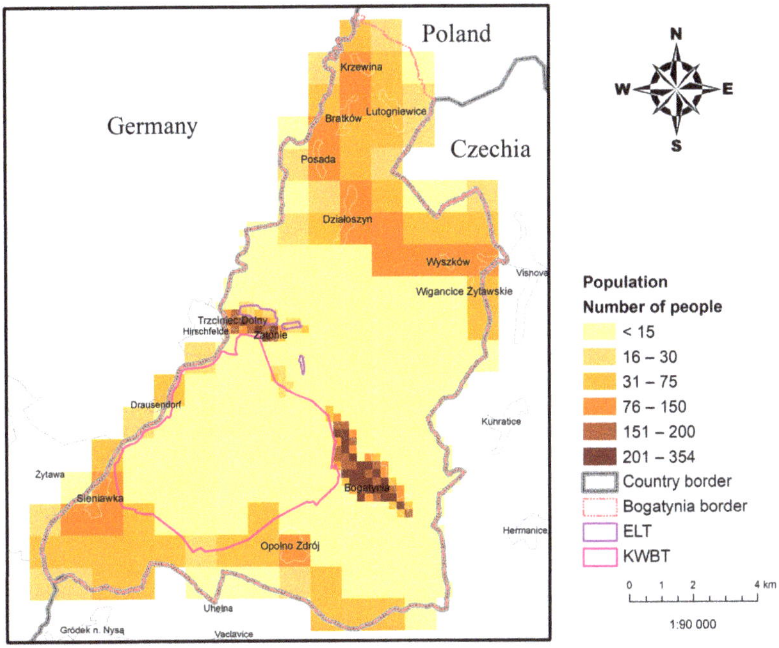

Figure 6. Map of the population density distribution (source: own study).

4. Results

4.1. Meteorological Conditions in the Base Year

On the basis of calculations made with the WRF/CALMET model, the analysis of meteorological conditions significantly affecting the dispersion of pollutants in 2018 was performed (Figures 7–15). Both spatial variability of selected parameters per year and monthly variability were discussed based on selected locations—Bogatynia (the largest city in the commune), Sieniawka (located in the southwest of the examined area) and Działoszyn (located in the north of the examined area).

Wind as a parameter shaping the rate and direction of the pollution spread is one of the most important parameters for the dispersion process. With complex terrain, it can be highly variable, and when dealing with such a specific object as a deep open-pit mine, it is necessary to include it in the model as a field variable in space. This is clearly marked when analyzing the prevailing wind directions, where a significant change in the air flow occurs in the area of the open pit and, while the wind is southern for the most part of the area, in the pit, it changes direction to the west or east. The significant dynamics of the dominant wind direction can also be seen in the depression in which Zittau and Hrádek nad Nisou are located, where in 2018, winds from the western sector prevailed.

The calculations show that the wind field in the analyzed area is characterized by variability of annual mean speeds in the range of 3.8 to 4.4 m/s. The highest speeds (above 4.2 m/s) characterize the northwest and southeast parts of the analyzed area (Figure 7). Reduced wind speeds are characteristic of the valley areas (around Bogatynia, as well as Zittau and Hrádek nad Nisou).

Based on the hourly wind speeds and directions, annual wind roses were prepared for selected grids within meteorological domains representing the following cities: Bogatynia, Sieniawka and Działoszyn (Figure 8). The obtained calculation results indicate that, in 2018 in Bogatynia, winds from the southern sector (19%) definitely dominated; the highest speeds were also recorded in this sector. In Sieniawka, the share of winds from the southern sector was also the largest (13%), but the share of winds from the NW and WNW sectors was also significant (around 10% per year). In Działoszyn, apart from southern winds (13%), there was a large share of SSW winds (11%). Only in Działoszyn, the share of winds from the direction in which KWBT is located (in this case, the southern sector) was significant. The highest amount of calm winds was recorded in Bogatynia (1.8%); in other cities, this value was around 1%.

In the case of the analysis of wind speed values, higher values of monthly mean wind speeds (except for the summer months, i.e., June–August) were observed in Bogatynia than in Działoszyn and Sieniawka (Figure 9). According to the distribution of monthly mean wind speeds in 2018, higher wind speeds (above 5 m/s) occur in the autumn and winter months (October, January, November, and December). The spring–summer period (May–August) was characterized by lower wind speeds (below 3.5 m/s).

On the other hand, the classification of wind speeds for selected grids from the studied area shows that the most frequently occurring were the winds with speeds in the 3–5 m/s range (30.5–38.1%), referred to as mild winds (Figure 10). Weak winds, i.e., 1.5–3 m/s, and winds with speeds above 5 m/s occurred with similar frequency in all towns (around 21%–27%). In Bogatynia, more than 1% of winds with very high speeds >10.8 m/s were observed. In 2018, wind speeds less than 1.5 m/s occurred in 9.9% of cases in Działoszyn, 11% in Sieniawka and 12.6% in Bogatynia.

According to the Climate Monitoring Bulletin of Poland, published annually by the Institute of Meteorology and Water Management, 2018 has been classified as an extremely warm year [40,41]. The analyses show that the annual mean air temperature in 2018 in the studied area varied from around 9 °C to over 10 °C, while in most of the area, it was around 9.5 °C. Temperatures above 10 °C occurred in the northern part of KWBT and in the vicinity of ELT and in the northwest of the area (Figure 11). The coldest month in 2018 and the only one with the average temperature below 0 °C was February (around −4 °C). March was also quite cool (monthly mean around 0 °C) (Figure 12). The month with the highest mean (>20 °C) was August. At the same time, from April to October, monthly mean temperatures were above 10 °C. The characteristics of monthly mean temperatures indicate that the Bogatynia and Działoszyn regions are thermally similar, while the Sieniawka area is slightly cooler.

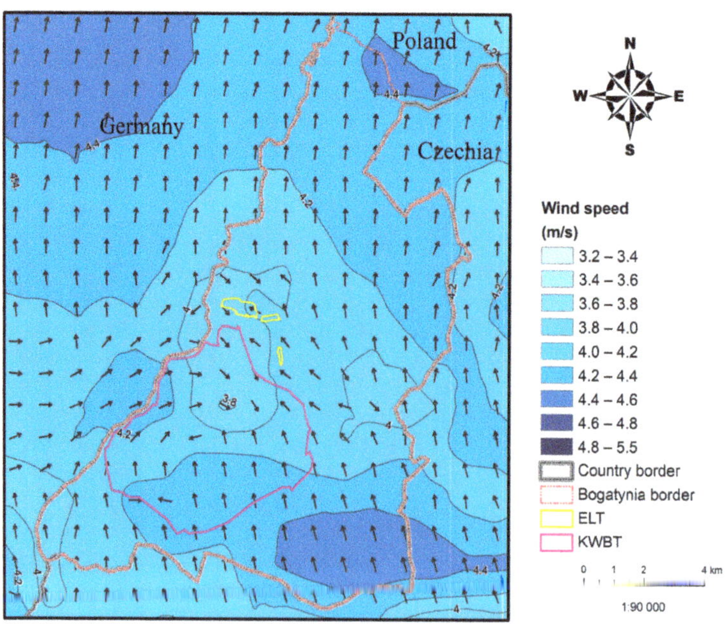

Figure 7. Spatial distribution of annual mean speeds and dominant wind direction determined by the WRF/CALMET model in 2018 (source: own study).

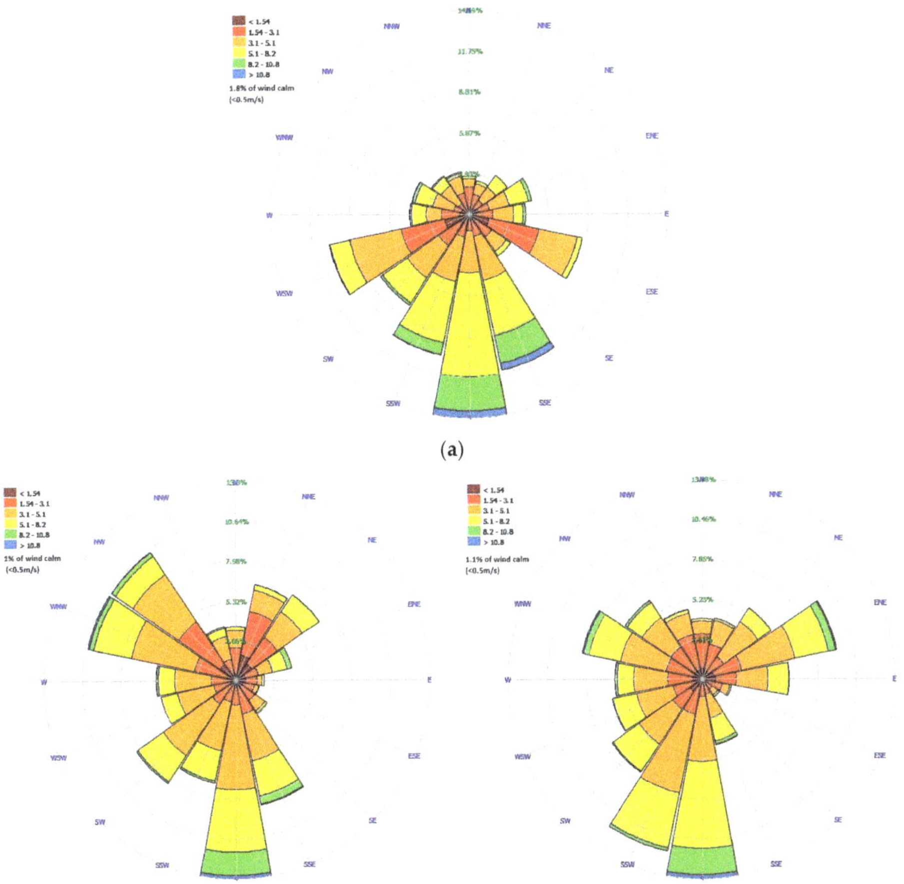

(a)

(b) (c)

Figure 8. The distribution of wind directions and wind speeds determined by the WRF/CALMET model in the meshes of the meteorological grid corresponding to the location of selected towns in 2018 (Source: Own study): (**a**) Bogatynia, (**b**) Sieniawka, (**c**) Działoszyn.

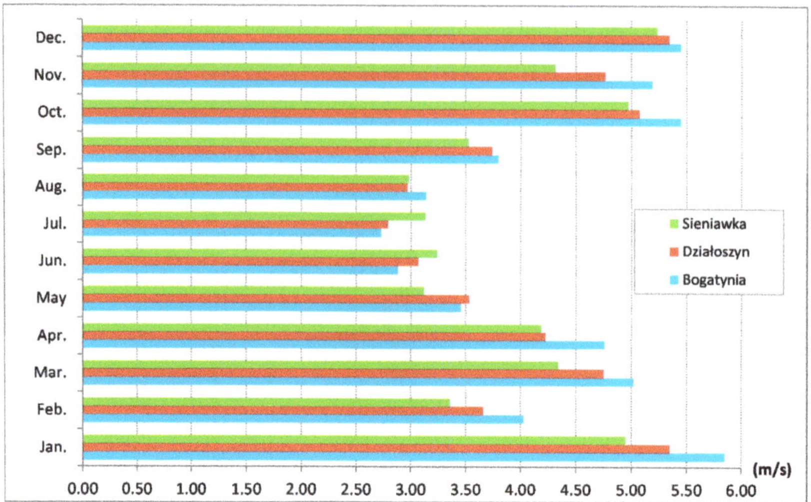

Figure 9. Monthly mean wind speeds determined by the WRF/CALMET model in selected towns in 2018 (source: own study).

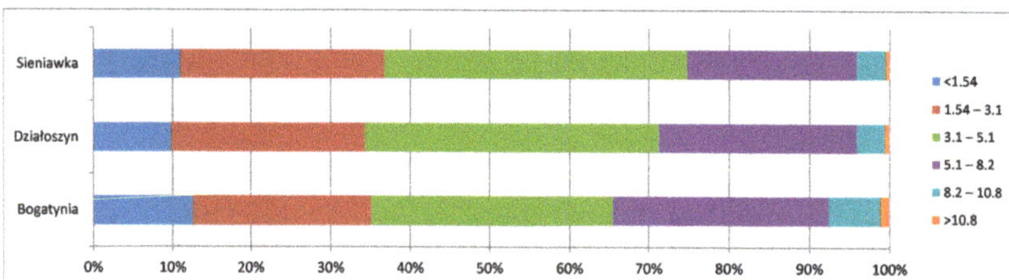

Figure 10. The frequency of wind speeds in specific ranges in selected towns in 2018 (source: own study).

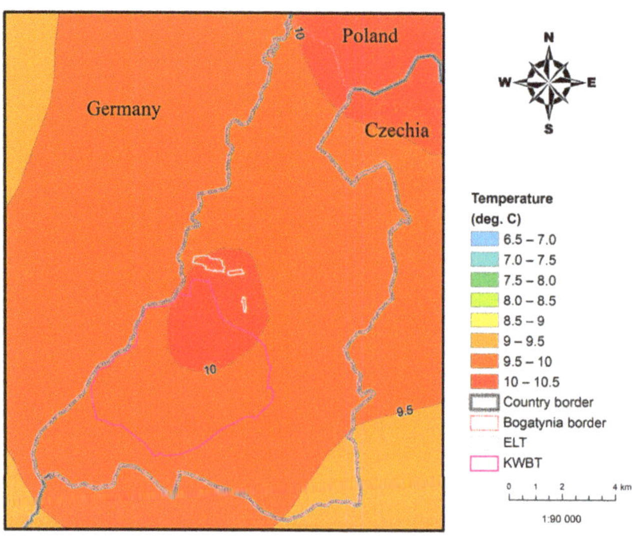

Figure 11. Spatial distribution of annual mean air temperature values determined by WRF/CALMET in 2018 (source: own study).

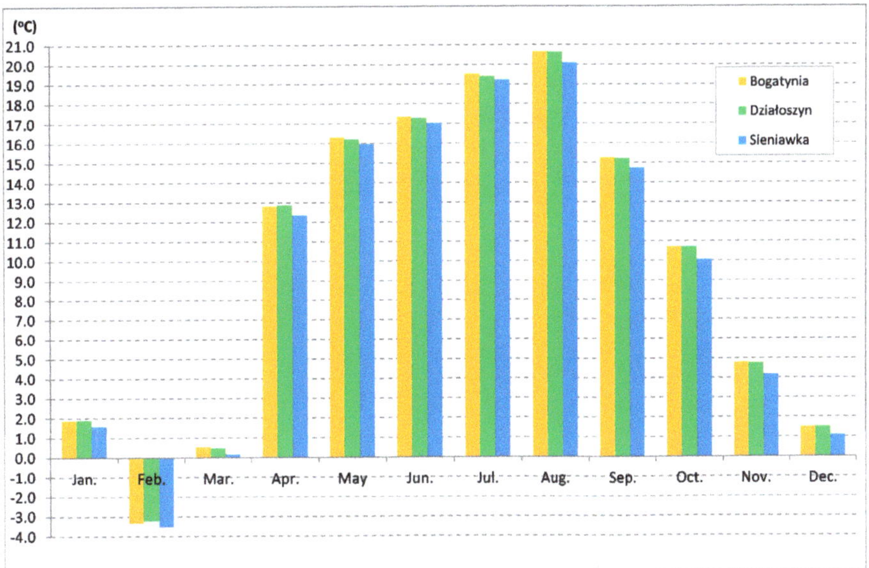

Figure 12. The course of the monthly mean air temperature determined by the WRF/CALMET model in selected towns in 2018 (source: own study).

The Pasquill atmospheric stability classes, which describe the vertical air turbulence associated with temperature gradient and wind speed, are very important parameters for the dispersion of pollution. The model adopts six stability classes (PGT1–PGT6). Classes 1 and 2 are unfavorable for the dispersion of pollutants due to the fact that the trail of exhaust gases rises and falls due to intense turbulence. Classes 5 and 6, in which inverse conditions occur, are very unfavorable; the pollutants remain in the given area at low altitudes because they have no conditions for dispersion. The incidence of individual classes was determined for the towns of Sieniawka, Działoszyn and Bogatynia (Figure 13). The calculations show that, in the vicinity of the towns in question, in 2018, the most common was the atmospheric stability class 4, which represents neutral conditions (around or over 50% of cases). Class 1, defined as extremely unstable conditions, was very rare (less than 1% of cases). However, unfavorable classes 5 and 6 occurred in a total of around 21–30% cases during the year, most often in Sieniawka.

The year of 2018 was a dry year, which is also confirmed by the spatial distribution of the annual total precipitation in the area (Figure 14). Such conditions adversely affect the rise of dust pollutants, which, in the case of a large-scale object such as a mine, may contribute to the occurrence of high concentrations. In 2018, the annual rainfall totals in the studied area ranged from around 600 (in the west in the area of the Nysa Łużycka Valley) to 800 mm (in the east in the area of the Izera Mountains). The analysis of the annual rainfall totals in selected locations indicates a relatively small variability: from around 590 mm in Sieniawka to around 620 mm in Bogatynia and Działoszyn. The analysis of the variability of precipitation in 2018 shows that the lowest rainfall occurred in February and November—below 10 mm—while the highest was measured in June and December (77–97 mm) (Figure 15). In Sieniawka, high rainfall was noted also in January, where the total was around 30 mm higher than in Bogatynia and Działoszyn. For the remaining period of the year, rainfall totals in Sieniawka were slightly lower than for Bogatynia and Działoszyn, especially in May and June. In the remaining months of the year, the differences in the total precipitation between individual towns did not exceed 10 mm.

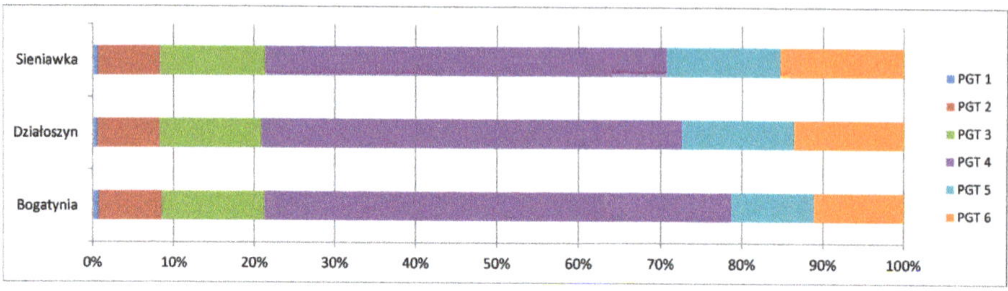

Figure 13. Share of the Pasquill atmospheric stability classes determined by the WRF/CALMET model in selected towns in 2018 (source: own study).

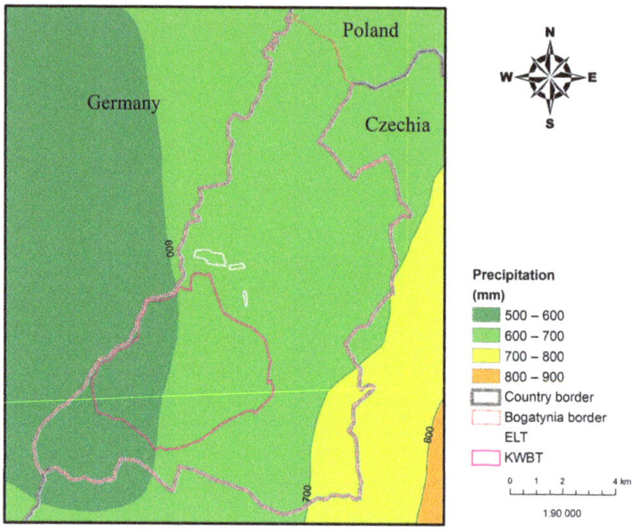

Figure 14. Spatial distribution of annual precipitation determined by the WRF/CALMET model in 2018 (source: own study).

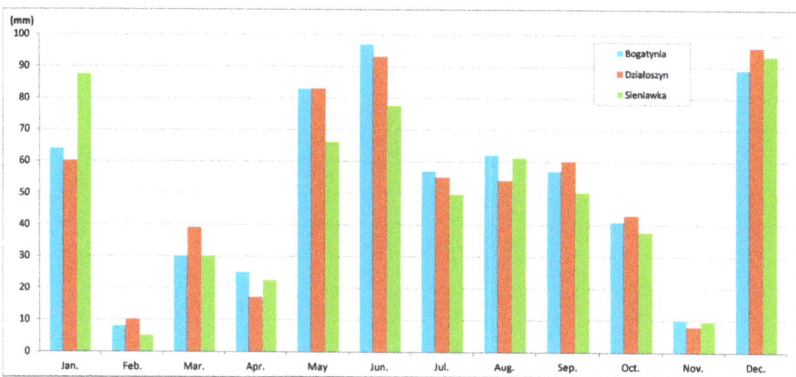

Figure 15. Monthly precipitation totals determined by the WRF/CALMET model in selected towns in 2018 (source: own study).

4.2. The Results of Air Quality Modeling

Human health protection is the main criterion for air quality assessment. This study focuses on the analysis of long-term effects; therefore, the values of annual mean concentrations of PM_{10} and $PM_{2.5}$ were assessed, with limit levels of 40 μg/m^3 and 25 μg/m^3, respectively. The assessment of changes resulting from the implementation of individual scenarios is presented as a relative difference.

Additionally, the shares of the main particulate matter emission sources in concentrations in two selected profiles of different emission nature were analyzed for scenarios 1 and 2. The first profile, around 40 km long, running from Działoszyn through Zatonie to Hrádek nad Nisou, clearly shows the impact of the energy complex. The second profile, around 26 km long, running from Olbersdorf through Zittau, Bogatynia and Hermanice to Frydlant, reflects the characteristics of the concentration field associated with emissions from household heating.

4.2.1. Annual Mean PM_{10} Concentrations

The obtained results show that, in the year of diagnosis, the annual mean PM_{10} concentration practically in the entire analyzed area did not exceed the limit value (Figure 16). The area located near the ash storage area, where maximum concentrations reached over 100 μg/m^3, was the exception. However, this area is located outside built-up areas, within a forest complex, so its range is limited. High concentrations also occur within the open pit, which is justified by the industrial characteristics of the area. Within the settlement area, the annual mean concentrations of PM_{10} are in the range of 18–26 μg/m^3 (45–65% of the permissible level). The highest concentrations in built-up areas occurred in Bogatynia and reached 34 μg/m^3 (85% of the permissible level). The lowest concentrations in the studied area occurred in its northwestern and southeastern regions (in elevated areas, marked in dark green).

The model calculations show that the implementation of corrective actions discussed in scenario 1 (Figure 16b) regarding the power complex facilities results in a maximum 70% reduction in concentrations nearby ash storage area. The greatest effectiveness is expected in the immediate vicinity of the ash stockyard (area marked with an arrow). Within a radius of 4 km from the main emission sources, a maximum reduction of 5% can be expected. The greatest effectiveness of actions in development areas is expected in the districts of Bogatynia–Zatonie (30–40%) and Trzciniec (10–15%), but also in Sieniawka (5–10%) and on the German side in the cities of Hirschfelde and Drausendorf and in eastern areas of Zittau (5–10%).

On the other hand, of the analyses carried out for activities indicated in scenario 2, they will be most effective in built-up areas (Figure 16c). In Bogatynia, as much as 20–40% decrease in annual mean PM_{10} concentrations is estimated. In larger cities, concentrations will fall by around 10–20%, and, in the remaining areas, the change will not exceed 5–10%.

The best effect was obtained for the compilation of scenarios 1 and 2, which is scenario 3 (Figure 16d), in which case at least a 10% decrease in concentrations can be expected basically in the entire analyzed area within the Polish borders. A small impact on the decrease in concentrations on the Czech and German side may, however, prove that the impact of both the energy complex and local emissions associated with the combustion of fuels in household heating devices is very limited.

Analyzing the share of emission sources in the annual mean PM_{10} concentrations in the profile between Działoszyn and Hrádek nad Nisou (Figure 17), it can be concluded that the inflow of pollutants from outside the computational domain has a very significant share, which is estimated at around 16 μg/m^3 basically along the entire length of the profile. The concentrations related to household heating, which are clearly marked only on the Polish side of the border, are also an important component of the profile. The closer the sources of the power complex, the more significant the increase in their share; however, their range of influence is very limited (a few km). In the immediate vicinity of the ash storage area, its share is similar to the share of the inflow. The impact of KWBT is practically limited to industrial areas and may be associated with local high concentrations, even exceeding the target values. At the same time, it is clear that individual operations (transshipment at a coal yard, dumping or mining) have a significant impact basically in the place of their performance.

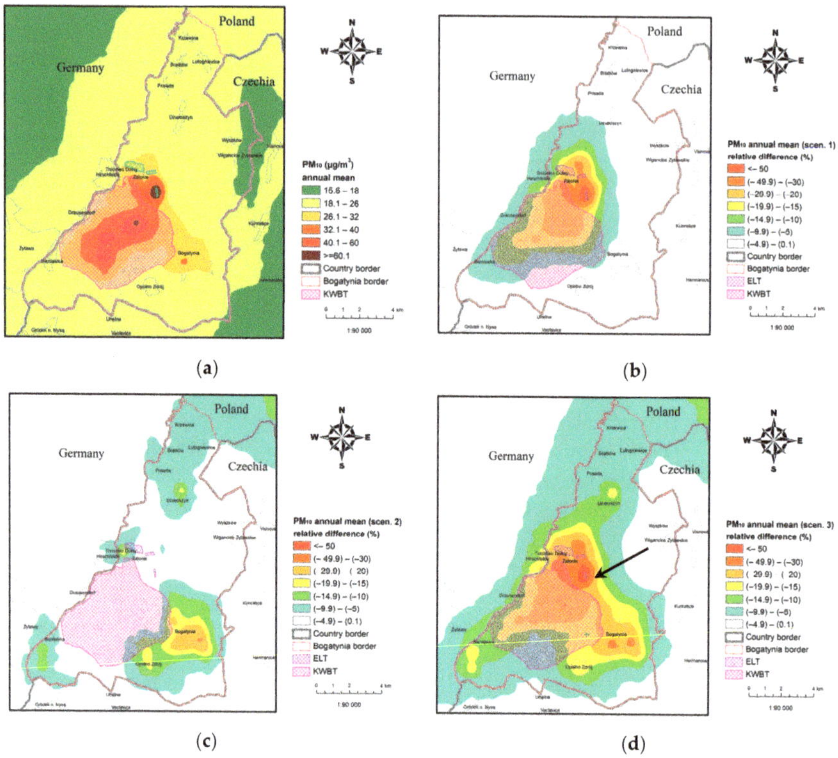

Figure 16. Annual mean PM_{10} concentrations in 2018: modeled values and relative differences for the analyzed scenarios (source: own study): (**a**) in 2018, (**b**) relative difference for scenario 1, (**c**) relative difference for scenario 2, (**d**) relative difference for scenario 3.

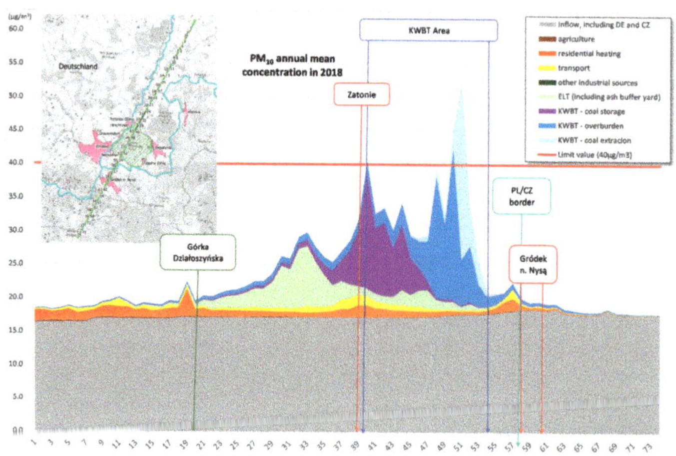

Figure 17. Profile of annual mean concentrations of PM_{10} between Działoszyn and Hrádek nad Nisou, considering the shares of individual source groups in 2018 (source: own study).

The implementation of the measures from scenario 1 will result in a very large decrease in concentrations, and so air quality standards will be met throughout the entire length of the analyzed profile (Figure 18). In this case, the most important factor shaping the air quality in the studied area will be the inflow of pollution. The impact of the power complex will be comparable to the current impact of emissions from household heating.

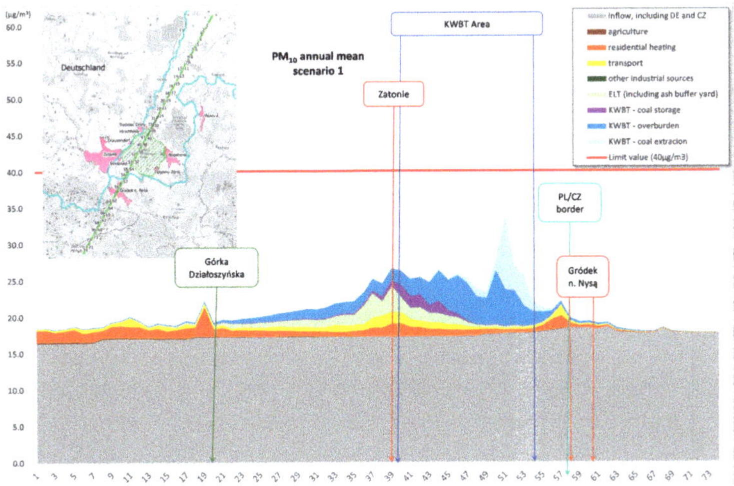

Figure 18. Profile of annual mean concentrations of PM_{10} between Działoszyn and Hrádek nad Nisou, considering the shares of individual source groups for scenario 1 (source: own study).

In the profile from Olbersdorf to Frydlant, the inflow of pollutants from outside of the studied area has the largest share in annual mean PM_{10} concentrations (Figure 19), similar to the profile analyzed earlier. However, in Poland, emissions related to household heating also have a very high share in concentration. This is particularly evident in Bogatynia, where concentrations from this type of emission can reach up to 12 $\mu g/m^3$. Transport is the third most important group of sources in the studied profile (maximum annual mean PM_{10} concentrations reach up to around 3 $\mu g/m^3$). The impact of the mine is relatively small, and it is significant only within the open pit (mining). At the same time, there are no exceedances of air quality standards virtually along the entire length of the profile.

The implementation of the anti-smog resolution will practically eliminate the impact of emissions from household heating, which will result in a significant reduction in the concentrations associated with it. This is clearly marked on the analyzed profile (Figure 20).

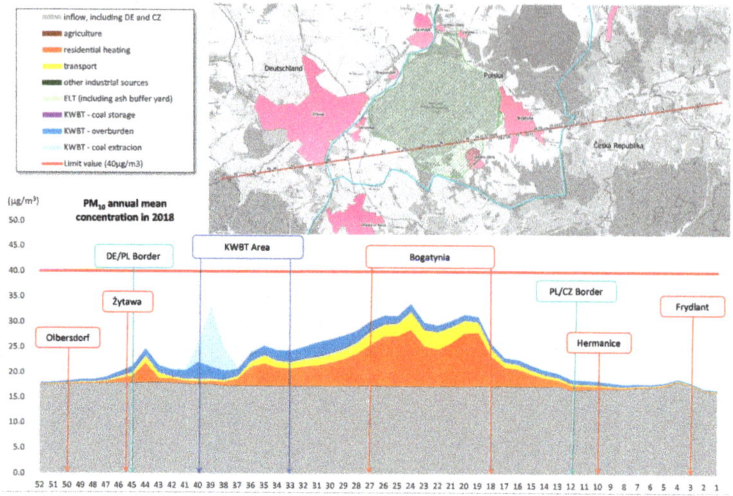

Figure 19. Profile of annual mean concentrations of PM_{10} between Działoszyn and Hrádek nad Nisou, considering the shares of individual source groups in 2018 (source: own study).

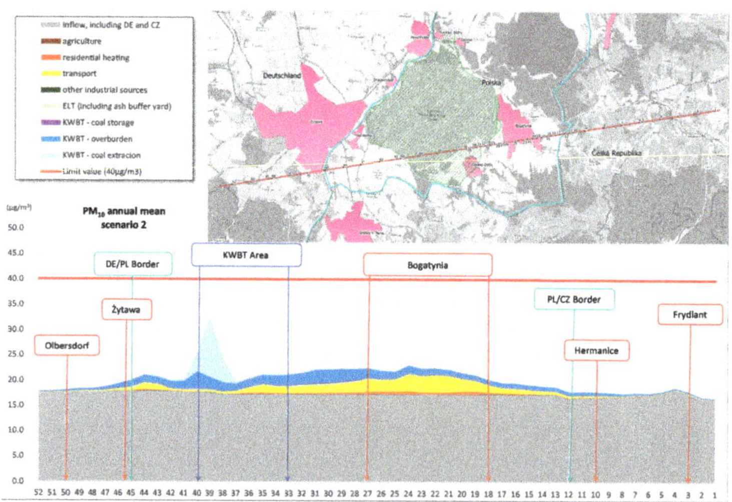

Figure 20. Profile of annual mean concentrations of PM_{10} between Olbersdorf and Frydlant, considering the shares of individual source groups in scenario 2 (source: own study).

4.2.2. Annual Mean $PM_{2.5}$ Concentrations

The results of the conducted model tests indicate that the lowest values of the annual mean $PM_{2.5}$ concentration occurred in the eastern part of the area, where they do not exceed 14 µg/m^3 (56% of the limit value) (Figure 21a). In the central part of the area, concentrations remain in the range of 14–16 µg/m^3. Outside the industrial area, the highest annual mean $PM_{2.5}$ concentrations occur in Bogatynia, where they reach around 23 µg/m^3 (92% of the limit value). In other locations in the studied area, higher annual mean concentrations of $PM_{2.5}$ were also recorded in Trzciniec Dolny (84% of the limit value), Zatonie (79% of the limit value), Sieniawka (76% of the limit value) and Działoszyn (70% of the limit value).

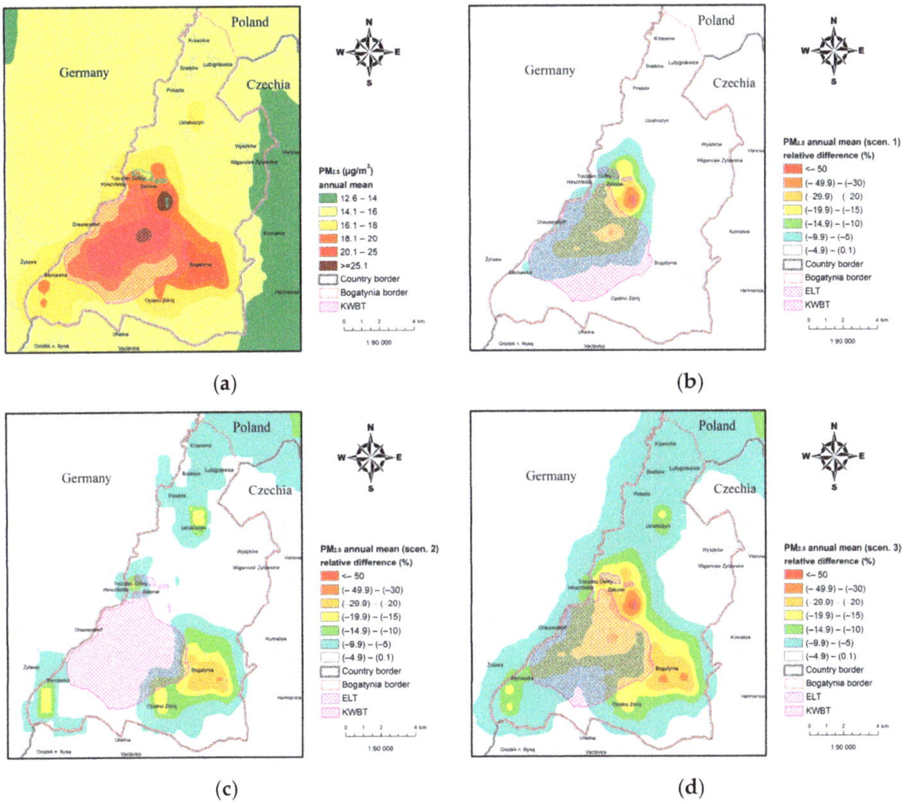

Figure 21. $PM_{2.5}$ annual mean concentrations in 2018: modeled values and relative differences for the analyzed scenarios (source: own study): (**a**) in 2018, (**b**) relative difference for scenario 1, (**c**) relative difference for scenario 2, (**d**) relative difference for scenario 3.

Minimizing measures implemented in scenario 1 will locally (nearby ash storage area) result in a maximum 75% reduction in annual mean $PM_{2.5}$ concentrations (Figure 21b). However, the range of maximum reductions is smaller than it was in the case of PM_{10}, which is due to the nature of the dust associated with the emitters of the power complex (in particular, KWBT and ash storage area). Emissions from this type of facility primarily concern mineral dust with larger fractions and lower volatility. The greatest efficiency is expected in the immediate vicinity of the ash storage area. Within a radius of 1.5 km from the main sources, a maximum reduction of 5% can be expected. Analyzing the development areas, the direct significant impact of the application of the measures resulting from scenario 1 can only be seen in Zatonie and Trzciniec.

The implementation of scenario 2 will result in similar reductions in $PM_{2.5}$ concentrations, as in the case of PM_{10}, but the reduction range is greater (Figure 21c).

As in the case of $PM_{2.5}$, the best effect was obtained for scenario 3: a 10% reduction in concentration values was obtained practically throughout the entire analyzed area within the Polish borders (Figure 21d).

In the profile between Działoszyn and Hrádek nad Nisou (Figure 22), there are no exceedances of air quality standards set for annual mean concentrations of $PM_{2.5}$. Analyzing the shares of emission sources, it can be stated that, also in the case of this pollution, the inflow of pollutants from outside the computational domain has a very significant share, which is estimated at around 13 µg/m^3 on the entire length of the profile. An important component of the profile is also concentrations related to household heating. As the sources of the power complex approach, their share increases, and the impact is much smaller than it was in the case of PM_{10}.

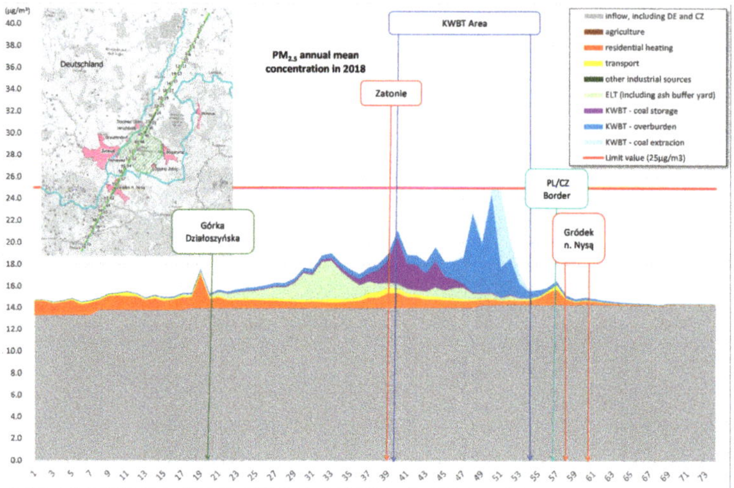

Figure 22. Profile of annual mean concentrations of $PM_{2.5}$ between Działoszyn and Hrádek nad Nisou, considering the shares of individual source groups in 2018 (source: own study).

The emission reduction related to the implementation of scenario 1 will significantly reduce the annual mean concentrations of $PM_{2.5}$ and the share of concentrations from local emissions will be lower than the concentrations from inflow (Figure 23). The impact of emissions related to the ash storage area will practically disappear and the impact of emissions from the mine will be limited to its area. The largest share of local emissions in concentrations will be associated with the impact of local heating sources.

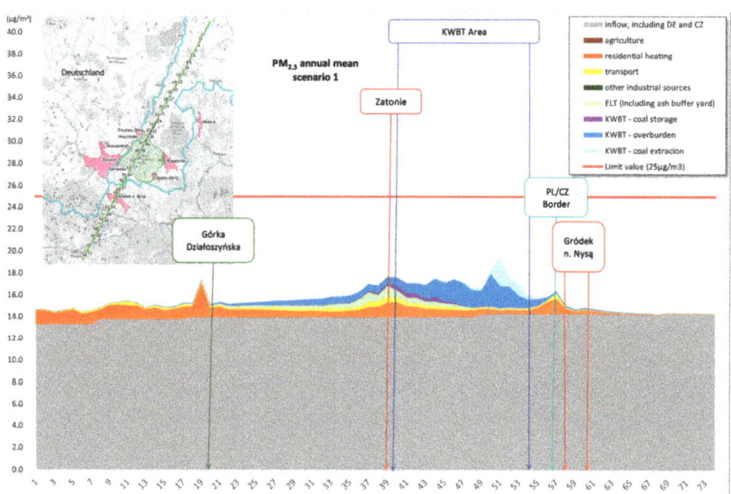

Figure 23. Profile of annual mean concentrations of $PM_{2.5}$ between Działoszyn and Hrádek nad Nisou, considering the shares of individual source groups for scenario 1 (source: own study).

In the profile between Olbersdorf and Frydlant, the annual mean concentration of $PM_{2.5}$ also has the most significant share of pollution (Figure 24). Locally, however, in Bogatynia, there is a very large share of emissions associated with household heating systems. The concentrations in Bogatynia are approaching the limit value, but they do not exceed it. Other local sources (also emissions related to KWBT) are much less important.

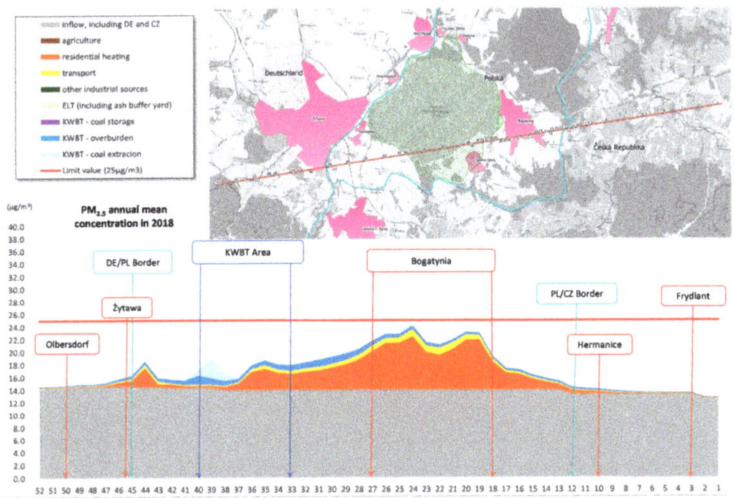

Figure 24. Profile of annual mean concentrations of $PM_{2.5}$ between Olbersdorf and Frydlant, considering the shares of individual source groups in 2018 (source: own study).

The implementation of the anti-smog resolution in the examined area will practically result in a very large reduction of local emissions and thus its impact on the formation of air quality in the examined area (Figure 25). The more significant impact of KWBT is practically limited to the open-pit area, but its share is similar to the share of emissions from local transport.

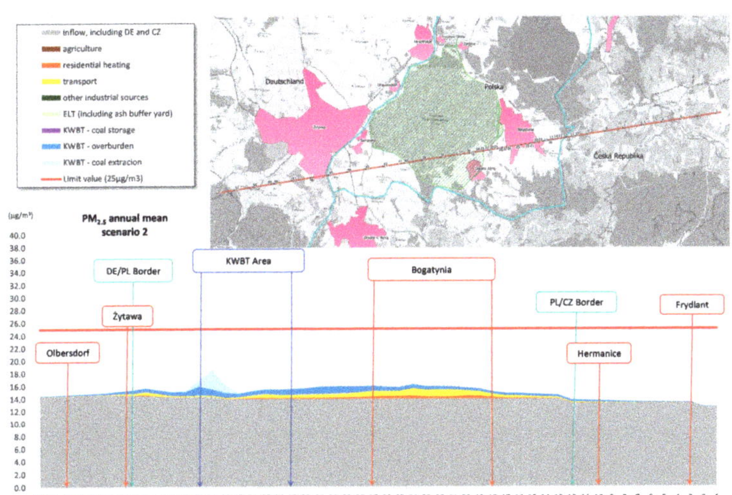

Figure 25. Profile of annual mean concentrations of $PM_{2.5}$ between Olbersdorf and Frydlant, considering the shares of individual source groups in scenario 2 (source: own study).

4.3. Assessment of the Effectiveness of the Implementation of Various Scenarios Based on the Results of Health Risk Analyses

The obtained results of health risk assessment analyses indicate the 6% general improvement in air quality from the implementation of scenario 1 (all activities specified for the mine) in the entire commune leads to the reduction of health effects by nearly 5%. This confirms that the mine itself has a negligible impact on the health of residents (current estimated health impact—base scenario—is approximately 7%; see Table 7).

Table 7. $PM_{2.5}$ concentration and related premature deaths calculated for emission scenarios.

Scenario	Annual Mean $PM_{2.5}$ Concentration		Estimated Total Number of Premature Deaths (95% CI)
	Arithmetic Average (95% CI)	Population Weighted Average	
	(μg/m³)	(μg/m³)	-
Baseline scenario	17.1 (10.4:23.9)	19.3	122.2 (78.8:163.6)
Including the mine's influence	1.01 (−2.1:4.1)	1.4	9.0 (5.8:12.0)
Scenario 1	(16.1 19.8)	18.3	116.2 (75.0:155.6)
Scenario 2	15.9 (9.7:22.0)	16.5	104.6 (67.5:140.1)
Scenario 3	14.9 (12.8: 16.9)	15.5	98.6 (63.6:132.0)

A completely different situation is observed in the case of scenario 2 (introduction of provisions resulting from the implementation of regional measures, i.e., anti-smog resolution). A noticeable decrease of over 7% in the level of pollution in the commune when implementing scenario 2 causes over a two-fold higher (14.4%) decrease in health risk (premature deaths).

Both abovementioned scenarios can be implemented independently and simultaneously by separate units based on their competence (mine authorities and local government). The implementation of both scenarios together (scenario 3) causes a significantly higher improvement in air quality and reduction in health risk in the studied area (e.g., changes in the values of the weighted population annual average concentration). After implementing the scenario 3 measures in the analyzed area, the 18% decrease in the number of premature deaths can be expected (estimated according to the baseline scenario) when the concentration of average annual particulate matter is reduced by just over 13%.

In each case (scenarios), the spatial distribution of health effects reduction changes is not regular (Figures 26–28). The analysis of the obtained results of premature death changes distribution for the areas of the commune shows little impact on the populated area for scenario 1, where in the vast majority of areas (nearly 73% of the area), the observed changes do not exceed 5% (Figure 26). Despite this, in this scenario, there are close to 1% areas with over 30% improvement in health. Generally, in 7% of areas, the number of premature deaths is reduced by at least 15%.

The situation looks much better with the implementation of scenario 2 (Figure 27). For this scenario, a decrease in the number of premature deaths associated with long-term exposure to $PM_{2.5}$ lower than 5% is already observed in nearly 58% of areas. An improvement in health—that is, at least a 15% reduction in premature deaths—has already been noticed in 12% of areas.

Implementation of scenarios 1 and 2 at the same time remains the most effective and health-promoting solution (for scenario 3, see Figure 28). Assessment results for scenario 3 show that only in 25% of the areas, the improvement in health is less than 5%, while a reduction in premature deaths of over 15% is observed almost in 33% of the commune's area.

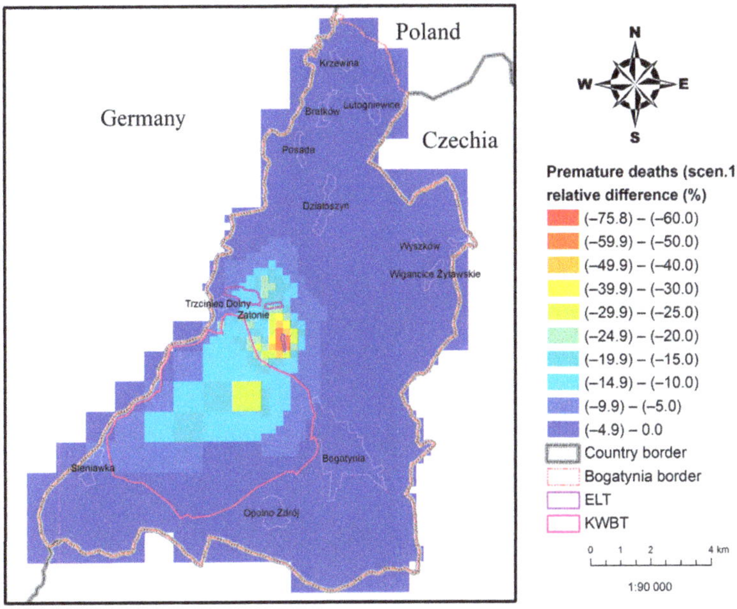

Figure 26. Spatial distribution of relative changes (reductions) in the number of premature deaths for scenario 1 relative to the baseline scenario.

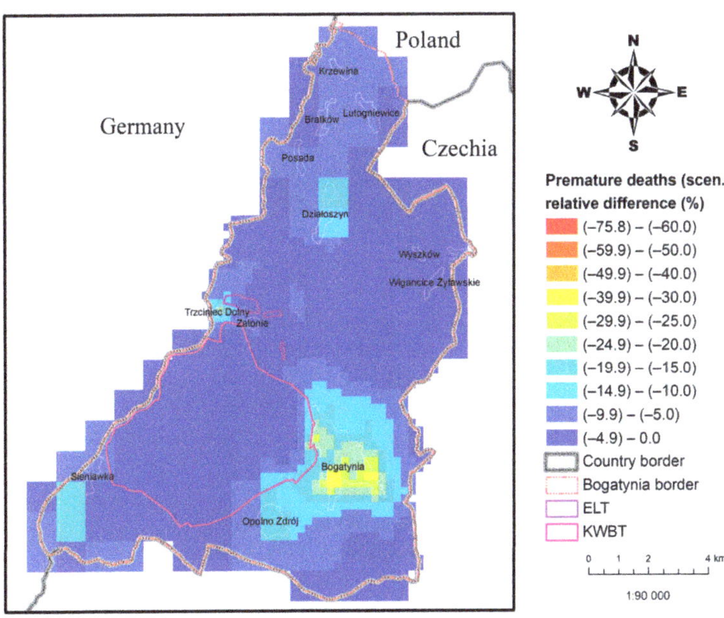

Figure 27. Spatial distribution of relative changes (reductions) in the number of premature deaths for scenario 2 relative to the baseline scenario.

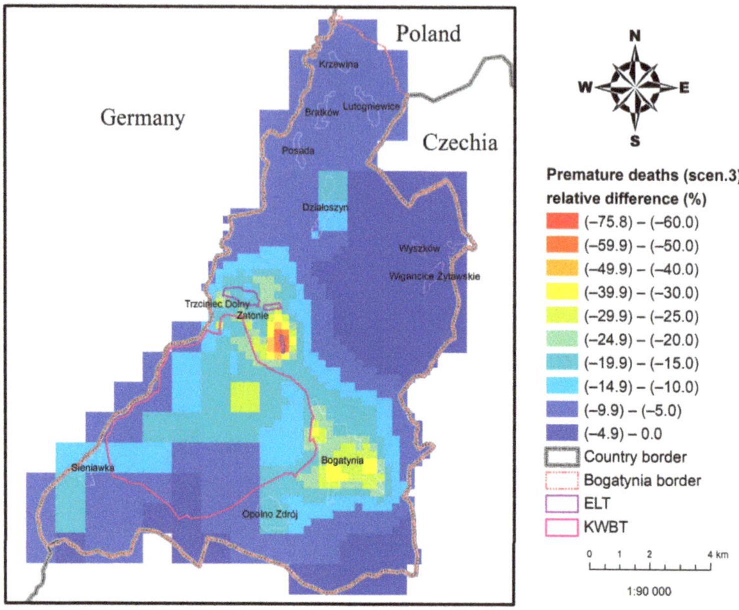

Figure 28. Spatial distribution of relative changes (reductions) in the number of premature deaths for scenario 3 relative to the baseline scenario.

5. Summary and Final Conclusions

Mathematical modeling is one of the tools whose application within the air quality management system is crucial. Commonly, mathematical models apply, among others, when developing plans and programs of corrective actions aimed at achieving and/or maintaining air quality at an appropriate level; analyses of the effectiveness of implementing specific solutions; analyses of the impact of air pollution on various elements of the environment, including human health; air quality forecasting, considering changes in the activity of emission sources and meteorological conditions, and providing the public and decision-makers with adequate information. In the case of modeling systems, it is important that the input data used in the analyses (including emission, meteorological and topographic data) are current and accurate. This has a significant impact on the results obtained by modeling and their credibility and representativeness. An important element of this type of analysis is the assessment of discrepancies in the results of model calculations, including the relative error rate of the modeling result in relation to the results of measurements made at measuring stations using devices compatible with or equivalent to the reference method specified for a given pollutant. This type of procedure was undertaken as part of the work in which the selected case study was the area of the commune in Poland, whose character was considered unique and complex due to the geographical location and diversity of the terrain and the functioning energy complex. Analyses were carried out considering the defined activities for three emission reduction scenarios: (1) scenario related to changes in emissions in the analyzed mine resulting from the minimizing measures indicated in the report on the mine's environmental impact, (2) scenario resulting from the "anti-smog" resolution in force in the Lower Silesian Voivodship and (3) scenario compiling the abovementioned scenarios. Additionally, to demonstrate the effectiveness of planned preventive and corrective actions taken in the analyzed area, a health risk analysis was performed. All analyses were made considering the changes in the distribution of pollutant concentrations within the boundaries of the commune.

The results of the conducted analyses indicated that the lowest values of the annual mean $PM_{2.5}$ concentration occurred in the eastern part of the studied area and did not exceed 14 μg/m^3 (56% of the limit value). The implementation of activities resulting from the considered scenario 1 will result in a reduction in annual mean concentrations of $PM_{2.5}$ (maximum 75% in industrial area), mainly due to the maximum reduction in activities carried out within the furnace ash storage area, as well as the construction of a new ash conveyor, thanks to which the storage area will be practically taken out of operation. The achieved range of maximum concentration reduction was smaller than in the case of PM10, which may be due to the nature of the dust associated with the emitters of the energy complex (in particular, KWBT and ash storage area). Emissions from this type of facility primarily concern mineral dust of larger fractions that are transported in the atmosphere to a much lesser extent. The greatest result is expected in the immediate vicinity of the ash storage area. Within a radius of 1.5 km from the main sources, a slight reduction in concentrations, not exceeding 5% of the current state, can be expected. For this reason, actions taken under this scenario do not have a major impact on the health of the surrounding residents.

The performed analyses have shown that the implementation of scenario 2 will result in similar levels of $PM_{2.5}$ concentration reductions as PM_{10}, but with a much vaster spatial range, which translates into a much higher impact on the population threat related to air quality, with an estimated 14% reduction.

The best effect was obtained for scenario 3, where, in total, the average 10% reduction in concentration values was obtained practically in the entire analyzed area within the borders of the community. Therefore, the implementation of both scenarios seems to be the most effective for limiting the health risk associated with the exposure of residents to particulate matter (estimated health effect reduction is almost 20%). The full implementation of scenario 3 shows that only in 25% of the area was the expected improvement of health lower than 5%. A significant decrease in health risk (more than 15%) was observed in as much as one third of the studied area.

Mathematical modeling as a tool should be disseminated, and the data used in the model should be available not upon special request but due to the obligation to provide access to information on the environmental impact of an installation, plant and/or the group of emitters in a given area.

Author Contributions: Conceptualization, I.S., M.P. and K.S.; methodology, I.S., M.P. and K.S.; software, M.P. and K.S.; validation, I.S., M.P. and K.S.; formal analysis, I.S., M.P., K.S., D.K., M.Z. and K.K.; investigation, I.S., M.P., K.S., D.K., M.Z. and K.K.; writing—original draft preparation, I.S., M.P., K.S., D.K., M.Z. and K.K.; writing—review and editing, I.S., M.P., K.S., D.K., M.Z. and K.K.; visualization, M.P. and K.S.; supervision, I.S. All authors have read and agreed to the published version of the manuscript.

Funding: This work was co-financed within the "Excellent Science" program of the Polish Ministry of Science and Higher Education.

Ministry of Science and Higher Education
Republic of Poland

Acknowledgments: The authors of the study thank for their cooperation: Anita Kuliś from ONE WAY Anita Kuliś (Zielonka, Poland), Rafał Skorupiński, as well as Milena Gola-Kozak and Dorota Sucholas from PGE Górnictwo i Energetyka Konwencjonalna S.A., Kopalnia Węgla Brunatnego Turów (Bogatynia, Poland).

Conflicts of Interest: The authors declare no conflict of interest. The funders had no role in the design of the study; in the collection, analyses, or interpretation of data; in the writing of the manuscript, or in the decision to publish the results.

References

1. CAFE Directive. Directive 2008/50/EC of the European Parliament and of the Council of 21 May 2008 on ambient air quality and cleaner air for Europe. *Off. J. Eur. Union* **2008**, *L152*, 1–44.
2. European Parliament and of the Council. Directive 2004/107/EC of the European Parliament and of the Council of 15 December 2004 relating to arsenic, cadmium, mercury, nickel and polycyclic aromatic hydrocarbons in ambient air. *Off. J. Eur. Union* **2008**, *L23*, 3–16.

3. Sejm of the Republic of Poland. Act of 27 April 2001—Environmental Protection Law (Announcement of the Marshal of the Sejm of the Republic of Poland of 19 July 2019 regarding the publication of a uniform text of the Act—Environmental Protection Law. *J. Laws* **2019**, *1396*, 1–155. (In Polish)
4. Polish Minister of the Environment. Regulation of the Polish Minister of the Environment of 24 August 2012 on the levels of certain substances in the air. *J. Laws* **2012**, *1031*, 1–9. (In Polish)
5. Air Quality in Europe—2019 Report, European Environment Agency, Copenhagen 2019. Available online: https://www.eea.europa.eu/publications/air-quality-in-europe-2019 (accessed on 6 May 2020).
6. Polish Minister of the Environment. Regulation of the Polish Minister of the Environment of 24 August 2012. of 8 October 2019 changing the regulation on the levels of certain substances in the air. *J. Laws* **2019**, *1931*, 1. (In Polish)
7. Kobus, D.; Iwanek, J.; Skotak, K. *Assessment of Air Quality in the Zones in Poland for 2018, Commissioned by the Polish Chief Inspectorate of Environmental Protection by the Institute of Environmental Protection—National Research Institute*; Polish Chief Inspectorate of Environmental Protection: Warsaw, Poland, 2018. (In Polish)
8. Sówka, I.; Kobus, D.; Skotak, K.; Zathey, M.W.; Merenda, B.; Paciorek, M. Assessment of the health risk related to air pollution in selected Polish health resorts. *J. Ecol. Eng.* **2019**, *20*, 132–145. [CrossRef]
9. Lower Silesian Regional Assembly. Resolution No. XLI/1407/17/of the Lower Silesian Regional Assembly of 30 November 2017 regarding the introduction of restrictions and bans on operation of fuel-burning installations in the Lower Silesian Voivodship, with the exception of the Municipality of Wrocław and health resorts. *Low. Sil. Voiv. J. Laws* **2017**, *5155*, 1–2. (In Polish)
10. Lower Silesian Regional Assembly. Resolution No. XLI/1406/17/of the Lower Silesian Regional Assembly of 30 November 2017 regarding the introduction of restrictions and bans on operation of fuel-burning installations in the area of health resorts in the Lower Silesian Voivodship. *Low. Sil. Voiv. J. Laws* **2017**, *5154*, 1–3. (In Polish)
11. Lower Silesian Regional Assembly. Resolution No. XLI/1405/17/of the Lower Silesian Regional Assembly of 30 November 2017 regarding the introduction of restrictions and bans on operation of fuel-burning installations in the area of the Municipality of Wrocław. *Low. Sil. Voiv. J. Laws* **2017**, *5153*, 1–3. (In Polish)
12. EU Commission. EU Commission Regulation 2015/1189 of 28 April 2015 on the implementation of Directive 2009/125/EC of the European Parliament and of the Council with regard to ecodesign requirements for solid fuel boilers. *Off. J. Eur. Union* **2015**, *L193*, 100–114.
13. Witamy na Stronie Internetowej KOBiZE. Available online: https://www.kobize.pl/ (accessed on 6 May 2020).
14. KOBIZE. *Krajowy Bilans Emisji SO2, NOX, CO, NH3, NMLZO, Pyłów, Metali Ciężkich i TZO za Lata 2015–2017 w Układzie Klasyfikacji SNAP*; Raport Syntetyczny: Warsaw, Poland, 2019. (In Polish)
15. Kobus, D.; Nych, A.; Sówka, I. Analysis of PM_{10} high concentration episodes in Warsaw, Krakow and Wroclaw in the years 2005–2017 with application of selected elements of information systems. *E3S Web Conf.* **2018**, *44*, 70. [CrossRef]
16. Fine Dust in the Atmosphere. Compendium of Knowledge about Air Pollution with Particulate Matter in Poland. Available online: https://www.gios.gov.pl/images/aktualnosci/Pyly_drobne_w_atmosferze.Kompendium_wiedzy.pdf (accessed on 6 May 2020).
17. Polish Minister of the Environment. Regulation of the Polish Minister of the Environment of 8 August 2018 on the assessment of the levels of certain substances in ambient air. *J. Laws* **2018**, *1119*, 1–25. (In Polish)
18. Minister of the Environment. *The Polish State Environmental Monitoring Program for the Years 2016–2020*; Chief Inspector of Environmental Protection: Warsaw, Poland, 2015. (In Polish)
19. The Application of Models under the European Union's Air Quality Directive: A Technical Reference Guide, EEA Technical Report No 10/2011. Available online: https://www.eea.europa.eu/publications/fairmode (accessed on 6 May 2020).
20. Blagojević, M.; Papić, M.; Vujičić, M.; Šućurović, M. Artificial neural network model for predicting air pollution. Case study of the Moravica district, Serbia. *Environ. Prot. Eng.* **2018**, *44*, 129–139.
21. Exponent Engineering and Scientific Consulting. Available online: http://www.src.com/ (accessed on 6 May 2020).
22. The Weather Research & Forecasting Model. Available online: https://www.mmm.ucar.edu/weather-research-and-forecasting-model (accessed on 6 May 2020).

23. PGE GIEK S.A. Oddział KWB Turów. *Kontynuacja Eksploatacji Złoża Węgla Brunatnego Turów, Raport o Oddziaływaniu na Środowisko (Report on the Environmental Impact, Codified Version)*; Reference Number WOOŚ. 4235.1.2015. MS; PGE GIEK S.A. Oddział KWB Turów: Bogatynia, Poland, 2019. (In Polish)
24. SRTM 90m Digital Elevation Database v4.1. Available online: https://cgiarcsi.community/data/srtm-90m-digital-elevation-database-v4-1/ (accessed on 6 May 2020).
25. CORINE Land Cover. Available online: https://land.copernicus.eu/pan-european/corine-land-cover (accessed on 6 May 2020).
26. ArcGIS 10 Tutorial. Available online: https://www.esri.com (accessed on 6 May 2020).
27. Global Forecasts of Aerosol. Available online: https://atmosphere.copernicus.eu/catalogue#/ (accessed on 6 May 2020).
28. Assessment of Air Quality in the Lower Silesian Voivodship in 2017, Provincial Environmental Protection Inspectorate in Wrocław. Available online: https://www.wroclaw.pios.gov.pl/pliki/powietrze/ocena_roczna_2017.pdf (accessed on 6 May 2020). (In Polish)
29. Uwaga! Informujemy, że Wszystkie Aktualne Oceny Stanu Środowiska od 2019 r. Znajdują sie na Stronie Głównego Inspektoratu Ochrony Środowiska. Available online: https://www.wroclaw.pios.gov.pl/index.php?dzial=monitoring&pod=powietrze&pod2=oceny (accessed on 6 May 2020).
30. Compilation of Air Pollutant Emission Factors, 13.2.4 Aggregate Handling And Storage Piles (US-EPA). Available online: https://www3.epa.gov/ttn/chief/ap42/ch13/final/c13s0204.pdf (accessed on 6 May 2020).
31. Cesky Hydrometeorologicky Ustav. Available online: http://portal.chmi.cz/historicka-data/ovzdusi (accessed on 6 May 2020).
32. Stoffeintrage Diffuser Quellen in Die Luft. Available online: https://diffuse-quellen.de/diffuse-quellen/map.html?type=luft (accessed on 6 May 2020).
33. *Health Risks of Air Pollution in Europe—HRAPIE Project Recommendations for Concentration–Response Functions for Cost–Benefit Analysis of Particulate Matter, Ozone and Nitrogen Dioxide*; World Health Organization: Geneva, Switzerland, 2013. Available online: https://www.euro.who.int/__data/assets/pdf_file/0006/238956/Health_risks_air_pollution_HRAPIE_project.pdf?ua=238951 (accessed on 6 May 2020).
34. *Review of Evidence on Health Aspects of Air Pollution (REVIHAAP)*; Technical Report; WHO Regional Office for Europe: Copenhagen, Denmark, 2013; Available online: https://www.euro.who.int/en/health-topics/environment-and-health/air-quality/publications/2013/review-of-evidence-on-health-aspects-of-air-pollution-revihaap-project-final-technical-report (accessed on 6 May 2020).
35. GUS. Available online: https://bdl.stat.gov.pl/BDL/start (accessed on 6 May 2020).
36. Geostatistics Portal. Available online: https://geo.stat.gov.pl/imap/ (accessed on 6 May 2020).
37. Historia Klasyfikacji NUTS. Available online: https://stat.gov.pl/statystyka-regionalna/jednostki-terytorialne/klasyfikacja-nuts/historia-klasyfikacji-nuts/ (accessed on 6 May 2020).
38. Dane z Państwowego Rejestru Granic i Powierzchni Jednostek Podziałów Terytorialnych Kraju—PRG. Available online: http://www.gugik.gov.pl/pzgik/dane-bez-oplat/dane-z-panstwowego-rejestru-granic-i-powierzchni-jednostek-podzialow-terytorialnych-kraju-prg (accessed on 6 May 2020).
39. Zmieniona Organizacja Pracy Rejestru TERYT. Available online: http://eteryt.stat.gov.pl/eTeryt/rejestr_teryt/aktualnosci/aktualnosci.aspx (accessed on 6 May 2020).
40. *Polish Climate Monitoring Newsletter*; Institute of Meteorology and Water Management—National Research Institute: Warsaw, Poland, 2018; Available online: https://klimat.imgw.pl/pl/biuletyn-monitoring (accessed on 6 May 2020).
41. Biuletyn Monitoringu Klimatu Polski PSHM IMGW-PIB. Available online: http://klimat.pogodynka.pl/pl/biuletyn-monitoring/ (accessed on 6 May 2020).

© 2020 by the authors. Licensee MDPI, Basel, Switzerland. This article is an open access article distributed under the terms and conditions of the Creative Commons Attribution (CC BY) license (http://creativecommons.org/licenses/by/4.0/).

energies

MDPI

Article

Domestic Organic Rankine Cycle-Based Cogeneration Systems as a Way to Reduce Dust Emissions in Municipal Heating

Piotr Kolasiński

Department of Thermodynamics and Renewable Energy Sources, Wrocław University of Science and Technology, Wybrzeże Wyspiańskiego 27, 50-370 Wrocław, Poland; piotr.kolasinski@pwr.edu.pl; Tel.: +48-71-320-23-39

Received: 9 June 2020; Accepted: 24 July 2020; Published: 2 August 2020

Abstract: Environmental issues are nowadays of great importance. In particular air and water quality should be kept at as high levels as possible. Energy conversion systems and devices which are applied for converting the chemical energy contained in different fuels into heat, electricity and cold in the industry and housing are sources of different gases and solid particle emissions. Medical data show $PM_{2.5}$ dust in particular is highly dangerous for human health. Therefore, limiting the number of low-quality fuel combustion processes is a key issue of modern energy policy. Statistical data show that domestic heating systems account for a large share of the total emissions of $PM_{2.5}$ and PM_{10} dust. For example in Poland in 2017, the share of households in the total annual emissions of $PM_{2.5}$ dust was equal to ca. 35.8%, while the share of $PM_{2.5}$ emission in industry (i.e., power generating plants, industrial power plants and technologies) was equal to only 23.6%. A possible way of solving this problem is by the successful replacement of old domestic furnaces by combined heat and power (CHP) or multigeneration boilers which can be used for heating the rooms and sanitary water and generating electricity and cold. Such systems can possibly contribute in the future to significant reductions of dust emissions and air pollution in urban and rural areas by limiting the number of low-quality fuel combustion processes. This article presents design considerations and experimental results related to a domestic micro-CHP unit which is based on organic Rankine cycle (ORC) technology. The main aim of the design works and experiments was therefore the analysis of the possibility of integrating the ORC system with a standard domestic central heating gas-fired boiler. The specially designed micro-ORC system was implemented in the laboratory and experiments were performed using this test stand. The main design aims of the test-stand were: low operating pressure, small working fluid flow, low price and compact dimensions. To meet these aims, volumetric machines were chosen as the expander and working fluid pump. The experimental results were positive and show that it is possible to integrate an ORC system with a standard domestic central heating gas boiler. For different heat source temperatures, the obtained expander power ranged from 109 W to 241 W and the thermodynamic cycle efficiency ranged from 4.3% to 8.8%. These positive research results were achieved partly thanks to the positive features of the different system subassemblies.

Keywords: air pollution; $PM_{2.5}$ dust; ORC; working fluid; selection method; volumetric expander; thermodynamic analysis

1. Introduction

One of the most important issues of present times is preventing excessive air pollution. Different industrial and domestic energy conversion systems and devices which are applied for converting the chemical energy contained in different fuels into heat, electricity and cold are sources of different gases and solid particle emissions. Especially in the countries (such as e.g., China, India, USA, RPA, Japan, Poland) whose power sector and industry are mainly based on the use of different fossil fuels (i.e., coal, coke, furnace oil or biomass) for the heat and electricity generation limiting these emissions is of great importance.

As the result of solid fuels combustion, in addition to the emission of gaseous pollutants (i.e., carbon, sulfur and nitrogen oxides), solid particles of different sizes are contained in the flue gases and emitted into the atmosphere in the form of dust. This problem became especially visible since many new measuring stations were installed around the world. Recently, the issue of micron particle sized (i.e., PM_{10} and $PM_{2.5}$) dust emissions has become particularly widely discussed in the media and the professional articles by many officials and researchers due to significant, negative impact of these dust on human health. $PM_{2.5}$ dust (i.e., dust featuring a particle size smaller than 2.5 μm) is especially hazardous to human health as its microparticles are able to penetrate the lungs, where they are dissolved in biological fluids and carried via the bloodstream through the whole body, causing different health issues (e.g., respiratory reactions). The importance of this issue is raised in numerous scientific studies on dust emissions into the atmosphere [1–12] and their impact on human health and the environment [13–17].

Statistical data on dust emission is published by many governmental and private agencies, for example by the statistical office of the European Union (Eurostat) and Statistics Poland (GUS). Table 1 summarizes statistical data on $PM_{2.5}$ dust emission in the EU-28 countries in 2017 [18].

Table 1. Statistical data on $PM_{2.5}$ dust emissions in EU-28 countries in 2017 [18].

Source of Dust Emission	Emission Tonnes
Energy production and distribution	49,716
Energy use in industry	108,887
Road transport	142,621
Non-road transport	32,085
Commercial, institutional and households	732,863
Industrial processes and product use	142,801
Agriculture	44,137
Waste	50,773
Other	485
Total sectors of emissions for the national territory	1,304,368

Table 2 summarizes the data on dust emissions in 2017 from various branches of the economy in Poland [19].

Table 2. The data on $PM_{2.5}$ dust emissions in 2017 from various branches of the economy in Poland [19].

Source of Dust Emission	Emission Tonnes
Power generating plants	10,400
Industrial power plants	35,500
Industrial technologies	34,700
Households	122,000
Other stationary sources (local boiler plants, trade, workshops, agriculture and others)	102,600
Mobile sources	35,500
Total	340,700

The data on the $PM_{2.5}$ emissions reported in Table 1 and in Table 2 show that in EU-28 countries and in Poland households have the largest share of $PM_{2.5}$ dust emissions. The households share of $PM_{2.5}$ dust emissions in Poland reached ca. 45% of the total emissions in 2017. Statistical data [18,19] show that other stationary installations are the second and transport is the third main source of PM_{10} and $PM_{2.5}$ dust emissions. For example in Poland, it was reported in 2017, [19] that emissions of these dusts from road transport, other vehicles and equipment (including rail transport) was equal to ca. 10% of total national emissions. It was also reported that a significant part of these emissions came from abrasion of tires and brakes and abrasion of road surfaces rather than fuel combustion [19]. The same data sources [18,19], show that the share of dust emission from the professional commercial and industrial installations which are commonly equipped with modern and efficient flue gas cleaning systems (such as for example electrostatic precipitators) is much smaller than in the case of other sectors of the economy.

Significant air pollution with dust is clearly visible in the autumn, winter and spring in rural and urban areas in which furnaces and boilers fed by low-quality fuels, such as wood, humid coal and other solid fuels are often used for domestic heating. The efficiency of these furnaces is low and usually ranges from 56% up to 70% [20], which additionally results in the unfavorable ratio of the amount of burned fuel to the heating effect, thereby increasing the amount of emitted flue gases. In large cities, the problem of dust emission is visible mostly in the areas of old, large housing estates consisting of buildings, which in most cases have not been properly insulated and whose residents use low-quality furnaces for heating the flats. Such settlements are, for example, Biskupin and Sępolno in Wrocław (Poland), which were built in 1930s and are composed of one floor brick blocks and single-family buildings.

Solid fuels which are usually combusted in domestic furnaces are featuring significant share of volatile matter. Table 3 reports the share of volatile matter (V^{daf}), and calorific values (Q_{wr}) for the solid fuels which are commonly combusted in the domestic furnaces [21]. Natural gas and heating oil were added to this Table for comparison.

Table 3. The share of volatile matter (V^{daf}), and calorific values (Q_{wr}) for the solid fuels which are commonly combusted in the domestic furnaces [21].

Fuel Type	V^{daf} %	Q_{wr} MJ/kg
Flame coal	30	25.0
Wheat	63–78	14.3–15.2
Waste wood	55–81	13.0
Natural gas	100	43.0
Heating oil	0	39.7–42.7

It was reported in [19] that in Poland, the average concentration of $PM_{2.5}$ dust in air was in 2018 equal to 22 μg/m^3. This concentration was 10% above the 20 μg/m^3 concentration which Poland should meet in 2015 and 4 μg/m^3 above the national $PM_{2.5}$ reduction target which was set by the government to 18 μg/m^3 and should be achieved by 2020. The highest values of the average concentration of $PM_{2.5}$ dust were observed in 2018 in Upper Silesian and Kraków agglomerations (31 μg/m^3), Rybnik-Jastrzębska agglomeration and Bielsko-Biała (30 μg/m^3) as well as in Częstochowa (27 μg/m^3) and Legnica (25 μg/m^3). These are agglomerations and cities having above 100,000 inhabitants. The highest values of emissions (for both PM_{10} and $PM_{2.5}$ dust) were recorded in 2017 in Central Europe (i.e., in Poland, Bulgaria, Croatia, Slovenia, Romania and Hungary) as well as in Italy and Cyprus.

In large agglomerations of cities, settlements and villages reducing the dust emissions from heating devices is not an easy task. One possible solution is connecting the houses, settlements and villages to the municipal heating network. However, sometimes it is not possible for various reasons (e.g., field restrictions, lack of space for heat exchangers in buildings, social issues, etc.). For the companies that are producers and distributors of network heat, such an investment is often not economically justified due to the above-mentioned problems, installation costs and a limited number (density) of potential heat recipients in a given area.

Currently, different municipal authorities in Poland and other countries are implementing co-financing programs focused on modernization of domestic heating systems by replacement of the old furnaces by modern heating devices (such as e.g., heat pumps or gas boilers). There are many devices of this type available on the market. These devices are featuring different technical parameters depending on the manufacturer and the design. For example, domestic gas boilers can be classified by the design of the combustion chamber [22]. Boilers with a closed combustion, boilers with an open combustion chamber and condensing boilers are applied in the domestic heating. Condensing boilers are featuring the highest energy conversion efficiency.

One of the possible methods of reducing the dust emission from domestic heating into the environment may be the application of combined heat and power (CHP) or multigeneration systems specially designed for domestic application. These systems can be installed as replacements of the currently applied old low-efficient boilers. This solution can in effect result in limiting the number of combustion processes and emissions of harmful substances into the atmosphere.

Domestic CHP system can be successfully designed and implemented with the application of technologies that are currently available on the market such as photovoltaic (PV) cells combined with heat pumps or heating systems based on electric heaters [23]. However, the power output of PV systems is usually limited by the availability of space required for their installation (e.g., the roof surface). The multigeneration system can be implemented for example by a coupling set of PV cells, heat pump and air conditioning unit [24]. Domestic boilers which are currently available on the market are fed by gas, oil or ecological solid fuels and are manufactured as heating devices, not CHPs. To the knowledge of the author there are currently no cogeneration or multi-generation domestic boilers commercially available. Domestic CHP or multigeneration boilers are promising alternative to other domestic power technologies, especially if the system power output is taken into account as the assessment criterion. Thanks to the high temperature of the heat source obtained by the fuel combustion it is possible to implement the vapour power plant cycle (i.e., Clausius Rankine cycle) in micro scale using specially designed machines and devices. Therefore, the system power per unit size can be higher than in the case of alternative technologies (e.g., PV cells) and the investment costs can be lower.

The successful implementation of domestic CHP systems or multigeneration boilers which can be used for heating the rooms and sanitary water and generating electricity and cold can possibly contribute in the future to significant reduction of the amount of low-quality furnaces which are currently used for heating purposes, and therefore significant reduction of the dust emission and air pollution in urban and rural areas. Moreover, CHP or multigeneration boilers can operate in combination with other devices (i.e., PV cells, heat pumps, air conditioners and others).

Research and development studies on this type of domestic power systems are carried out in different research and development units around the world [25], including the Department of Thermodynamics and Renewable Energy Sources of Wrocław University of Science and Technology. The integration of a gas central heating boiler with a power generating unit is usually done by redesigning the boiler into a home CHP micro power plant. This way a CHP unit featuring the electrical power output of ca. 2 kW_e and thermal power output of 24 kW_t can be implemented. Depending on the size of the system and consumer's energy needs such systems can be sufficient to cover the energy needs of smaller properties or can be treated as additional energy systems which can be used in combination with other technologies such as for example fuel cells or PV cells. The operating principle of these power generation systems is the same as in the case of classic steam power plants which are operating according to Clausius-Rankine (C-R) cycle. Different working fluids (i.e., water and low boiling substances) can be applied in these systems. Often, organic Rankine cycle (ORC) technology (i.e., a power system which uses low-boiling working fluid instead of water and operates according to the modified cycle of a classic steam power plant) is selected to be applied in power generating unit. The ORC system must be properly designed and miniaturized for this purpose, which is a significant scientific, design and construction challenge because the thermal phenomena and issues related to large and micro scale power systems are different. Proper selection of the working fluid and expander to such a system is of great importance [26]. Due to the low power of the domestic system, low flow rates of working fluid and low range of operating pressures are expected in the system. Therefore, in addition to axial or radial micro-turbines, it is possible to use much simpler and cheaper volumetric expanders (such as multi-vane, scroll, screw, lobe or piston expanders). These issues are treated in more detail in the following part of this paper together with design considerations and results of experiments which were proceeded at the Wrocław University of Science and Technology using the in-house designed and made prototype of domestic CHP ORC plant.

2. Design Considerations and Comparison of Different Domestic CHP Systems

Domestic CHP systems can be based on different technologies. Gas microturbines, fuel cells, photovoltaic cells, Stirling engines, small gas engines, steam micro power plants and ORC micro power plants are currently considered as possible technologies that can be applied in domestic energy conversion systems. Research on these types of cogeneration systems is carried out by various researchers around the world and these technologies are currently at different stages of development. PV cells are for example successfully applied in domestic conditions, however they are further being intensively developed. Present works are aimed at maximizing the power obtained from the PV panel per unit surface and energy conversion efficiency. New materials (such as e.g., perovskites) are considered as promising for application in PV panels [23]. Issues related to research and development works related to the application of fuel cells in domestic conditions are presented in [27,28]. Natural gas is considered as fuel for domestic fuel cells. Recently, research and development works have been carried out on the domestic CHP systems based on gas microturbines. The design of the domestic CHP system employing microturbine is similar to the design of the domestic gas boilers. Such systems are prototyped and presented in [29]. Small Stirling engines [30] are also developed to meet domestic operational conditions. Since some time domestic vapour power plants implementing Clausius-Rankine cycle and using different working fluids are developed and tested [31]. They are treated in more details in other sections of this article.

Domestic CHP systems should be simple in the design, easy to use and install, fully automated and their investment and service costs should be low. Not in the case of all of the above described systems these features can be satisfied. Table 4 shows a comparison of investment costs (per kW of power output) for different energy conversion technologies that can be applied in/as domestic CHP systems.

Table 4. A comparison of investment costs (per kW of power output) for different energy conversion technologies that can be applied in/as domestic CHP systems.

Technology Type	Investment Costs EUR/kW	Overall Efficiency %	Electrical Efficiency %	Electricity/Heat Supply Ratio kW_e/kW_t	Part-Load Flexibility	Ref.
PV cell	1700	-	14–16	5/-	Yes	[32]
Fuel cell	890	70–92	37–55	0.75/0.9–30.8	No	[33]
Heat pump	350	4–5	-	-/30	Yes	[34]
ORC system	1000	90	10	2.5–10/11–44	Yes	[25]
Micro CHP gas turbine	360	90	26	30/30	No	[35]
Stirling engine	2200	90	15	2–10/4–35	Yes	[36]

It can be seen from Table 4 that among the mentioned technologies ORC systems are characterized by positively low investment costs in comparison to competitive technologies.

ORC systems are modified classical steam power plants. Unlike classical steam power plants, they are driven by heat sources featuring lower thermal parameters, power and output therefore they are featuring lower power, lower efficiency and smaller external dimensions. Moreover, they are using a low-boiling working fluid instead of water. However, the set of thermodynamic processes proceeding in the ORC system and operating principle are the same as in the case of the classical Clausius-Rankine system. The main components of the ORC system are two heat exchangers (an evaporator and a condenser), a liquid working fluid reservoir, a liquid working fluid pump and an expander coupled to a electric generator. In order to improve the efficiency of the system an additional heat exchanger (i.e., regenerator) is often applied. Figure 1 shows the scheme of the ORC system with regenerator. Figure 2 visualizes the thermodynamic processes of the simple organic Rankine cycle in the T-s plane.

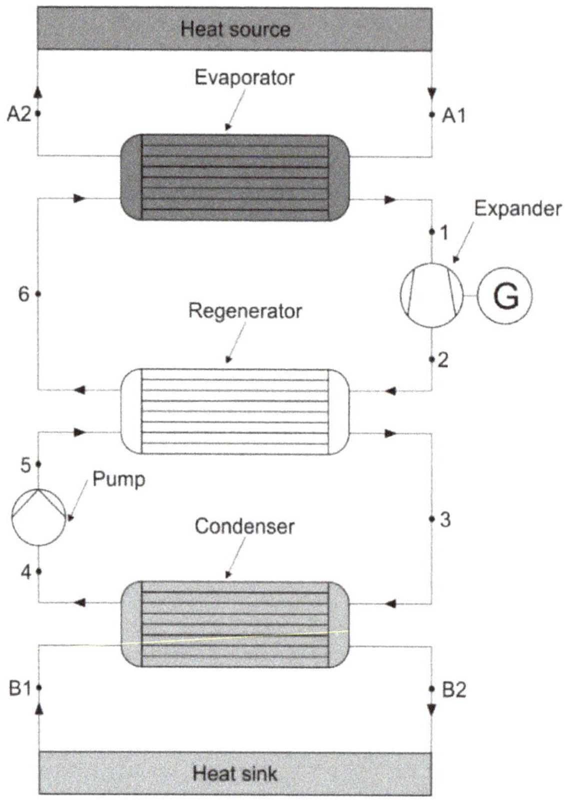

Figure 1. Scheme of the ORC system with regenerator.

The most important design choices of the ORC systems are the working fluid and expander selection. Different low-boiling substances are applied as working fluids in ORC systems. These can be for example refrigerants or silicone oils. By the shape and course of the saturation curve, the working fluids are classified as dry, wet and isentropic [26,37,38]. The selection of the working fluid for the ORC system is often challenging as environmental, thermodynamic and design considerations should be taken into consideration. Different methods for working fluid selection are reported in literature [39–50]. The working fluid should be non-toxic and featuring low values of global warming potential (GWP) and ozone depletion potential (ODP) indicators. In terms of thermodynamic properties, the working fluid should be selected to fit the thermal characteristics of the heat source (i.e., its temperature, thermal power, heat capacity and mass flow rate of the heating medium). The heat source thermal parameters and the type of the applied working fluid have the influence on the ORC system design and configuration. For example, ORCs with direct evaporation of the low-boiling working fluid (whose layout is presented in Figure 3) or using an intermediate (e.g., thermal oil) heat transfer working fluid (which layout is presented in Figure 4) are implemented depending on the heat source temperature and the type of the applied low-boiling working fluid. In the case of domestic scale power systems, the thermal power of the heat source is usually low or medium (typically ranging from 10 up to 50 kW_t). Therefore, if the ORC system is considered to be powered by a heat source featuring such a range of thermal power the mass flow rate of the low-boiling working fluid should be kept at low value. In addition, to provide operational safety of the ORC system, the applied working fluid should feature low values of pressure within the temperature range of the heat source. The list of the selected working fluids which can be used in domestic ORCs is presented in Table 5.

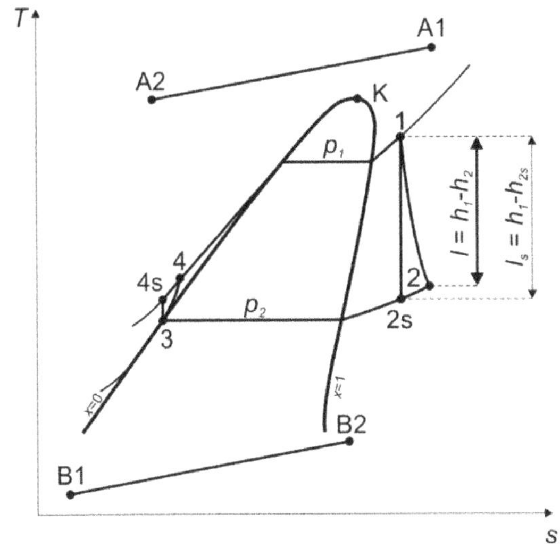

Figure 2. Thermodynamic processes of the simple organic Rankine cycle in the T-s plane.

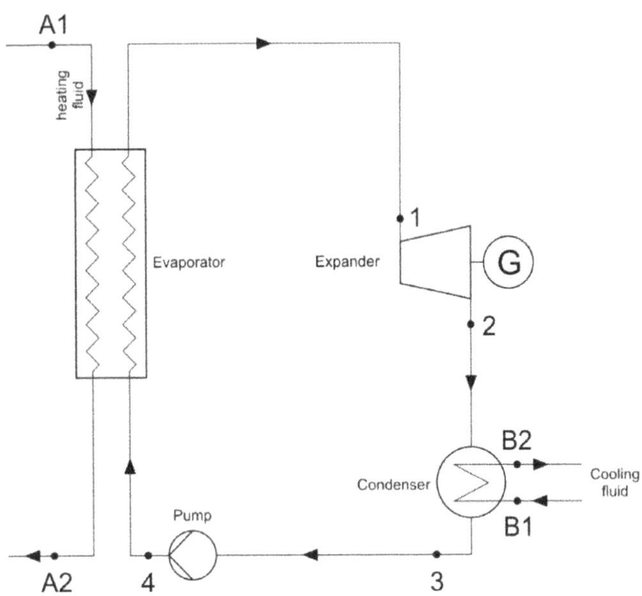

Figure 3. Scheme of the ORC system with direct evaporation of the working fluid.

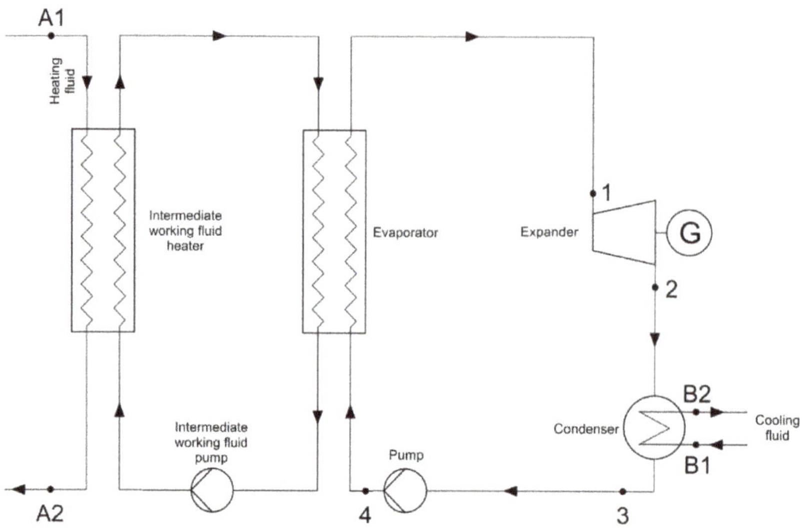

Figure 4. Scheme of the ORC system with intermediate working fluid loop.

Table 5. The working fluids suitable for domestic Organic Rankine Cycle (ORC) systems [37–50].

No.	Working Fluid	Triple Point Temperature	Normal Boiling Point Temperature	Critical Point Parameters			ODP	GWP	Working Fluid Class [39]
		t_{trp} °C	t_{nbp} °C	t_{cr} °C	p_{cr} MPa	ρ_{cr} kg/m^3			
1	R113	−36.22	47.59	214.06	3.39	560.00	0.8	4800	ANZCM
2	R114	−92.52	3.59	145.68	3.25	579.97	1.0	3.9	AZCM
3	R123	−107.15	27.82	183.68	3.66	550.00	0.02	77	ACNMZ
4	R124	−199.15	−11.96	122.28	3.62	560.00	0.02	620	ACNZM
5	R1234ze	−104.53	−18.95	109.37	3.63	489.24	0	6	ACNZM
6	R134a	−103.30	−26.07	101.06	4.06	512.00	0	1300	ACZ
7	R152a	−118.59	−24.02	113.26	4.51	368.00	0	120	ACZ
8	R227ea	−128.60	−16.34	101.75	2.93	594.25	0	3220	ANCMZ
9	R236fa	−93.63	−1.44	124.92	3.20	551.30	0	9810	ACNMZ
10	R365mfc	−34.15	40.15	186.85	3.22	473.84	0	825	ANZCM
11	R245ca	−81.65	25.13	174.42	3.39	523.59	0.12	N/A	ANCMZ
12	R245fa	−102.10	15.14	154.01	3.65	516.08	0	1030	ACNMZ
13	R601a	−160.50	27.83	187.2	3.38	236.00	0	20	ANCMZ
14	R141b	−103.47	32.05	204.35	4.21	458.56	0.12	725	ACNMZ
15	R142b	−130.43	−9.12	137.11	4.05	446.00	0.07	2310	ACNMZ
16	R236ea	−103.15	6.19	139.29	3.5	563.00	0	1330	ANZCM
17	R600a	−159.42	−11.75	134.66	3.63	225.5	0	3	ACNMZ
18	RC318	−39.80	−5.97	115.23	2.78	620.00	0	10,300	AZCM
19	R1234yf	−53.15	−29.45	94.7	3.38	475.55	0	4	ACNZM
20	R290	−187.63	−42.11	96.7	4.25	220.48	0	20	ACZ

R113 is banned by the Montréal Protocol. R143a and R123 will be phased out soon. However, the author decided to include these working fluids in the comparison as the reference substances which were considered to be applied or were applied in ORCs. The literature shows [51] that the electric power output of different domestic CHP systems usually ranges between 1 and 10 kW$_e$. Considering the advisable features of domestic CHP ORC systems they should be assembled of components featuring simple design and low investment and service costs. Brazed or welded plate heat exchangers meet these requirements. These heat exchangers are simple in design and low weight thanks to the application of specially formed plates that are welded together. The specially formed plate increases the heat exchange surface and leads to the intensification of the heat exchange phenomena. This way it is possible to achieve high thermal power in relation to the heat exchanger dimensions. Plate heat exchangers can be applied for clean gases and liquids. In the case of dusty working fluids, the intermediate heat transfer substance have to be applied. Plate heat exchangers feature small external dimensions in relation to their power and are cheap compared to the other types of the heat exchangers. What is more, the heat transfer rate in plate heat exchanger is high. Shell-and-tube heat exchangers can be also applied in domestic ORCs. Compared to plate heat exchangers, shell-and-tube heat exchangers are larger, heavier and more complex to manufacture as they require the use of many welded joints. The heat transfer rate in shell-and-tube heat exchangers is lower compared to plate heat exchangers and a larger heat transfer area is needed. Therefore these heat exchangers feature significantly larger external dimension, mass and investment costs compared to plate heat exchangers. In general, the ORC systems can utilize two types of expansion machines. First type are axial or radial (inflow or outflow) turbines [52]. The others are volumetric expanders (e.g., piston, screw, scroll, lobe and multi-vane machines) [53]. Only turbines which are featuring small external dimensions and a power output of few kW (i.e., microturbines) can be applied in domestic ORC systems [52]. However, microturbines are featuring very high rotational speeds and clearance losses become large for small scale turbines, which leads to the need of using complicated and high-precision manufacturing technologies and tools for their fabrication. Therefore, their investment costs are high. For these reasons, in the opinion of the author, their applicability to domestic systems is currently limited. Compared to turbines, volumetric expanders are having positive features as they are operating at much lower rotational speeds, lower working fluid flow rates, and higher pressure drops that can be achieved in one stage [54]. Therefore, they fit better to small scale ORCs. Piston [55–61], screw [62–66], scroll [67–73], lobe [74,75] and multi-vane [25] expanders can be applied in domestic ORCs. These expanders are at different stage of development. Scroll expanders are applied in many prototypes of the domestic ORC systems. Their mechanical power output ranges from 1 to 10 kW [53], while the pressure expansion ratio of a scroll expander typically ranges between 2 and 4.5. Piston and linear piston expanders feature the highest pressure expansion ratios among the volumetric expanders applied in ORCs. Therefore, these expanders are a good option for high-pressure ORC systems. Their power output ranges from 0.5 to 15 kW [52,53]. The value of pressure expansion ratio of a piston expander can reach 200. Screw expanders are insensitive to wet gas operating conditions, therefore these machines can be applied in systems utilizing floating heat sources. The pressure expansion ratio of a screw expander typically ranges between 10 and 15 and power output ranges from 1 to 8 kW [72]. Rotary lobe expanders are currently under research and development [74,75]. Thanks to their positive features (insensitivity to wet gas operating conditions, compact and simple design, high power output in relation to external dimensions) they seem to be good option for different industrial and domestic ORCs. Multi-vane expanders can operate in wet gas conditions, their design is simple and compact, they feature small gas consumption and low rotational speed and are relatively cheap. Therefore these expanders are particularly promising for application in domestic ORCs. Their possible application in such systems is currently being investigated by many researchers. The pressure expansion ratio of a multi-vane expander typically ranges between 3 and 5 and power output ranges from 1 to 10 kW [25]. Multi-vane expanders combine in one design the best features of turbines and volumetric machines. Working fluid pumps that can be applied in domestic ORC systems should be characterized by low energy consumption, high reliability and simplicity of design. Centrifugal and volumetric pumps are the options that can be applied in these systems. Centrifugal pumps are featuring high mass flow rates, therefore their application is better justified in the case of high-power industrial ORC systems utilizing turbines as the expanders rather than domestic systems. In micro-power ORCs, volumetric pumps are a better choice, as they are featuring smaller energy consumption, smaller mass flow rates, dimensions, lower power and rotational speed. Positive displacement pumps, such as for example multi-vane pumps [76] are currently applied in prototypes. ORC system subassemblies should be hermetically sealed and connected using similar techniques as are used in refrigeration and air conditioning units, (e.g., soldering with silver). The automatic control system should provide the safe operation of the system. Additionally it should give the opportunity of measurement and observation of system's operating parameters as well as signaling errors, emergency conditions and necessity of inspections.

Additionally, in order to efficiently use the generated heat in the summer, the micro-absorption cooling unit may be integrated with a CHP ORC system. This way, the multi-generation system can be implemented. According to the author's knowledge, no research is currently proceeding on the integration of absorption cooling unit with domestic ORC systems, therefore this opportunity should be considered as one of the possible ways of further research on domestic ORCs.

3. Description of the Experimental Test-Stand and the Experiment Results

At the Laboratory of Thermodynamics and Thermal Properties of Materials of the Department of Thermodynamics and Renewable Energy Sources at the Wroclaw University of Science and Technology an experimental test-system composed of ORC test-stand and domestic gas boiler was designed and commissioned. Thanks to the special design of this test-setup and its flexibility to different experimental conditions it is possible to conduct a wide range of experiments related to micro CHP ORC systems. For example, it is possible to quickly change the expander or working fluid. The key assumptions for the design of this test stand were defined in order to meet the advisable features of domestic ORCs described in the first part of this article. In particular, the following design requirements were set: small dimensions and compact design, safety, low investment costs, low weight, low power needs, low operating pressure and small mass flow rates of the working fluids. These requirements were achieved thanks to the application of specially selected and designed subassemblies in the test-stand, i.e., plate heat exchangers (evaporator and condenser) and multi-vane volumetric machines (pump and expander). In addition, the working fluid featuring low pressure in the range of temperatures of domestic heat sources (such as e.g., solar heat, hot water, etc.) was applied in the test-stand.

3.1. Description of the Test-Stand

Central heating boilers fed by different fuels, e.g., natural gas, are commonly used for individual heating in many properties. The possibility of integration of a electricity generating system in the boiler seems to be interesting option. This way boiler can be transformed into a CHP system, which, in addition to heat, can generate electricity that can be consumed by the boiler user or fed to the grid. This solution can contribute to increase of energy generation efficiency and share of distributed energy systems and therefore to increased energy safety of end energy consumers. The ORC technology can be used for this purpose. In the following part of this paper in order to check the possibility of integrating the ORC system with the domestic gas boiler a standard, commercially available gas central heating boiler was selected and used to power the ORC test-stand. This is a closed combustion chamber boiler featuring the thermal power of 24 kW. The automatic controller which is applied in the boiler enables smooth regulation (per 1 °C) of the central heating water temperature in the range of 40–85 °C and temperature of the sanitary water in the range of 30–60 °C. Central heating water was used to supply the ORC test-stand as it is visualized in the scheme of the test-system (see, Figure 5). Figure 6 shows the general view of the test-system, while Figure 7 shows a detailed view of the ORC test-stand. The test-system operating principle is described in the following.

The central heating water from the gas boiler is pumped by a water pump (which is installed in the boiler) to the inlet of a plate heat exchanger i.e., evaporator of the ORC system (featuring the height of 622 mm, length of 165 mm and width of 197 mm). The direction of water and low-boiling working fluid flow in the evaporator is countercurrent. Thanks to the possibility of smooth regulation of the temperature of the central heating water, it is possible to conduct experimental tests for different temperatures of the heat source. Therefore, it is possible to simulate in laboratory conditions various domestic heat sources which can be applied for powering the ORC system. The implementation of the Clausius-Rankine cycle in the range of the heat source temperature provided by the applied gas boiler was possible thanks to the application of R123 (2,2-dichloro-1,1,1-trifluoroethane) as the working fluid in the ORC test-setup. This working fluid features a low boiling temperature (ca. 27 °C) and low pressures in the analyzed range of the heat source temperature.

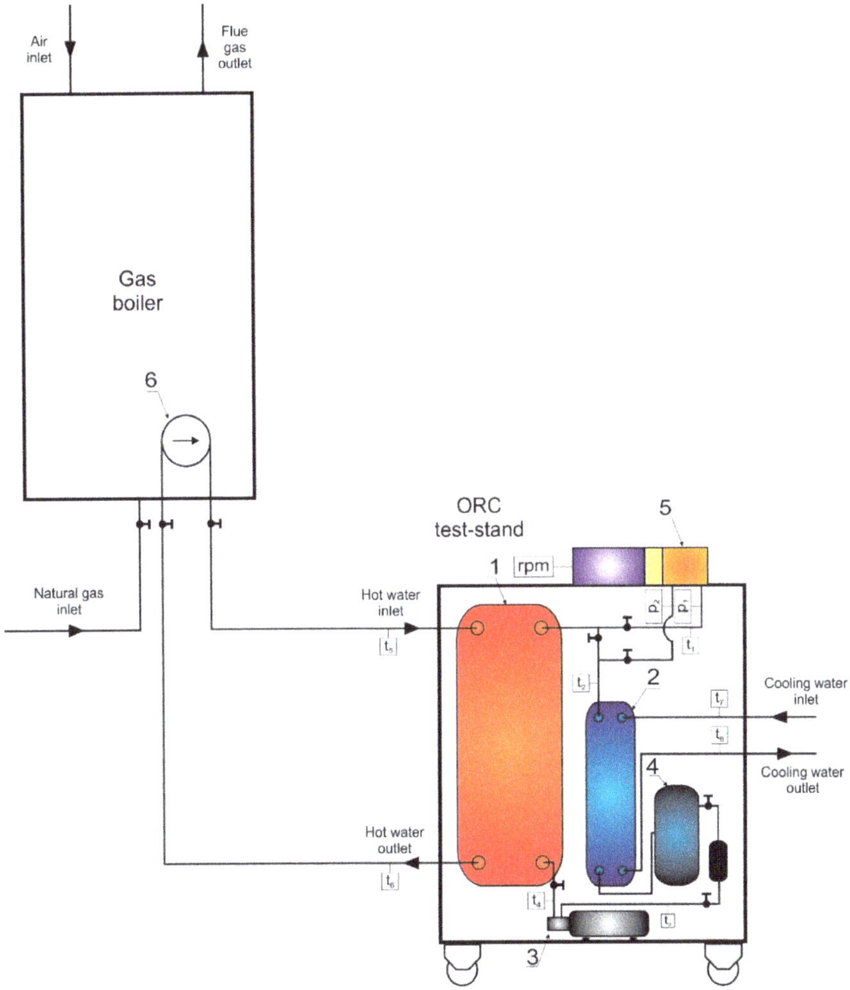

Figure 5. Scheme of the test-system.

A positive displacement multi-vane pump (featuring the power of 185 W) was applied in the ORC test-stand to force the flow of liquid low-boiling working fluid. By using this type of working fluid pump, it was possible to conduct the experiments under small and very small flow rates of low-boiling working fluid (typically during experiments the working fluid flow rate is regulated in the range of 50–250 dm^3/h). In addition, this type of pump gives the ability to easily adjust operating parameters (i.e., rotational speed, working fluid flow rate and pressure) and is characterized by low rotational speed, simple and easy-to-service design and a low price. What is more it is insensitive to operation in two-phase flow conditions. The working fluid pump forces the flow of the working fluid from the liquid reservoir (featuring a volume of 3.9 dm^3) through pipelines to all of the test-stand components. On the outlet pipeline of the liquid reservoir the working fluid drier, filter and a sight-glass are installed.

Figure 6. General view of the test-system (1—gas boiler; 2—ORC test-stand).

The working fluid vapour which is obtained at the outlet of the evaporator features high values of thermodynamic parameters (i.e., pressure and temperature) and it flows through the pipelines and control valves to the inlet of the expander. In the ORC test-stand a specially modified multi-vane air motor was applied as the expander (featuring the power of 330 W). The modifications of the expander design were necessary to adapt the machine to low-boiling working fluid operating conditions. These modifications were described in more details in earlier publications of the author [77,78]. This type of the expander (similarly to the applied multi-vane pump), features a simple design, low rotational speed and low price and is insensitive to two-phase operating conditions. It can be easily hermetically sealed and coupled to the generator. Detailed results of a comprehensive experimental and numerical research related to this expander and possibilities of its optimization are reported in [79].

After expansion in the expander, the thermodynamic parameters of the working fluid (i.e., pressure and temperature) decrease and the working fluid flows through the pipeline to the inlet of a plate condenser. A bypass pipeline with a shut-off valve is installed between the outlet of the evaporator inlet of the condenser and the expander inlet. Thanks to its opening it is possible to direct the gas flow from the outlet of the evaporator directly to inlet of the condenser bypassing the expander. The bypass pipeline is used during starting (heating up) and stopping (cooling down) the test-stand. The plate heat exchanger (featuring the height of 456 mm, length of 47 mm and width of 80 mm) was applied as the condenser in the ORC test-setup. Condenser is cooled by the cold water from the municipal pipelines. The flow rate of the cooling water is regulated by means of set of control valves. Therefore it possible to simulate different domestic heat sinks in laboratory conditions. The direction of the cooling water and low-boiling working fluid flow in the condenser is countercurrent. After condensation in the condenser, the working fluid is forced back by the pump to the liquid reservoir. The Clausius-Rankine cycle is therefore completed with this point.

Figure 7. Detailed view of the ORC test-stand (1—evaporator; 2—condenser; 3—pump; 4—working fluid reservoir; 5—expander and generator).

The test-stand is equipped with different measuring sensors. Temperature measurements are carried out in 10 measuring points using T-type thermocouples (temperature measuring points are marked with t letters in Figure 5). Thermocouples are connected to a microprocessor recorder with LCD screen which enables continuous observation and recording of the measured values. Pressure is measured using pressure gauges (pressure measuring points are marked with p letters in Figure 5) while the rotational speed of the expander is measured using a laser tachometer.

An automatic control system is not applied in the ORC test-stand due to its limitations (automatic controller is however applied in the gas boiler). All operating parameters of the ORC system are manually controlled, therefore different experimental conditions (including simulation of failures and emergency situations) can be tested.

3.2. The Experimental Description

The aim of the experiment presented in this article was observation of the associated operation of a gas boiler and the ORC test-stand under various experimental conditions and recording the operating parameters.

Before starting the experiments, the test-setup was prepared for tests according to the following procedure. Initially, all of the shut-off valves that are installed in ORC test-stand were opened except the shut-off valves installed at the inlet and at the outlet of the expander. Then, the low-boiling working fluid pump was started and the flow of low-boiling working fluid was forced from the liquid reservoir tank through the pipelines to the inlet of the evaporator. Subsequently, using the control valve on the low-boiling medium pump, the flow of the low-boiling working fluid was set (constant flow of the working fluid was kept for each experimental series). Then, the condenser cooling water shut-off valve was opened and the flow of the cooling water was set constant using the control valve.

After preparing the ORC test-stand, using the boiler controller, the temperature of the central heating water (i.e., the temperature of the heat source) was set and the boiler was switched on.

After reaching the set temperature of the heat source, the bypass between evaporator and condenser was closed and the flow of the working medium through the expander was opened using shut-off valves installed on the inlet and outlet ports of the expander. When the gas flow through the expander was opened, the operation of the expander and variation of working fluid parameters in the system were observed. At the time when operating parameters stabilized (i.e., ORC system reached steady state operation) operating parameters were measured.

After the experimental series was completed for one setting of the heat source temperature and measurements were recorded, the shut-off valves at the inlet and the outlet of the expander were closed again, the boiler was switched off and the stand was cooled to ambient parameters using cooling water. The same experimental procedure was then carried out for different settings of the heat source temperature.

4. Experimental Results and Discussion

The experiments were carried out for variable temperature of the heat source, which was controlled between 45 °C and 85 °C with a step of 10 °C. During the experiments thermodynamic parameters of the working fluids (i.e., R123, heating water and cooling water) were recorded and operation of the test-system was put under the observation. In this experimental series the low-boiling working fluid flow rate was set to 35 dm^3/h, the heating water flow rate was set to 150 dm^3/h and cooling water flow rate was set to 70 dm^3/h.

The collected experimental results are summarized in Table 6, while Table 7 reports the values of indicated and real power output of the expander (calculated from the relation $N_{iex} = m_{WF} \times (h_1 - h_{2s})$ and $N_{rex} = m_{WF} \times (h_1 - h_2)$ correspondingly, where m_{WF} is the mass flow rate of the working fluid), expander internal efficiency (calculated from the relation $\eta_{iex} = (h_1 - h_2)/(h_1 - h_{2s})$), power of the evaporator (calculated from the relation $Q_{EV} = m_{WF} \times (h_1 - h_4)$), power of the condenser (calculated from the relation $Q_{CN} = m_{WF} \times (h_2 - h_3)$) and thermodynamic cycle efficiency (calculated from the relation $\eta_{iex} = N_{rex}/Q_{EV}$). These operational parameters were calculated using the values of the enthalpy (h) of the working fluid in measuring points which were obtained using CoolProp computer software [80].

Table 6. Thermodynamic parameters of R123 and heating and cooling water in measuring points and rotational speed of the expander.

No.	t_{hs} (°C)	p_1 (bara)	t_1 (°C)	p_2 (bara)	t_2 (°C)	t_3 (°C)	t_4 (°C)	t_5 (°C)	t_6 (°C)	t_7 (°C)	t_8 (°C)	n_{ex} (rev/min)
1	45	1.5	38.9	1.6	24.1	17.4	17.9	39.5	38.4	14.7	15.6	700
2	55	1.85	45.2	1.6	26.0	17.6	18.0	46.9	46.4	14.8	16.9	3280
3	65	2.2	50.9	1.6	28.1	17.9	18.3	52.7	51.6	15.0	19.1	3300
4	75	2.9	60.7	1.7	31.4	18.1	18.4	60.9	59.8	15.1	17.5	2900
5	85	3.4	66.7	1.7	33.4	18.0	18.6	71.5	65.9	15.1	18.3	2950

Table 7. The values of indicated and real power output of the expander, expander internal efficiency, power of the evaporator, power of the condenser and thermodynamic cycle efficiency.

No.	σ_p (-)	N_{iex} (W)	N_{rex} (W)	η_{iex} (%)	Q_{EV} (W)	Q_{CN} (W)	η_{ORC} (%)
1	2.17	164	109	66.4	2525	2417	4.3
2	2.67	207	141	67.8	2572	2433	5.5
3	3.14	244	167	68.5	2613	2447	6.4
4	4.13	307	213	69.3	2687	2476	7.9
5	4.87	346	241	69.6	2735	2496	8.8

The experimental results are visualized in Figures 8–10. Figure 8a shows the variation of the expander isentropic power (N_{iex}) and expander power output (N_{rex}) vs. the heat source temperature (t_{hs}) while Figure 8b shows the variation of the expander isentropic power (N_{iex}) and expander power output (N_{rex}) vs. the pressure expansion ratio (σ_p). Figure 9a shows the variation of the expander isentropic efficiency (η_{iex}) vs. the heat source temperature (t_{hs}) while Figure 9b shows the variation of the expander isentropic efficiency (η_{iex}) vs. the pressure expansion ratio (σ_p). Figure 10a shows the variation of the thermodynamic cycle efficiency (η_{ORC}) vs. the heat source temperature (t_{hs}) while Figure 10b shows the variation of the thermodynamic cycle efficiency (η_{ORC}) vs. the pressure expansion ratio (σ_p).

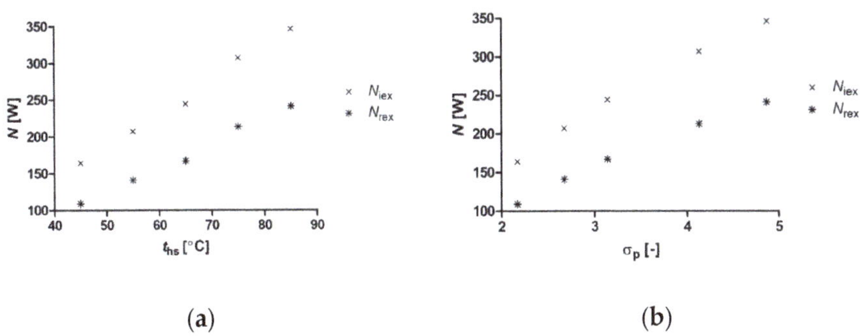

Figure 8. The variation of expander isentropic power (N_{iex}) and expander power output (N_{rex}): (**a**) $N = f(t_{hs})$; (**b**) $N = f(\sigma_p)$.

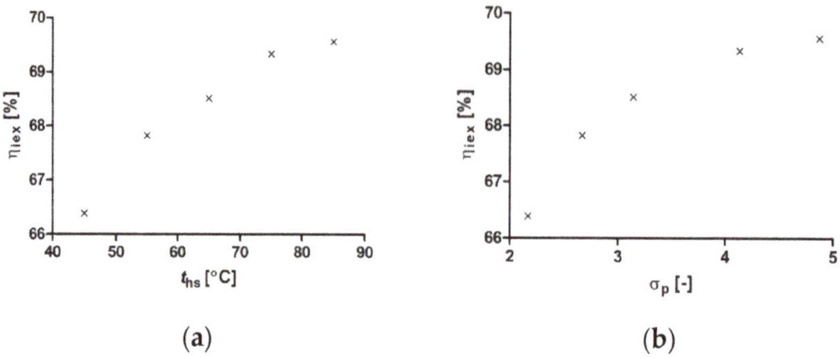

Figure 9. The variation of expander isentropic efficiency (η_{iex}): (**a**) η_{iex} = f(t_{hs}); (**b**) η_{iex} = f(σ_p).

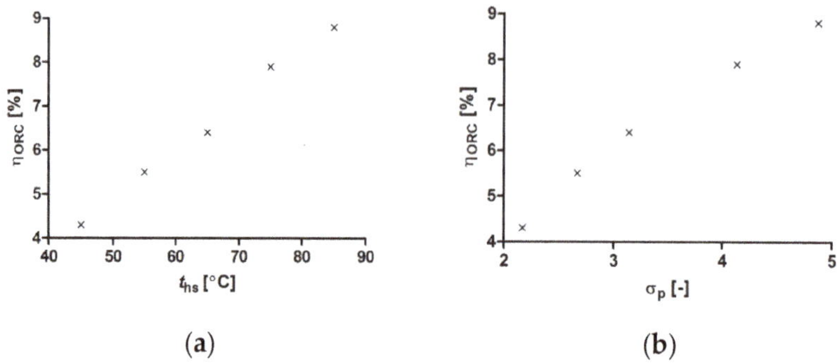

Figure 10. The variation of thermodynamic cycle efficiency (η_{ORC}): (**a**) η_{ORC} = f(t_{hs}); (**b**) η_{ORC} = f(σ_p).

The experiment was conducted in the heat source temperature range of 45–85 °C. Figures 8–10 show the increasing trend of all of the ORC test-setup operational indicators which is caused by the growing heat source temperature (t_{hs}). Results visualized in Figure 8 show that the multi-vane expander power output was varying during the experiment in the range of 109–241 W. The increasing expander power output which is observed for increasing temperature of the heat source is caused by increasing temperature and pressure of vapour at the inlet to the expander. The experiment therefore proves, that the multi-vane expander power output depends mostly on the pressure expansion ratio (not the working fluid flow rate which was kept constant during the experiments). Results visualized in Figure 9 show that the multi-vane expander internal efficiency was varying during the experiment in the range of 66.4–69.6%, while the thermodynamic cycle efficiency was increasing with increasing heat source temperature (t_{hs}), increasing expander power output (N_{rex}) and pressure expansion ratio (σ_p) and was varying in the range of 4.3–8.8% as it is visualized in Figure 10. Experimental data which are reported in Table 6 show, that the decrease of heating water temperature between the inlet and the outlet of the ORC evaporator ranges between Δt = 0.5–5.6 °C depending on the experimental conditions. These low drops of water temperature results from the difference in heat capacity of heating water and low-boiling fluid and were observed by the author during his earlier experiments on ORC systems [54,77–79]. The value of the heating water temperature drop is particularly important in case of domestic CHP ORC systems in which the water flowing out from the evaporator will be then transferred to the inlet of the central heating system. Therefore, lower the temperature drops are higher is the thermal comfort and better are the operating conditions of central heating installation. Low drops of water temperature which were observed during described experiments are promising from the point of view of the possibility of integrating the ORC system with a gas boiler as such small drops of temperature they will be practically imperceptible to the user.

Literature reports [81] that the heat demand in standard houses ranges between 70 and 100 kWh/m^2, and in older buildings exceeds 120 kWh/m^2 per year. Table 8 reports the comparison of the $PM_{2.5}$ and PM_{10} dust emission of domestic boilers fed by different fuels [82]. Therefore, if technologies listed in Table 8 are considered as the heating boilers, the annual emission of dust per 1 m^2 of property area will reach the values reported in Table 8 as E_s (for standard house) and E_o (for old house). The analysis of the data reported in Table 8 shows that coal-fired boilers are the largest source of dust emission. For example in Wrocław (Poland) the literature reports [19,83] that the number of low-quality coal-fired boilers and furnaces is equal to ca. 20,000. These furnaces are used mostly for heating old houses and old flats. Application of gas-fired CHP ORC systems as the replacements of the coal-fired boilers in these properties can therefore led to clean heating and power generation and significant reduction of the dust emitted into the environment. The average area of the flat in Poland is 70.4 m^2 [84]. If such an area is considered to be heated by coal-fired boiler, then for the earlier mentioned number of coal-fired boilers the avoided dust emission in Wrocław thanks to the application of the CHP ORC units can be as high as ca. 47.3 t/a. Additionally ca. 40,000 kW$_e$ of electricity might be generated if single CHP ORC unit electric power is equal to 2 kW$_e$.

Table 8. The comparison of the $PM_{2.5}$ and PM_{10} dust emission from domestic boilers fed by different fuels [82].

Bolier Type	E mg/kWh$_t$	E_s mg/m^2·a	E_o mg/m^2·a
Oil-fired bolier	40	2800–4000	4800
Gas-fired bolier	5	350–500	600
Coal-fired bolier	280	19,600–28,000	33,600
Wood-fired boiler (manually operated)	140	9800–14000	16,800
Wood-fired boiler (automatically controlled)	65	4550–6500	7800
Pellet-fired bolier	40	2800–4000	4800
Wood chips-fired boiler	52	3640–5200	6240
CHP unit	0	0	0

5. Summary and Conclusions

This paper presents the results of experimental study on possible integration of domestic gas boilers with micro-ORC systems and conversion of such a boiler into a domestic CHP unit. Such systems are promising options for application as replacements for standard low-efficient boilers fed by low-quality solid fuels which are commonly applied for domestic heating. The design issues of such systems were discussed and the selection of heat exchangers, expander and working fluid were found to be most important issues that need to be solved during the implementation of such systems. The experimental test-setup was designed in order to fit to the stated design requirements (i.e., simple design, safety and low costs). The multi-vane expander was applied in the experimental ORC test-stand. The experimental series were performed using the test-system in order to record the operational conditions. The experiments succeeded. The power output of the expander of was ranging from 109 to 241 W, and the thermodynamic cycle efficiency was ranging from 4.3 to 8.8%. Operating parameters were found to be dependent on the heat source temperature and working fluid pressure at the inlet to the expander. The decrease of heating water temperature between the inlet and the outlet of the ORC evaporator was found to range between Δt = 0.5–5.6 °C depending on the experimental conditions. Therefore, the experimental results show that there is a possibility of integrating the ORC system with a standard domestic central heating gas boiler. This positive research results were achieved partly thanks to the positive features of the applied multi-vane pump and multi-vane expander as well as plate heat exchangers, which are characterized by positively high values of the heat transfer coefficients, high thermal power in relation to the dimensions and high efficiency. The simple design of the multi-vane expander and its additional positive features including the lack of the need for additional complicated and high-cost automatic adjustment systems, lubrication systems and rotational speed reduction devices.

During the tests, the operation of the test-system was observed and there was no emergency or safety hazards noticed. Further development of the design supported by the results of further experimental research is needed and can successfully lead to the implementation of these systems on a large scale. In effect, this may lead to the increased energy generation efficiency (energy will be generated and consumed at the same place without the necessity of transmission), decentralization of energy generation, and what is the most important, significant reduction of the number of low-quality and low-efficient furnaces used in domestic heating. This way, a reduction in the number of low-quality solid fuel combustion processes and therefore a reduction of solid pollutant emissions (such as $PM_{2.5}$ and PM_{10} dust) into the atmosphere can be achieved. In the case of large cities, like Wrocław (Poland) the avoided dust emission can be as high as 47.3 t/a.

Experiments were performed using a micro-power multi-vane expander. Further works should be focused on increasing the power output and efficiency of the expander and CHP system. Then an expander featuring bigger power output will be applied in the test-stand by the author in his further works. Another further research topic is the integration of the micro-absorption cooling unit in a CHP ORC system in order to efficiently use the heat generated during the summer.

Funding: This research received no external funding.

Acknowledgments: The author would like to thank Izabela Sówka and Sławomir Pietrowicz for the invitation to publish this article.

Conflicts of Interest: The author declares no conflict of interest.

References

1. Kumar, P.; Pirjola, L.; Ketzel, M.; Harrison, R.M. Nanoparticle emissions from 11 non-vehicle exhaust sources—A review. *Atmos. Environ.* **2013**, *67*, 252–277. [CrossRef]
2. Commodo, M.; Sgro, L.A.; D'Anna, A.; Minutolo, P. Size Distribution of Nanoparticles Generated by a Heating Stove Burning Wood Pellets. *Int. J. Environ. Qual.* **2012**, *8*, 21–26.
3. Saikia, B.K.; Saikia, J.; Rabha, S.; Silva, L.F.O.; Finkelman, R. Ambient nanoparticles/nanominerals and hazardous elements from coal combustion activity: Implications on energy challenges and health hazards. *Geosci. Front.* **2018**, *9*, 863–875. [CrossRef]
4. Nussbaumer, T.; Doberer, A.; Klippel, N.; Bühler, R.; Vock, W. Influence of Ignition and Operation Type on Particle Emissions from Residential Wood Combustion. In Proceedings of the 16th European Biomass Conference and Exhibition, Valencia, Spain, 2–6 June 2008.
5. Annesi-Maesano, I. The Air of Europe: Where are we going? *Eur. Respir. Rev.* **2017**, *26*, 170024. [CrossRef] [PubMed]
6. Collective work. *Air Quality in Europe—2019 Report*; European Environment Agency: Copenhagen, Denmark, 2019.
7. Hata, M.; Chomanee, J.; Thongyen, T.; Bao, L.; Tekasakul, S.; Tekasakul, P.; Otani, Y.; Furuuchi, M. Characteristics of nanoparticles emitted from burning of biomass fuels. *J. Environ. Sci.* **2014**, *26*, 1913–1920. [CrossRef] [PubMed]
8. Johansson, L.S.; Tullin, C.; Leckner, B.; Sjövall, P. Particle emissions from biomass combustion in small combustors. *Biomass Bioenergy* **2003**, *25*, 435–446. [CrossRef]
9. Burtscher, H. The origin and Production of Nanoparticles in Environment and Industry. In *Nanoparticles in Medicine and Environment: Inhalation and Health Effects*; Springer: Dordrecht, The Netherlands, 2009.
10. Triantafyllou, A.G. Levels and trend of suspended particles around large lignite power stations. *Environ. Monit. Assess.* **2003**, *89*, 15–34. [CrossRef]
11. Shanthakumar, S.; Singh, D.N.; Phadke, R.C. Flue gas conditioning for reducing suspended particulate matter from thermal power stations. *Prog. Energy Combust. Sci.* **2008**, *34*, 685–695. [CrossRef]
12. Zhiqiang, Q.; Siegmann, K.; Keller, A.; Matter, U.; Scherrer, L.; Siegmann, H.C. Nanoparticle air pollution in major cities and its origin. *Atmos. Environ.* **2000**, *34*, 443–451. [CrossRef]

13. World Health Organization. Air Pollution. Available online: www.who.int/news-room/fact-sheets/detail/ambient-(outdoor)-air-quality-and-health (accessed on 8 June 2020).
14. European Environment Agency. Cutting air Pollution in Europe would Prevent Early Deaths, Improve Productivity and Curb Climate Change. Available online: https://www.eea.europa.eu/highlights/cutting-air-pollution-in-europe (accessed on 8 June 2020).
15. Calderón-Garcidueñas, L.; González-Maciel, A.; Mukherjee, P.S.; Reynoso-Robles, R.; Pérez-Guillé, B.; Gayosso-Chávez, C.; Torres-Jardón, R.; Cross, J.V.; Ahmed, I.A.M.; Karloukovski, V.V.; et al. Combustion- and friction-derived magnetic air pollution nanoparticles in human hearts. *Environ. Res.* **2019**, *176*, 108567. [CrossRef]
16. Calderón-Garcidueñas, L.; González-Maciel, A.; Kulesza, R.J.; González-González, L.O.; Reynoso-Robles, R.; Mukherjee, P.S.; Torres-Jardón, R. Air Pollution, Combustion and Friction Derived Nanoparticles, and Alzheimer's Disease in Urban Children and Young Adults. *J. Alzheimers Dis.* **2019**, *70*, 343–360. [CrossRef] [PubMed]
17. Jayaraj, R.L.; Rodriguez, E.A.; Wang, Y.; Block, M.L. Outdoor Ambient Air Pollution and Neurodegenerative Diseases: The Neuroinflammation Hypothesis. *Curr. Environ. Health Rep.* **2017**, *4*, 166–179. [CrossRef] [PubMed]
18. Eurostat. Air Pollution Statistics—Emission Inventories. Available online: https://ec.europa.eu/eurostat/statistics-explained/index.php/Air_pollution_statistics_-_emission_inventories#Fine_particulate_matter (accessed on 8 June 2020).
19. Collective work. *Environment 2019*; Główny Urząd Statystyczny: Warsaw, Poland, 2019.
20. US Department of Energy. Furnaces and Boilers. Available online: https://www.energy.gov/energysaver/home-heating-systems/furnaces-and-boilers (accessed on 8 June 2020).
21. Kruczek, S. *Kotły*; Wrocław University of Technology Publishing: Wrocław, Poland, 2001.
22. Kowalski, C. *Kotły Gazowe Centralnego Ogrzewania*; WNT: Warszawa, Poland, 1994.
23. Herrando, M.; Markides, C.N.; Hellgardt, K. A UK-based assessment of hybrid PV and solar-thermal systems for domestic heating and power: System performance. *Appl. Energy* **2014**, *122*, 288–309. [CrossRef]
24. Chen, H.; Riffat, S.B.; Fu, Y. Experimental study on a hybrid photovoltaic/heat pump system. *Appl. Therm. Eng.* **2011**, *31*, 4132–4138. [CrossRef]
25. Kolasiński, P. Application of the Multi-Vane Expanders in ORC Systems—A Review on the Experimental and Modeling Research Activities. *Energies* **2019**, *12*, 2975. [CrossRef]
26. Bao, J.; Zhao, L. A review of working fluid and expander selections for organic Rankine cycle. *Renew. Sustain. Energy Rev.* **2013**, *24*, 325–342. [CrossRef]
27. Elmer, T.; Worall, M.; Wu, S.; Riffat, S.B. Fuel cell technology for domestic built environment applications: State of-the-art review. *Renew. Sustain. Energy Rev.* **2015**, *42*, 913–931. [CrossRef]
28. Staffell, I.; Ingram, A.; Kendall, K. Energy and carbon payback times for solid oxide fuel cell based domestic CHP. *Int. J. Hydrog. Energy* **2012**, *37*, 2509–2523. [CrossRef]
29. Visser, W.P.J.; Shakariyants, S.A.; Oostveen, M. Development of a 3 kW microturbine for CHP applications. *J. Eng. Gas Turbines Power* **2011**, *133*, 042301. [CrossRef]
30. Conroy, G.; Duffy, A.; Ayompe, L.M. Economic, energy and GHG emissions performance evaluation of a WhisperGen Mk IV Stirling engine µ-CHP unit in a domestic dwelling. *Energy Convers. Manag.* **2014**, *81*, 465–474. [CrossRef]
31. Wajs, J.; Mikielewicz, D.; Bajor, M.; Kneba, Z. Experimental investigation of domestic micro-CHP based on the gas boiler fitted with ORC module. *Arch. Thermodyn.* **2016**, *37*, 79–93. [CrossRef]
32. Liu, X.; O'Rear, E.G.; Tyner, W.E.; Pekny, J.F. Purchasing vs. leasing: A benefit-cost analysis of residential solar PV panel use in California. *Renew. Energy* **2014**, *66*, 770–774. [CrossRef]
33. Staffell, I.; Green, R. The cost of domestic fuel cell micro-CHP systems. *Int. J. Hydrog. Energy* **2013**, *38*, 1088–1102. [CrossRef]
34. Poppi, S.; Sommerfeldt, N.; Bales, C.; Madani, H.; Lundqvist, P. Techno-economic review of solar heat pump systems for residential heating applications. *Renew. Sustain. Energy Rev.* **2018**, *81*, 22–32. [CrossRef]
35. Do Nascimento, M.; Rodrigues, L.; dos Santos, E.; Gomes, E.; Dias, F.; Velásques, E.; Carrillo, R. Micro Gas Turbine Engine: A Review. In *Progress in Gas Turbine Performance*; IntechOpen: London, UK, 2014.
36. Proctor, C. Cool Energy Turns Waste Heat into Power with 200-Year-old Technology. Denver Business Journal, 3 July 2014. Available online: https://www.bizjournals.com/denver/blog/earth_to_power/2014/07/cool-energy-turns-waste-heat-into-power-with-200.html (accessed on 8 June 2020).

37. Lai, N.A.; Wendland, M.; Fischer, J. Working fluid for high-temperature organic Rankine cycles. *Energy* **2011**, *36*, 199–211. [CrossRef]
38. He, C.; Liu, C.; Gao, H.; Xie, H.; Li, Y.; Wu, S.; Xu, J. The Optimal Evaporation Temperature and Working Fluids for Subcritical Organic Rankine Cycle. *Energy* **2012**, *38*, 136–143. [CrossRef]
39. Györke, G.; Deiters, U.K.; Groniewsky, A.; Lassu, I.; Imre, A.R. Novel classification of pure working fluids for Organic Rankine Cycle. *Energy* **2018**, *145*, 288–300. [CrossRef]
40. Imre, A.R.; Kustán, R.; Groniewsky, A. Thermodynamic Selection of the Optimal Working Fluid for Organic Rankine Cycles. *Energies* **2019**, *12*, 2028. [CrossRef]
41. Zhang, X.; Zhang, Y.; Cao, M.; Wang, J.; Wu, Y.; Ma, C. Working Fluid Selection for Organic Rankine Cycle Using Single-Screw Expander. *Energies* **2019**, *12*, 3197. [CrossRef]
42. White, J.A.; Velasco, S. Approximating the Temperature–Entropy Saturation Curve of ORC Working Fluids From the Ideal Gas Isobaric Heat Capacity. *Energies* **2019**, *12*, 3266. [CrossRef]
43. Invernizzi, C.M.; Ayub, A.; Di Marcoberardino, G.; Iora, P. Pure and Hydrocarbon Binary Mixtures as Possible Alternatives Working Fluids to the Usual Organic Rankine Cycles Biomass Conversion Systems. *Energies* **2019**, *12*, 4140. [CrossRef]
44. Hung, T.C.; Wang, S.K.; Kuo, C.H.; Pei, B.S.; Tsai, K.F. A study of organic working fluids on system efficiency of an ORC using low-grade energy sources. *Energy* **2010**, *35*, 1403–1411. [CrossRef]
45. Wang, E.H.; Zhang, H.G.; Fan, B.Y.; Ouyang, M.G.; Zhao, Y.; Mu, Q.H. Study of working fluid selection of organic Rankine cycle (ORC) for engine waste heat recovery. *Energy* **2011**, *36*, 3406–3418. [CrossRef]
46. Setiawan, D.; Subrata, I.D.M.; Purwanto, Y.A.; Tambunan, A.H. Evaluation of Working Fluids for Organic Rankine Cycle Based on Exergy Analysis. *IOP Conf. Ser. Earth Environ. Sci.* **2018**, *147*, 12035. [CrossRef]
47. Siddiqi, M.A.; Atakan, B. Investigation of the Criteria for Fluid Selection in Rankine Cycles for Waste Heat Recovery. *Int. J. Thermodyn.* **2011**, *14*, 117–123.
48. Dai, X.; Shi, L.; Qian, W. Review of the Working Fluid Thermal Stability for Organic Rankine Cycles. *J. Therm. Sci.* **2019**, *14*, 597–607. [CrossRef]
49. Kolasiński, P. The Method of the Working Fluid Selection for Organic Rankine Cycle (ORC) System with Volumetric Expander. In Proceedings of the 3rd International Seminar on ORC Power Systems, Brussels, Belgium, 12–14 October 2015.
50. Kolasiński, P. The Method of the Working Fluid Selection for Organic Rankine Cycle (ORC) Systems Employing Volumetric Expanders. *Energies* **2020**, *13*, 573. [CrossRef]
51. Harrison, J. Stirling Engine Systems for Small and Micro Combined Heat and Power (CHP) Applications. In *Small and Micro Combined Heat and Power (CHP) Systems*; Woodhead Publishing: Cambridge, UK, 2011.
52. Alshammari, F.; Usman, M.; Pesyridis, A. Expanders for Organic Rankine Cycle Technology. In *Organic Rankine Cycle Technology for Heat Recovery*; IntechOpen: London, UK, 2018.
53. Lemort, V.; Legros, A. Positive Displacement Expanders for Organic Rankine Cycle Systems. In *Organic Rankine Cycle (ORC) Power Systems: Technologies and Applications*; Woodhead Publishing: Cambridge, UK, 2017.
54. Gnutek, Z.; Kolasiński, P. The application of rotary vane expanders in ORC systems—Thermodynamic description and experimental results. *J. Eng. Gas Turbines Power* **2013**, *135*, 61901. [CrossRef]
55. Usman, M.; Pesyridis, A.; Cockerill, S.; Howard, T. Development and Testing of a Free Piston Linear Expander for Organic Rankine Cycle Based Waste Heat Recovery Application. In Proceedings of the 5th International Seminar on ORC Power Systems, Athens, Greece, 9–11 September 2019.
56. Latz, G.; Erlandsson, O.; Skåre, T.; Contet, A.; Andersson, S.; Munch, K. Performance Analysis of a Reciprocating Piston Expander and a Plate Type Exhaust Gas Recirculation Boiler in a Water-Based Rankine Cycle for Heat Recovery from a Heavy Duty Diesel Engine. *Energies* **2016**, *9*, 495. [CrossRef]
57. Bianchi, M.; Branchini, L.; Casari, N.; De Pascale, A.; Melino, F.; Ottaviano, S.; Pinelli, M.; Spina, P.R.; Suman, A. Experimental analysis of a micro-ORC driven by piston expander for low-grade heat recovery. *Appl. Therm. Eng.* **2019**, *148*, 1278–1291. [CrossRef]
58. Wronski, J.; Imran, M.; Skovrup, M.J.; Haglind, F. Experimental and numerical analysis of a reciprocating piston expander with variable valve timing for small-scale organic Rankine cycle power systems. *Appl. Energy* **2019**, *247*, 103–116. [CrossRef]
59. Oudkerk, J.F.; Dickes, R.; Dumont, O.; Lemort, V. Experimental performance of a piston expander in a small-scale organic Rankine cycle. *IOP Conf. Ser. Mater. Sci. Eng.* **2015**, *90*, 12066. [CrossRef]

60. Glavatskaya, Y.; Podevin, P.; Lemort, V.; Shonda, O.; Descombes, G. Reciprocating Expander for an Exhaust Heat Recovery Rankine Cycle for a Passenger Car Application. *Energies* **2012**, *5*, 1751–1765. [CrossRef]
61. Li, G.; Zhang, H.; Yang, F.; Song, S.; Chang, Y.; Yu, F.; Wang, J.; Yao, B. Preliminary Development of a Free Piston Expander–Linear Generator for Small-Scale Organic Rankine Cycle (ORC) Waste Heat Recovery System. *Energies* **2016**, *9*, 300. [CrossRef]
62. Hsu, S.-W.; Chiang, H.-W.D.; Yen, C.-W. Experimental Investigation of the Performance of a Hermetic Screw-Expander Organic Rankine Cycle. *Energies* **2014**, *7*, 6172–6185. [CrossRef]
63. Zhang, Y.-Q.; Wu, Y.-T.; Xia, G.-D.; Ma, C.-F.; Ji, W.-N.; Liu, S.-W.; Yang, K.; Yang, F.-B. Development and experimental study on organic Rankine cycle system with single-screw expander for waste heat recovery from exhaust of diesel engine. *Energy* **2014**, *77*, 499–508. [CrossRef]
64. Tang, H.; Wu, H.; Wang, X.; Xing, Z. Performance study of a twin-screw expander used in a geothermal organic Rankine cycle power generator. *Energy* **2015**, *90*, 631–642. [CrossRef]
65. Öhman, H.; Lundqvist, P. Screw Expanders in ORC Applications, Review and A New Perspective. In Proceedings of the 3rd International Seminar on ORC Power Systems, Brussels, Belgium, 12–14 October 2015.
66. Kim, Y.M.; Shin, D.G.; Kim, C.G. Optimization of Design Pressure Ratio of Positive Displacement Expander for Vehicle Engine Waste Heat Recovery. *Energies* **2014**, *7*, 6105–6117. [CrossRef]
67. Kaczmarczyk, T.; Ihnatowicz, E.; Żywica, G.; Kiciński, J. Experimental investigation of the ORC system in a cogenerative domestic power plant with a scroll expanders. *Open Eng.* **2015**, *5*, 411–420. [CrossRef]
68. Gao, P.; Jiang, L.; Wang, L.W.; Wang, R.Z.; Song, F.P. Simulation and experiments on an ORC system with different scroll expanders based on energy and exergy analysis. *Appl. Therm. Eng.* **2015**, *75*, 880–888. [CrossRef]
69. Jradi, M.; Li, J.; Liu, H.; Riffat, S. Micro-scale ORC-based combined heat and power system using a novel scroll expander. *Int. J. Low Carbon Technol.* **2014**, *9*, 91–99. [CrossRef]
70. Dumont, O.; Parthoens, A.; Dickes, R.; Lemort, V. Experimental investigation and optimal performance assessment of four volumetric expanders (scroll, screw, piston and roots) tested in a small-scale organic Rankine cycle system. *Energy* **2018**, *165*, 1119–1127. [CrossRef]
71. Song, P.; Wei, M.; Liu, Z.; Zhao, B. Effects of suction port arrangements on a scroll expander for a small scale ORC system based on CFD approach. *Appl. Energy* **2015**, *150*, 274–285. [CrossRef]
72. Emhardt, S.; Tian, G.; Chew, J. A review of scroll expander geometries and their performance. *Appl. Therm. Eng.* **2018**, *141*, 1020–1034. [CrossRef]
73. Kim, D.; Chung, H.J.; Jeon, Y.; Jang, D.S.; Kim, Y. Optimization of the injection-port geometries of a vapor injection scroll compressor based on SCOP under various climatic conditions. *Energy* **2017**, *135*, 442–454. [CrossRef]
74. Kurkus-Gruszecka, M.; Krawczyk, P. Comparison of Two Single Stage Low-Pressure Rotary Lobe Expander Geometries in Terms of Operation. *Energies* **2019**, *12*, 4512. [CrossRef]
75. Kurkus-Gruszecka, M.; Krawczyk, P.; Badyda, K. CFD modelling of a fluid flow in a rotary lobe expander. In Proceedings of the International Conference of Numerical Analysis and Applied Mathematics (ICNAAM 2018), Athens, Greece, 13–18 September 2018.
76. Bianchi, G.; Fatigati, F.; Murgia, S.; Cipollone, R. Design and analysis of a sliding vane pump for waste heat to power conversion systems using organic fluids. *Appl. Therm. Eng.* **2017**, *124*, 1038–1048. [CrossRef]
77. Kolasiński, P. The Influence of the Heat Source Temperature on the Multivane Expander Output Power in an Organic Rankine Cycle (ORC) System. *Energies* **2015**, *8*, 3351–3369. [CrossRef]
78. Kolasiński, P.; Błasiak, P.; Rak, J. Experimental and Numerical Analyses on the Rotary Vane Expander Operating Conditions in a Micro Organic Rankine Cycle System. *Energies* **2016**, *9*, 606. [CrossRef]
79. Rak, J.; Błasiak, P.; Kolasiński, P. Influence of the Applied Working Fluid and the Arrangement of the Steering Edges on Multi-Vane Expander Performance in Micro ORC System. *Energies* **2018**, *11*, 892.
80. Bell, I.H.; Wronski, J.; Quoilin, S.; Lemort, V. Pure and Pseudo-pure Fluid Thermophysical Property Evaluation and the Open-Source Thermophysical Property Library CoolProp. *Ind. Eng. Chem. Res.* **2014**, *53*, 2498–2508. [CrossRef] [PubMed]
81. Zangheri, P.; Armani, R.; Pietrobon, M.; Pagliano, L.; Boneta, M.; Müller, A. *Heating and Cooling Energy Demand and Loads for Building Types in Different Countries of the EU*; Politecnico di Milano: Milano, Italy, 2014.
82. Quintero, R.; Genty, A.; Vieitez, E.; Wolf, O. *Development of Green Public Procurement Criteria for Water-Based Heaters*; Publications Office of the European Union: Luxembourg, 2014.

83. Baborska-Narożny, M.; Szulgowska-Zgrzywa, M.; Mokrzecka, M.; Chmielewska, A.; Fidorow-Kaprawy, N.; Stefanowicz, E.; Piechurski, K.; Laska, M. Climate justice: Air quality and transitions from solid fuel heating. *Build. Cities* **2020**, *1*, 120–140. [CrossRef]
84. Collective Work. *Housing Conditions in Poland in 2017*; Główny Urząd Statystyczny: Warsaw, Poland, 2018.

© 2020 by the author. Licensee MDPI, Basel, Switzerland. This article is an open access article distributed under the terms and conditions of the Creative Commons Attribution (CC BY) license (http://creativecommons.org/licenses/by/4.0/).

energies

MDPI

Article

Theoretical and Economic Evaluation of Low-Cost Deep Eutectic Solvents for Effective Biogas Upgrading to Bio-Methane

Edyta Słupek, Patrycja Makoś * and Jacek Gębicki

Department of Process Engineering and Chemical Technology, Faculty of Chemistry, Gdansk University of Technology, G. Narutowicza St. 11/12, 80-233 Gdansk, Poland; edyta.slupek@pg.edu.pl (E.S.); jacek.gebicki@pg.edu.pl (J.G.)

* Correspondence: patrycja.makos@pg.edu.pl; Tel.: +48-508997100

Received: 29 May 2020; Accepted: 28 June 2020; Published: 1 July 2020

Abstract: This paper presents the theoretical screening of 23 low-cost deep eutectic solvents (DESs) as absorbents for effective removal of the main impurities from biogas streams using a conductor-like screening model for real solvents (COSMO-RS). Based on thermodynamic parameters, i.e., the activity coefficient, excess enthalpy, and Henry's constant, two DESs composed of choline chloride: urea in a 1:2 molar ratio (ChCl:U 1:2), and choline chloride: oxalic acid in a 1:2 molar ratio (ChCl:OA 1:2) were selected as the most effective absorbents. The σ-profile and σ-potential were used in order to explain the mechanism of the absorptive removal of CO_2, H_2S, and siloxanes from a biogas stream. In addition, an economic analysis was prepared to demonstrate the competitiveness of new DESs in the sorbents market. The unit cost of 1 m^3 of pure bio-methane was estimated to be in the range of 0.35–0.37 EUR, which is comparable to currently used technologies.

Keywords: biogas; deep eutectic solvents; upgrading; absorption; COSMO-RS; economic analysis

1. Introduction

Due to European Union (EU) energy policies to promote the utilization of renewable resources, there has been a significant increase in biogas plants and the level of biogas production [1,2]. The number of biogas plant installations and the amount of produced bio-methane in recent years is presented in detail in Figure 1. Biogas can be produced by anaerobic digestion from different waste materials (i.e., manure and food residue, wastewater sludge, or industrial by-products) or landfill gas. Biogas mainly consists of methane (50–70%) and contaminants including carbon dioxide, water, nitrogen, oxygen, hydrogen sulfide, ammonia, and numerous organic compounds (i.e., siloxanes) [3–5]. The presence of these contaminants prevents the use of biogas as an alternative transport fuel or natural gas substitute. Among the biogas impurities, carbon dioxide, hydrogen sulfide and siloxanes are the most problematic [6,7].

Carbon dioxide is present in high concentrations in biogas and it acts as a ballast; this significantly reduces the quality of biogas because it reduces the caloric power of biogas in proportion to its concentration. Biogas should contain more than 90% pure methane, depending on its further application. During the biogas combustion process, hydrogen sulfide reacts with water, forming sulfuric acid, which corrodes the surface in the combustion chamber [8], while the siloxanes are converted into silicon dioxide (SiO_2), which can be deposited into the cylinder, impeller, valves, piston rings, liners, spark plugs, and turbochargers. Accumulation of hard deposits of SiO_2 reduces the life span of the turbines and engine efficiency, which results in detonation in the combustion chambers and an increase in the exhaust gas emissions due to unburned fuel. This also results in higher plant maintenance costs. In addition, the presence of certain groups of trace compounds in biogas can cause the emission of toxic by-products into the atmosphere. The occurrence of these contaminants is a major barrier to the use of biogas as a renewable energy source.

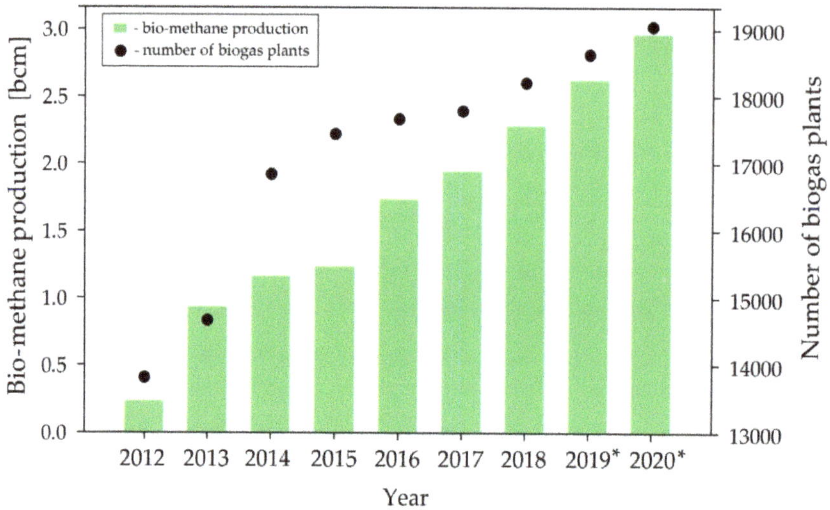

Figure 1. Number of biogas plants and bio-methane production from 2012 to 2020 in the European Union (bcm—billion cubic meters; 2019* and 2020* are estimated values).

Currently, there are several technologies for removing CO_2, H_2S, and siloxanes from biogas including adsorption, refrigeration with condensation, membrane technologies, biological methods, and absorption [9–13]. Among these technologies, physical absorption is one of the most popular. This process consists of transferring contaminants from a gas phase to an absorbent. Different types of absorbents such as water, organic compounds, and oils are used [14–16]. However, there are a few disadvantages associated with conventional organic absorbents, which can lead to equipment corrosion and harmful effects on the environment. Therefore, in the past decade, ionic liquids (ILs) have been proposed as a potential alternative for conventional absorbents for CO_2 [17–20] and H_2S [17,21–23] removal from different type of gas streams. Despite the attractive physicochemical properties of ILs (i.e., good thermal stability, non-volatile properties, and high absorption capacity [24,25]), they not found practical industrial application due to their high viscosity, potential toxicity, high cost and complicated synthesis processes [26]. Due to the limitations of both conventional solvents and ILs, alternative solutions are still in demand. Nowadays, one of the most promising group of green absorbents is deep eutectic solvents (DESs). DESs are synthesized by the direct mixing of two ingredients—hydrogen bond acceptor (HBA) with a hydrogen bond donor (HBD). DES mixtures are characterized by a lower melting point compared to the individual components [27]. In addition, DESs are characterized by specific physicochemical properties such as their non-volatility, non-flammability, high absorption capacity, non-toxic character, and high thermal stability [28,29]. A comparison of the properties of DES with other absorbents is presented in Table 1.

Table 1. Comparison of physicochemical properties of absorbents [28–37].

Properties	Water	Organic Amine	ILs	DESs
The synthesis	No	No	Multi-step synthesis	Easy
Applicability	Single function	Single function	Multifunction	Multifunction
Tunability	No	No	High	High
Thermal stability	Low	Low	Tunable, but generally high	Tunable, but generally high
Boiling Points	100 °C	111–350 °C	>250 °C	Higher than other solvents (214–1774 °C)
Environmentally friendly	Yes	No	Not all	Yes
Toxicity	No	Yes	Often increase toxicity for aquatic systems	Acceptable toxicity profiles
Corrosive nature	High	High	Low	Low
Biodegradability	Readily	Readily	Difficult	Readily
Density	Low	Medium	Tunable, but generally higher than other solvents	Tunable, but generally lower than ILs
Viscosity	Low	Medium	Tunable, but generally higher than other solvents	Tunable, but generally lower than ILs
Surface tension	High	Low	Generally lower than water and higher than organic amine	Low
Vapor pressure	High	High	Low	Low
Flammability	No	Yes	No	No
Nature	Neutral	Basic	Basic/neutral/acid	Basic/neutral/acid
Type of absorption	Physical	Chemical/Physical	Physical	Physical
Absorption capacity	Medium	Medium	High	High
Biodegradable	Yes	No	Poor	Yes
Cost	Low	Moderate	High	Low

Because of their unique properties, DESs are now successfully used as extractants [38–41] and absorption solvents [42–46] for the purification of gas and liquid streams [46–49]. Of the available DESs, solvents composed of quaternary ammonium salts are considered to be the most promising absorbents. DESs can also be synthesized from natural compounds, which makes them so-called "green solvents" due to the lack of or very low toxicity and their biodegradability [50]. Due to the high thermal stability of DESs, they can be regenerated repeatedly without loss of absorption capacity and the regeneration step requires less energy compared to other popular absorbents. Hence, the use of DESs as absorption solvents in the biogas upgrading process are considered as environmentally friendly technologies for the production of green bio-energy.

The application of upgraded biogas for the production of energy is considered as one of the most efficient methods for reducing greenhouse gas emissions to the atmosphere. For this reason, 23 deep eutectic solvents composed of quaternary ammonium salts and low-cost organic components were examined as potential absorbents for the removal of siloxanes, CO_2, and H_2S from a model biogas stream. A conductor-like screening model for real solvents (COSMO-RS) was used for the pre-selection of DESs. The selection of DESs with the highest dissolution potential for all impurities was made on the basis of the activity coefficient, excess enthalpy, and Henry's constant values. The absorption mechanism for the removal of the main impurities (CO_2, H_2S, siloxanes) was explained based on σ-profiles and σ-potential analysis. In addition, an economic analysis of the biogas upgrading processes was prepared. To the best of our knowledge, this is the first economic analysis report dedicated to biogas upgrading processes that use DESs.

2. Materials and Methods

2.1. Procedures

2.1.1. Computational Studies

In this investigation, COSMO-RS calculations were carried out using ADF COSMO-RS software (SCM, Netherlands). The geometry optimization of all DESs were performed using the continuum solvation COSMO model at the BVP86/TZVP level of theory. This level of theory was selected due to proven high efficiency and low computational costs [51]. The list of 23 DESs is presented in Table 2. The main thermodynamic parameters, i.e., the activity coefficient, excess enthalpy, and Henry's constant were calculated based on previous studies [52,53]. The parameters were determined for model biogas composed of 64.9% of CH_4, 31% of CO_2, 3% of H_2O and 1.04 of H_2S, and 0.02% of hexamethyldisiloxane (L2), octamethyltrisiloxane (L3), and octamethylcyclotetrasiloxaan (D4), which represents the typical composition of biogas from wastewater treatment plants and landfills [8,54].

Table 2. List of deep eutectic solvents (DESs) used for the conductor-like screening model for real solvents (COSMO-RS) calculation.

No.	HBA	HBD	HBA:HBD Molar Ratio	Abbreviation
1	Choline chloride	ethylene glycol	1:3	ChCl:EG (1:3)
2	Choline chloride	glycerol	1:3	ChCl:Gly (1:3)
3	Choline chloride	levulinic acid	1:3	ChCl:Lev (1:3)
4	Choline chloride	lactic acid	1:2	ChCl:LA (1:2)
5	Choline chloride	butyric acid	1:2	ChCl:Bu (1:2)
6	Choline chloride	phenol	1:2	ChCl:Ph (1:2)
7	Choline chloride	urea	1:2	ChCl:U (1:2)
8	Choline chloride	diethylene glycol	1:2	ChCl:DEG (1:2)
9	Choline chloride	oxalic acid	1:2	ChCl:OA (1:2)
10	Choline chloride	methacrylic acid	1:2	ChCl:MthA (1:2)
11	Choline chloride	propylene glycol	1:2	ChCl:PG (1:2)
12	Tetrabutylammonium chloride	ethylene glycol	1:3	TBACl:EG (1:3)
13	Tetrabutylammonium chloride	glycerol	1:3	TBACl:Gly (1:3)
14	Tetrabutylammonium chloride	levulinic acid	1:3	TBACl:Lev (1:3)
15	Tetrabutylammonium chloride	lactic acid	1:2	TBACl:LA (1:2)
16	Tetrabutylammonium chloride	butyric acid	1:2	TBACl:Bu (1:2)
17	Tetrabutylammonium chloride	phenol	1:2	TBACl:Ph (1:2)
18	Tetrapropylammonium bromide	ethylene glycol	1:3	TEABr:EG (1:3)
19	Tetrapropylammonium bromide	glycerol	1:3	TEABr:Gly (1:3)
20	Tetrapropylammonium bromide	levulinic acid	1:3	TEABr:Lev (1:3)
21	Tetrapropylammonium bromide	lactic acid	1:2	TEABr:LA (1:2)
22	Tetrapropylammonium bromide	butyric acid	1:2	TEABr:Bu (1:2)
23	Tetrapropylammonium bromide	phenol	1:2	TBABr:Ph (1:2)

Henry's constant (K_H) was applied to systems in thermodynamic equilibrium. The K_H links the solubility of solute impurities (i) to its partial pressure above the mixture (p_i^{vap}). K_H was calculated using Equation (1).

$$K_H = \frac{1}{\gamma_i p_i^{vap}} \quad (1)$$

where γ_i is the infinite dilute activity coefficient of impurities (i), and p_i^{vap} is the vapor pressure of impurities (i).

The activity coefficient was calculated using Equation (2) and Equation (3).

$$\ln(\gamma_i) = \frac{\mu_i^{solv} - \mu_i^{pure}}{RT} \quad (2)$$

where μ_i^p is the chemical potential of pure impurities (i), μ_i^j is the chemical potential of impurities in the liquid phase, T is the temperature (K), and the universal gas constant R = 8.314 J/mol.

The excess enthalpy of mixtures H^E (kJ/mol) was calculated based on Gibbs-Helmholtz using Equation (3).

$$H^E = -T^2 \frac{\partial\left(\frac{G^E}{T}\right)}{\partial T} \quad (3)$$

where T is the temperature (K), and G^E is the excess Gibbs free energy (kJ/mol).

2.1.2. Biogas Upgrading Technology Description

The scheme for the biogas upgrading technology described in this paper is presented in Figure 2. The physicochemical properties of DESs are similar to the most commonly used absorbents (i.e., amine or water), therefore, DESs can be applied in existing and currently used absorption installations. In order to better compare the benefits of DESs application in the absorption process, the size of the installations (absorption and desorption column, compressor, pump, blower, dryer, and heat exchangers) and the process streams (inlet biogas stream 813 m^3/h and inlet air stream 403 m^3/h) was adopted from previous studies [55,56].

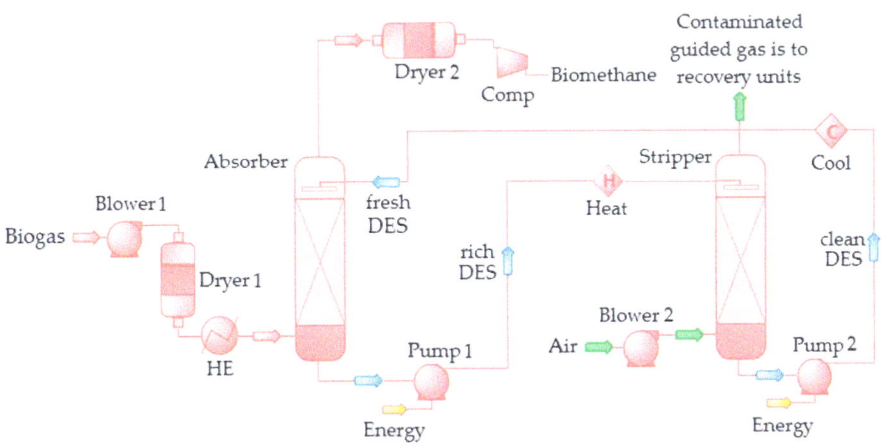

Figure 2. Scheme for the biogas upgrading technology [55,56].

In the first stage, biogas stream is introduced into Dryer 1. Then, the biogas is passed directly to the heat exchanger, after which biogas is directed into the bottom part of the absorber (813 m^3/h), which operates at a temperature of 20 °C, and pressure of 100 kPa. The biogas stream is introduced at the bottom of the absorber. The DES is introduced at the top of the column. The biogas and DES move through a counter-flow scrubbing column. In the column, the biogas comes into contact with a DES to dissolve the main impurities (L2, L3, D4, CO_2, and H_2S). This is a process of mass transfer of pollutants from the biogas phase to the liquid DES phase. The upgraded bio-methane is downloaded from the top of the absorber, drained again (Dryer 2), and compressed. The obtained renewable bio-methane can be directly injected into the distribution gas grid at 700 kPa. The biogas purification system also contains the stripper column, which is operated under a temperature of 115–125 °C and pressure of 140–170 kPa. Saturated DES from the absorption column is directed into the stripper column where

DES is purged with an inlet air stream (403 m^3/h). Most of the impurities (L2, L3, D4, CO_2, and H_2S) are liberated into a concentrated air stream that exits at the top of the stripper column. The impurities stream is directed to the H_2S, CO_2, L2, L3, and D4 recovery system. The regenerated DES is cooled and returned to the absorber column.

2.1.3. Cost and Economic Analysis

The cost simulations included an estimation of the total annual cost (*TAC*) of the biogas upgrading process. *TAC* included the annual capital investment cost (*ACIC*), and the annual operation and maintenance cost (OC and MC).

The *ACIC* was estimated based on the method of Scholz et al. [57] according to Equation (4).

$$ACIC = TCIC\frac{i(1+i)^n}{(1+i)^n - 1} \tag{4}$$

where *ACIC* is the annual capital investment cost, *TCIC* is the total capital investment cost, *i* is the interest rate (9%), and *n* is the depreciation period (15 years).

The *TCIC* was mainly estimated as the percentage value of the equipment cost (*EC*) [55]. The *EC* was estimated by Guthrie's method [58], according to Equation (5).

$$EC = PEC\left(f_{mp} + f_m - 1\right) \tag{5}$$

where *EC* is the equipment cost, *PEC is* the bare purchased equipment cost, f_{mp} the material and pressure correction factor, and f_m is the module factor, which depends on the size equipment. The values of f_{mp} and f_m were adopted according to the procedure proposed by Scholz et al. [57]. The *EC* of the absorption column, stripper column, blowers, pumps, compressors, and heat exchangers was adopted from other studies [56]. The list of basic parameters for maintenance and operation cost, which consist of operating supply cost, research, and development (R&D) costs, personnel labor cost, utility costs (i.e., electricity cost for heating and cooling, absorbent exchange cost) is presented in Table 3.

Table 3. Parameters for operation costs.

Parameter	Units	Costs	Ref.
Electricity	EUR/kWh	0.1	[59]
Heat	EUR/kWh	0.046	[57]
Personnel	EUR/h	38.88	[56]
Choline chloride	EUR/t	4550	[60]
Urea	EUR/t	218.4	[61]
Oxalic acid	EUR/t	455	[62]

The last step of the cost analysis was the estimation of the risk and economic benefits of the project. The financial assessment of the investment was carried out on the basis of the expected energy production, and total costs of the plant. The unit cost (*UC*) of 1 m^3 biogas purification was calculated according to Equation (6) [63].

$$UC = \left(\frac{\left(\frac{TCIC}{n}\right) + ((TCIC * i) + TAC)}{APB}\right) \tag{6}$$

where *UC* is the unit cost of 1 m^3 bio-methane, *i* is the interest rate (9%), *n* is the depreciation period (15 years), *APB* is the annual production of bio-methane [m^3], and *TAC* is the total annual cost.

The annual amount of cubic meters of upgraded biogas stream was determined according to Equation (7).

$$APB = BF \cdot \%\, CH_4 \cdot ML \tag{7}$$

where BF is the biogas flow, % CH_4 is the percentage of methane in biogas, and ML is the methane loss.

3. Results and Discussion

3.1. COSMO-RS Prediction—Pre-Selection of DESs

The preselection of DESs that are characterized by high solubility of siloxanes, H_2S, CO_2, H_2O, and CH_4 was made based on the Henry's constants, activity coefficients, and excess enthalpy of mixtures that were predicted according to the COSMO-RS method. Water was omitted in the calculations because it was assumed to be removed before the biogas enters the absorption column. All parameters were determined at 20 °C and 100 kPa. The calculation results are presented in Table 4.

The activity coefficient is a thermodynamic parameter that is associated with the affinity of siloxanes, H_2S, CO_2, and CH_4 to DESs. This parameter indicates the differences in strength among DESs and impurities, which are a result of the dominant interactions. Usually, the activity coefficient values are given as $\ln(1/\gamma)$, hence these are rather negative (Table 4) [64]. The higher negative values of logarithmic activity coefficients indicate greater solubility of siloxanes, H_2S, and CO_2 in DESs. The second main thermodynamic parameter is the excess enthalpy of mixtures (H^E). H^E is a sensitive measure of the intermolecular interactions between DESs and impurities. The results of H^E calculated for all DES-impurities models are presented in Table 4. The DES, which is characterized by a higher dissolution capacity of CO_2, H_2S, and siloxanes has lower values of H^E (higher negative). The third parameter is the Henry's Law constant (K_H). The K_H describes the ratio at the equilibrium of the concentration of impurities in the gas phase to the concentration of impurities in the DES phase, and it combines vapor pressure and solubility, which can be used to estimate the likelihood that a substance will be exchanged between the gas phase and a DES. Lower K_H indicates a higher concentration of impurities in the DES phase than in the gas phase.

ChCl:U (1:2) and ChCl:OA (1:2) showed lower values for all thermodynamic parameters, relative to all impurities. Slightly higher values were obtained for the rest of the DESs composed of choline chloride such as HBA. This indicates that this type of HBA in DES structures has a major influence on absorption efficiency. This is in line with the conclusions obtained in previous studies [46]. This can be caused by several factors, including HBA alkyl chain length, different charge density on the ammonium, as well as asymmetry in ChCl ammonium with a hydroxyl group in the longest branch, and theoretically, a type of counter-ion (Cl^- or Br^-). However, the obtained results indicate that this type of counter-ion in HBA only has a slight effect on the ability of DESs to dissolve all impurities. The use of DESs containing ChCl as HBA in the absorption process is preferred because they are characterized by less viscosity compared to DESs composed of quaternary ammonium salts with long alkyl chain length [65].

Principal components analysis (PCA) was used to obtain a better interpretation of all results (the activity coefficient, excess enthalpy, and Henry's constant). The PCA plot is presented in Figure 3. The numbers on the diagram correspond to the DESs numbers in Table 4. The results indicate that DESs can be divided into three groups. The first group is marked with a yellow circle and contains two DESs (ChCl:U 1:2 and ChCl:OA 1:2) that have the greatest dissolution potential for all impurities. The second group, marked with a green circle, includes DESs that have the potential to effectively absorb siloxanes, but they have low CO_2 and H_2S dissolution potential. These DESs may have potential use for selective siloxane removal, but their solubility is insufficient in applications that require the comprehensive removal of impurities from biogas. The last group includes DESs that have the lowest absorption potential for all of the tested compounds.

Table 4. The logarithmic activity coefficient of siloxanes H_2S, and CO_2 model at infinite dilution, excess enthalpy of mixtures and Henry's constant of siloxanes calculated by COSMO-RS at 20 °C and 101325 Pa.

No.	DES	Activity Coefficient					H^E [kJ/mol]					K_H [mol/L atm]				
		L2	L3	D4	H_2S	CO_2	L2	L3	D4	H_2S	CO_2	L2	L3 [*10^3]	D4 [*10^5]	H_2S	CO_2
1	ChCl:EG (1:3)	−5.28	−6.81	−6.57	−0.30	−1.09	−5.58	−5.57	−5.57	−5.62	−5.61	50.8	1.25	3.12	1.79	0.089
2	ChCl:Gly (1:3)	−4.86	−6.30	−6.03	−0.08	−0.83	−5.63	−5.62	−5.62	−5.68	−5.66	76.2	2.22	5.67	1.80	0.093
3	ChCl:Lev (1:3)	−3.68	−4.77	−4.42	0.16	−0.54	−5.68	−5.68	−5.68	−5.73	−5.71	151.3	5.67	15.49	1.90	0.104
4	ChCl:LA (1:2)	−5.49	−7.12	−6.81	−0.39	−1.14	−7.45	−7.44	−7.44	−7.48	−7.46	33.0	0.74	1.95	1.36	0.070
5	ChCl:Bu (1:2)	−3.13	−4.04	−3.66	0.01	−0.54	−7.52	−7.51	−7.52	−7.53	−7.52	383.8	18.57	52.76	1.89	0.119
6	ChCl:Ph (1:2)	−3.24	−4.19	−3.49	−0.14	−0.63	−7.77	−7.77	−7.82	−7.33	−7.76	366.3	17.29	69.11	1.59	0.107
7	ChCl:U (1:2)	−8.11	−10.52	−10.22	−1.15	−2.05	−7.29	−7.28	−7.30	−7.32	−7.31	1.9	0.02	0.05	0.83	0.037
8	ChCl:DEG (1:2)	−4.30	−5.57	−5.32	−0.11	−0.75	−7.65	−7.64	−7.63	−7.69	−7.67	97.5	3.17	7.94	1.49	0.086
9	ChCl:OA (1:2)	−8.11	−10.55	−9.43	−1.01	−1.88	−8.09	−8.08	−8.18	−8.12	−8.10	3.8	0.04	0.24	0.82	0.038
10	ChCl:MthA (1:2)	−3.45	−4.45	−3.98	−0.07	−0.62	−7.50	−7.50	−7.51	−7.52	−7.50	332.3	15.20	45.62	1.82	0.114
11	ChCl:PG (1:2)	−3.96	−5.10	−4.80	−0.01	−0.66	−7.77	−7.77	−7.76	−7.81	−7.79	203.7	7.94	20.31	2.06	0.118
12	TBACl:EG (1:3)	−3.39	−4.36	−4.12	0.18	−0.43	−4.13	−4.13	−4.13	−5.62	−5.61	367.9	18.13	45.69	2.04	0.120
13	TBACl:Lev (1:3)	−2.30	−2.97	−2.64	0.47	−0.09	−4.13	−4.13	−4.13	−5.73	−5.71	640.4	39.82	107.48	2.02	0.127
14	TBACl:LA (1:2)	−2.97	−3.83	−3.53	0.43	−0.11	−5.35	−5.35	−5.35	−5.38	−5.35	601.2	36.54	98.86	1.96	0.125
15	TBACl:Bu (1:2)	−1.42	−1.81	−1.49	0.38	−0.08	−5.49	−5.49	−5.50	−5.50	5.48	1413.2	109.25	293.16	2.10	0.146
16	TBACl:Ph (1:2)	−1.32	−1.69	−1.17	0.15	−0.18	−5.87	−5.88	−5.91	−5.84	−5.82	1579.3	125.21	423.11	1.63	0.128
17	TBABr:EG (1:3)	−3.46	−4.45	−4.18	0.12	−0.44	−3.85	−3.85	−3.85	−5.32	−5.30	334.4	16.01	41.41	1.90	0.118
18	TBABr:Gly (1:3)	−3.28	−4.25	−3.95	0.21	−0.34	−3.67	−3.66	−3.66	−5.37	−5.36	321.7	15.68	41.48	1.77	0.112
19	TBABr:Lev (1:3)	−2.35	−3.04	−2.70	0.41	−0.10	−3.85	−3.85	−3.85	−5.43	−5.42	593.7	36.09	98.92	1.89	0.125
20	TBABr:LA (1:2)	−3.03	−3.92	−3.59	0.03	−0.50	−4.97	−4.53	−4.98	−5.00	−4.98	261.8	11.87	32.05	1.56	0.101
21	TBABr:Bu (1:2)	−1.44	−1.84	−1.48	0.32	−0.08	−5.10	−5.10	−4.75	−5.11	−5.09	1386.6	106.58	295.07	1.97	0.145
22	TBABr:Ph (1:2)	−1.32	−1.70	−1.10	0.13	−0.18	−5.43	−5.44	−5.48	−5.39	−5.38	1567.8	124.03	449.89	1.60	0.129

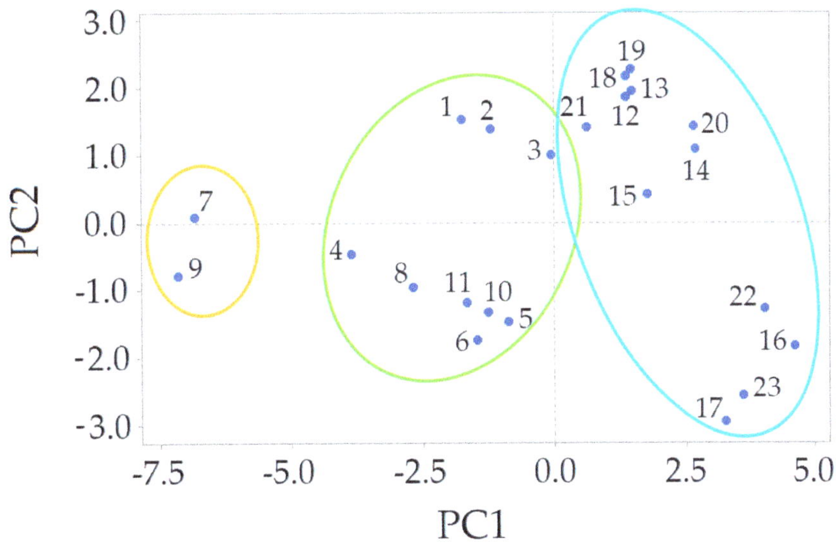

Figure 3. Principal components analysis (PCA) plot of all thermodynamic data including the activity coefficient, excess enthalpy and Henry's constant.

Based on the obtained thermodynamic results, only the DESs that showed the greatest dissolution potential for all impurities were adopted for further consideration (ChCl:U 1:2 and ChCl:OA 1:2). In practice, most of the obtained results using COSMO-RS are slightly overestimated, and this fact was more pronounced for temperatures far from room temperature. Due to the fact that all calculations were made for 20 °C, it can be concluded that the obtained results are very reliable, because the COSMO-RS model ensures acceptable accuracy (about 5%) with regard to experimental results [66,67].

3.2. Molecular Interactions

After geometric optimization, the absorption efficiency of DESs can be interpreted by molecular interactions. The geometric optimized structures of DESs are presented in Figure 4. Based on molecule-specific characteristics, the charge-related σ-profiles and σ-potential were successfully used to interpret the complex molecular interactions, according to previous studies [68–70].

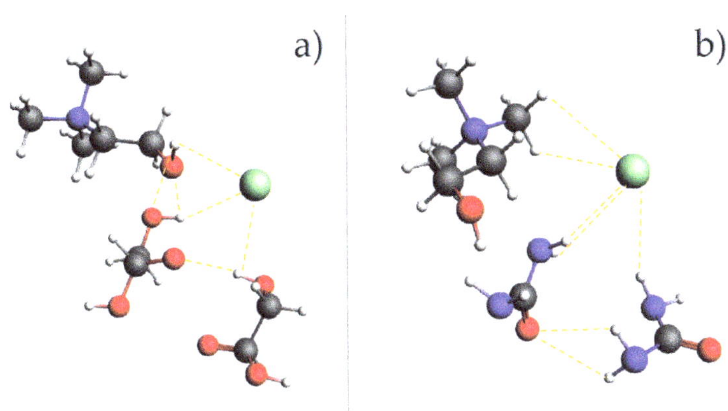

Figure 4. Optimized structures of (**a**) ChCl:OA (1:2), (**b**) ChCl:UA (1:2).

3.2.1. σ-Profiles

The σ-profile is the most important molecule-specific property because it indicates the probability distribution of the surface area of molecules that have charge density. The σ-profiles of all contaminants and DESs are presented in Figure 5. In the diagram, the range of surface area over charge density is between −0.025 and 0.025 eÅ^{-2}. This range can be divided into three segments, i.e., the non-polar region (−0.0084 eÅ^{-2} < σ < 0.0084 eÅ^{-2}), the hydrogen bond acceptor (HBA) region (−0.025 eÅ^{-2} < σ < 0.0084 eÅ^{-2}), and the hydrogen bond donor region (HBD) (0.0084 eÅ^{-2} < σ < 0.025 eÅ^{-2}). The HBA and HBD regions indicate the potential of the studied molecules to form strong hydrogen bonds. The results indicate that the σ-profile of ChCl:U (1:2) and ChCl:OA (1:2) almost overlap each other. This means that both DESs have similar properties with regard to molecular interaction. In both DESs, much larger peaks can be observed around negative values, compared to peaks around positive values, which shows more presence of HBA than HBD. The peaks of all siloxanes assume a similar shape and most of the areas are located in the non-polar area (−0.0084 eÅ^{-2} < σ < 0.0084 eÅ^{-2}), and there are small fragments of siloxane peaks in the HBD region. The opposite results can be observed for carbon dioxide, which may be a hydrogen bond acceptor to a small extent. Similar small areas of hydrogen sulfide peaks are found in the HBA and HBD parts. The σ-profile results show that the siloxanes have more negative activity coefficient values compared to CO_2 and H_2S.

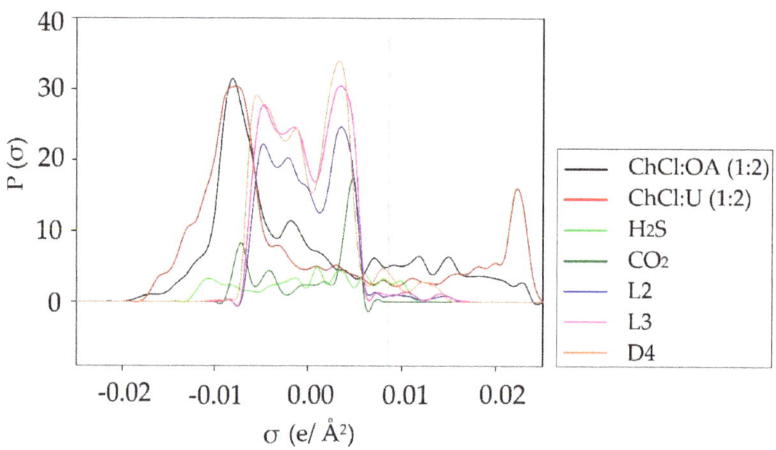

Figure 5. σ-profiles of ChCl:OA (1:2), ChCl:U (1:2), H_2S, CO_2, and siloxanes (L2, L3, D4).

3.2.2. σ-Potential

The σ-potential describes the affinity of the DESs to biogas impurities (CO_2, H_2S, L2, L3, D4) (Figure 6). The σ-potential diagram can also be divided into the same three fragments as in the σ-profile. The higher negative value of μ(σ) [kcal/molÅ] indicates stronger interaction between compounds. On the other hand, the higher positive values of μ(σ) suggest stronger repulsive interactions.

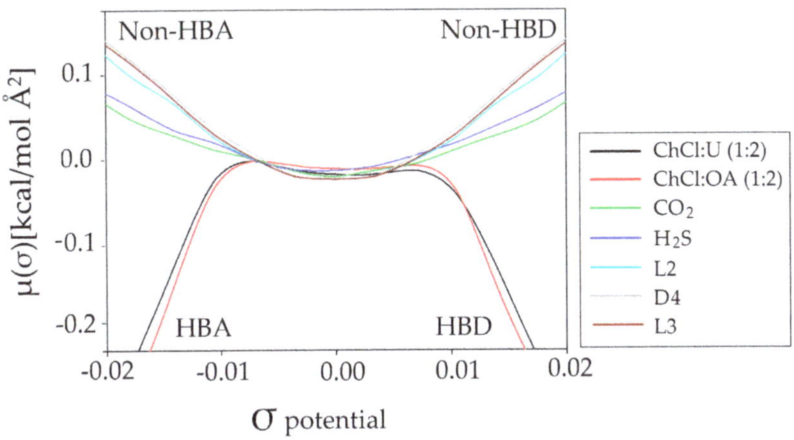

Figure 6. σ-potential of ChCl:U (1:2), ChCl:OA (1:2), CO_2, H_2S, L2, L3, and D4.

The graphic results show that all contaminants of the model biogas have parabolic curves of σ-potential. The negative values of μ(σ) in the non-polar segment indicate the non-polar nature of CO_2, H_2S, and siloxanes. The σ-potential of ChCl:U (1:2) and ChCl:OA (1:2) show negative values in the HBD, HBA, and non-polar region. This indicates that both DESs will tend to interact with hydrogen bond acceptor and donor surfaces and nonpolar molecules. The positive values of σ-potential in the HBA and HBD region of all impurities suggest that electrostatic interactions are probably the main driving force of the absorption process. In addition, the high negative value in the non-polar region of the DESs suggests a strong affinity to all biogas impurities. In addition, the similar σ-potential shape of both DESs suggests similar dissolution capabilities for all of the impurities.

3.3. Economic Evaluation

The main factor that determines the success of an investment is the economic cost [71]. The capital and running costs of biogas upgrading technology depend primarily on the size of the installation, type of technology, type of installed devices (their number and power), degree of technological advancement (degree of modernity and automation), system configuration, etc. Therefore, these costs are a function of many factors. The described technology for biogas upgrading assumes that the resulting bio-methane product will meet the quality standards of natural gas [72]. This enables the bio-methane to be introduced into natural gas installations. This is very important from an economic point of view because bio-methane does not require a specially dedicated infrastructure, which increases investment costs.

In order to better compare the cost of applying DESs, the size of installations and process streams were adopted from previous studies [55,56]. Based on an assumed biogas flow rate (813 m^3/h), estimated annual DESs consumption, and assumed biogas composition (CH_4 (64.9%; 31.0% CO_2; 3.0% H_2O; 1.04% H_2S, and 0,02% of L2, L3, and D4 [8,54]) the amount of raw biogas (7.13 Mm^3) supplied for installation per year was calculated. In addition, methane losses of 5% during the biogas upgrading process were assumed based on COSMO-RS theoretical calculations. The annual bio-methane production was calculated as 4.27 Mm^3 per year. Based on the solubility of individual biogas components in DESs, the saturation time of absorbents was calculated using the COSMO-RS model (Table 5). In order to obtain reliable information about the cost of 1 m^3 of pure bio-methane, the complete cost analysis including the total investment, operating, and maintenance costs was calculated.

Table 5. List of individual impurities and their solubility in DESs.

Type of Impurities	Impurities Concentration	Flow of Individual Impurities	Molar Mass of Impurities	ChCl:U (1:2)		ChCl:OA (1:2)	
				Solubility	Saturation Time	Solubility	Saturation Time
	[%]	[m^3/h]	[g/mol]	[mol/L DES]	[h]	[mol/L DES]	[h]
CO_2	31.0	252.03	44.01	11.53	12.89	11.49	12.84
L2	0.02	0.16	162.38	1.42	21.89	0.24	3.73
L3	0.02	0.16	236.53	0.99	20.76	0.10	2.01
D4	0.02	0.16	296.62	1.09	24.51	0.14	3.12
H_2S	1.04	8.46	34.10	32.93	1150.32	146.63	122.15

3.3.1. Investment Cost

The literature review indicated that the process scale of the biogas upgrading technology is the most important factor in the total capital investment cost (*TCIC*) calculations [73,74].

In this study, an absorption capacity of 427 m^3/h was obtained for the assumed flow rate of raw biogas, absorption and desorption column dimensions, and 8600 operating hours per year (Table 6). The assumed process parameters enabled the estimation of the individual equipment cost (*EC*) according to Equation (6). The *EC* costs (Table 6) include *EC* for the upgrading biogas section but do not include the biogas production sections. The values presented in Table 6 are average amounts from previous works [55,56]. Nevertheless, to minimize the risk of overly optimistic calculations, fluctuations in the market price of individual materials, i.e., steel and electronic components in the years from 2015–2020 were included [75].

Table 6. Estimated costs of equipment of biogas upgrading technology.

Equipment	Description	Equipment Cost (*EC*) ± SD [EUR]
Blower	Introduces biogas into the absorber	42,000 ± 3360
Absorber column	Column diameter: 1 m Column height: 15 m	50,000 ± 4000
Stripper column	Column diameter: 1 m Column height: 15 m	50,000 ± 4000
Centrifugal Pump	Pump Power	46,000 ± 3680
Heat Exchanger	Heat Exchanger	38,000 ± 3040
Centrifugal Compressor	Compressor Power	36,000 ± 2880
Dryer	Biogas water collection and disposal	25,000 ± 2000
Unlisted Equipment		300,000 ± 24,000
Total equipment cost (TEC)		587,000 ± 46,960

SD—standard deviation.

A total *EC* cost estimate was necessary to calculate the total capital investment cost (*TCIC*). *TCIC* was estimated mainly on the basis of the value of equipment cost (*EC*) [55]. In addition, statistical data for absorption technologies and laboratory processes scaling data were used for the estimation of the *TCIC* [55]. The general *TCIC* analysis for ChCl:U (1:2) and ChCl:OA (1:2) is presented in Table 7. The calculated *TCIC* for absorption using DESs was in the range of 3,152,088–3,164,929 EUR. The obtained *TCIC* is comparable to the *TCIC* of amine scrubber (3,166,000 EUR), pressure swing adsorption (3,140,000 EUR), and membrane separation (3,033,000 EUR) calculated for installations with a capacity of 500 m^3/h bio-methane. A much lower *TCIC* was obtained for the water scrubber (2,794,000 EUR) [76].

Table 7. The general estimate the total capital investment cost (*TCIC*) for DES.

Parameter	Cost ± SD [EUR] for ChCl:U (1:2)	Cost ± SD [EUR] for ChCl:OA (1:2)
Direct Cost (DC)		
Total equipment cost (TEC)	587,000 ± 46,960	587,000 ± 46,960
Installation instrumentation and control	610,480 ± 48,838	610,480 ± 48,838
Electrical and heat power	64,570 ± 5166	69,950± 5596
Building and building services and equipment installation	381,550 ± 30,524	381,550 ± 30,524
Yard improvement	58,700 ± 4696	58,700 ± 4696
External services	410,900 ± 32,872	410,900 ± 32,872
Total direct cost (TCD)	2,113,200 ± 169,056	2,118,580 ± 169,486
Indirect Cost (IC)		
Engineering and construction site	434,380 ± 34,750	434,380 ± 34,750
Law cost	23,480 ± 1878	23,480 ± 1878
Contractor's fee	129,140 ± 10,331	129,140 ± 10,331
Incidents	258,280 ± 20,662	258,280 ± 20,662
Total indirect cost (TCI)	845,280 ± 67,622	845,280 ± 67,622
Other Cost (OC)		
Floating capital	126,792 ± 10,143	126,792 ± 10,143
DES batch	66,816 ± 5345	74,277 ± 5942
Total other cost (TOC)	193,608 ± 15,489	201,069 ± 16,086
Total capital investment cost (TCIC)	3,152,088 ± 252,167	3,164,929 ± 253,194

SD—standard deviation.

3.3.2. Operation and Maintenance Cost

The annual fixed operating costs (FC) included the operation and maintenance cost (OC and MC) of biogas upgrading plants. The OC and MC included the costs of maintenance, operating, labor, and taxation, which are presented in Table 8. The cost of DESs was calculated for the scrubber volume (2.35 m^3), which was doubled in order to maintain the continuity of the process.

Table 8. General estimate of the operation cost (OC) and maintenance cost (MC) for DES.

Parameter	Cost ± SD [EUR] of ChCl:U (1:2)	Cost ± SD [EUR] of ChCl:OA (1:2)
Fixed Cost (FC)		
Regional taxes and insurance	46,066 ± 3685	46,066 ± 3685
Total fixed cost (TFC)	46,066 ± 3685	46,066 ± 3685
Direct Production Cost (DPC)		
Maintenance (M)	69,099 ± 5528	69,099 ± 5528
Salary for the operator (1500 man-hour/year) (SO) (10 Personnel)	58,320 ± 4666	58,320 ± 4666
Supervision (S)	8748 ± 700	8748 ± 700
Operating materials	10,365 ± 830	10,365 ± 830
Changes in electricity cost in the laboratory	20,425 ± 1634	20,425 ± 1634
Total direct production cost (TDPC)	166,957 ± 13,357	166,957 ± 13,357
General Expenses (GE)		
Administrative cost	1313 ± 105	1313 ± 105
Distribution, marketing and R&D cost	39,710 ± 3177	39,710 ± 3177
Total general Expenses (TGE)	41,023 ± 3282	41,023 ± 3282
DES Cost (DESC)		
DES replacement cost	334,080 ± 26,726	445,662 ± 35,653
Depreciation expense	3339 ± 267	339 ± 267
Total DES Cost (TDESC)	337,419 ± 26,994	449,001 ± 35,920
Total operation and maintenance cost (TOC and MC)	591,465 ± 47,317	703,047 ± 56,244

SD—standard deviation.

Due to the different absorption capacity of DES and regeneration cycles, the energy consumption in the absorption processes was different. Based on previous studies, it was assumed that ChCl:U (1:2) and ChCl:OA (1:2) can be regenerated 73 and 60 times, respectively, without loss of absorption capacity. From an economic and industrial point of view, recycling and reuse of DES after the absorption process is highly desirable because it reduces annual operating costs and the amount of waste. Numerous regeneration cycles can be achieved due to highly reversible absorption, which mainly depends on the structure and thermal stability of DESs. HBDs play the main role in the thermal stability of DESs, which depends mainly on the weak intermolecular interaction. The decomposition temperature of urea in ChCl:U is about 172.40 °C [77], while the decomposition temperature of oxalic acid in ChCl:OA is about 134.84 °C [78]. Both temperatures are higher than the temperature required for regeneration, which is enough to ensure long absorption–desorption cycles. However, the ChCl:OA structure and its lower decomposition temperature result in a slightly lower number of regeneration cycles. After a number of regeneration cycles, DESs must be replaced to further ensure the high quality of bio-methane. The other costs of OC and MC was estimated based on the literature [74,79] and using percentage factors of *TCIC*. The costs in Table 8 (FC, DPC, GE) are averaged values for selected European Union countries, i.e., Sweden, Germany, France, Norway, and Poland for which standard deviations have been determined. The one-time cost of replacing the absorbent is 66,816 EUR and 74,277 EUR for ChCl: U and ChCl: OA, respectively. Due to the 5-fold (ChCl: U) and 6-fold (ChCl: OA) exchange of absorbents to ensure the high quality of bio-methane, the total cost of replacement is 334,080 and 445,662 EUR for ChCl:U and ChCl:OA, respectively.

The total OC and MC cost for ChCl:OA (703,047 EUR) is comparable with amine scrubber (688,000 EUR) and membrane separation (662,000 EUR), while the total OC and MC cost for ChCl:U (591,465 EUR) is more comparable with water scrubber (513,000 EUR) and pressure swing adsorption (557,000 EUR) [76].

3.3.3. Economic Comparison of the Overall Biogas Upgrading Process

It is difficult to clearly estimate the costs of individual technologies due to the differences in the cost of components, materials and utilities, and local conditions. Therefore, is important to consider the total annual cost (*TAC*) of the biogas upgrading process, which was 982,510 ± 78,601 EUR (ChCl:U) and 1,095,685 ± 87,654 EUR (ChCl:OA) in the economic analysis. The *TAC* cost for ChCl:U was very similar to the *TAC* for pressure swing adsorption (970,000 EUR), while the *TAC* obtained for ChCl:OA was very similar to the *TAC* for amine scrubber (1,104,000 EUR) and membrane separation (1,061,000 EUR). The lowest *TAC* is for the water scrubber (880,000 EUR). Based on the above calculations, the unit cost of 1 m^3 of pure bio-methane was determined by means of Equation (4). The obtained unit cost of 1 m^3 of bio-methane was 0.35 ± 0.03 EUR/m^3 and 0.37 ± 0.03 EUR/m^3 for the physical absorption process using ChCl:U (1:2) and ChCl:OA (1:2), respectively. The unit cost for various biogas treatment technologies can be ordered as follows: amine scrubber > membrane separation > ChCl:OA (1:2) > ChCl:U (1:2) > PSA > water scrubbing [76] (Table 9). The values include the average standard deviation (8%), which was adopted based on the above calculations. Table 9 contains only the total *TAC* and UC values without standard deviations due to the lack of data from other studies. The main advantage of the innovative method based on DES is the cost of biogas upgrading compared to the most commonly used absorbents.

Table 9. Comparison of economic analysis.

Purification Methods	Total Annual Cost (*TAC*)	Unit Cost of Bio-Methane (UC)	Ref.
Amine scrubber	1,104,000 EUR	0.39 EUR/m^3	[56]
Membrane separation	1,061,000 EUR	0.38 EUR/m^3	[56]
ChCl:OA scrubber	1,095,685 EUR	0.37 EUR/m^3	This study
ChCl:U scrubber	982,510 EUR	0.35 EUR/m^3	This study
PSA	970,000 EUR	0.35 EUR/m^3	[56]
Water scrubbing	880,000 EUR	0.33 EUR/m^3	[56]

The application of traditional absorbents (water, amine), requires further biogas refinement operations, which involves additional costs, while the use of DES ensures that high-quality bio-methane is obtained in a one-step process. The obtained results indicate that biogas upgrading technology by means of DESs is a competitive technology for all currently used methods in the industry.

4. Conclusions

The study presents low-cost deep eutectic solvents (DES) as potential new sorption materials that enable one-step, effective biogas upgrading. This is a significant advantage compared to the currently used sorbents that are dedicated to removing only selected groups of impurities, which does not guarantee that the biogas will be of sufficient quality. The use of developed sorbents under absorption conditions results in biogas with high-methane gas parameters that meet the parameters for gas injected into the transmission network and transport fuel. In addition, the use of new sorbents based on DESs are highly advantageous from an economic and ecological point of view because the sorbents are synthesized from inexpensive, easily available materials that can be regenerated many times without loss of absorption capacity.

In this study, 23 low-cost DESs composed of quaternary ammonium salts and organic components were investigated. Based on the basic thermodynamic properties, i.e., the activity coefficient, excess enthalpy, and Henry's constant, two DESs (ChCl:U (1:2) and ChCl:OA (1:2)) were selected because they showed the highest dissolution potential of the siloxanes, CO_2, and H_2S. The high affinity of both DESs to all of the main biogas contaminations was confirmed by means of σ-profiles and σ-potential analysis. It was shown that the electrostatic interactions between biogas impurities and DESs are the main driving force of the absorption process. For the best DESs, economic analysis simulation was conducted in order to evaluate and compare ChCl:U (1:2) and ChCl:OA (1:2) to each other and to currently available industrial absorbents. The unit cost of DESs depend mainly on the DES structure, which is responsible for its absorption capacity, and regeneration cycles. The unit cost of obtaining 1 m^3 of high-quality bio-methane using DESs absorption is comparable to the costs of currently used technologies. However, the proposed biogas upgrading technology offers the possibility of removing CO_2, H_2S, and siloxanes in one step. This is a significant advantage compared to other commonly used technologies that only remove individual impurities. The obtained results show the great potential of DESs to improve biogas to high-quality bio-methane with properties comparable to natural gas. Such bio-methane could be injected into the natural gas network or used as an alternative to compressed natural gas fuel. However, further experimental research is needed to confirm the obtained results.

Author Contributions: Conceptualization, P.M., and E.S.; methodology, P.M., and E.S.; investigation, P.M., and E.S.; data curation, P.M., and E.S.; writing—original draft preparation, P.M., E.S., J.G.; writing—review and editing, P.M.; visualization, P.M., and E.S.; supervision, J.G. All authors have read and agreed to the published version of the manuscript.

Funding: This research received no external funding.

Conflicts of Interest: The authors declare no conflicts of interest. The funders had no role in the design of the study; in the collection, analyses, or interpretation of data; in the writing of the manuscript, or in the decision to publish the results.

References

1. European Biogas Association (EBA). *European Biogas Association Annual Report 2019*; EBA: Brussels, Belgium, 2019.
2. European Biogas Association (EBA). *Statistical Report 2018: Annual Report*; EBA: Brussels, Belgium, 2018; Volume 68.
3. Persson, M.; Jonsson, O.; Wellinger, A. Biogas Upgrading To Vehicle Fuel Standards and Grid Injection. *IEA Bioenergy Task* **2007**, *37*, 1–34.
4. Andrés, C.; De Guardia, A.; Couvert, A.; Wolbert, D.; Le, S.; Soutrel, I.; Nunes, G. Odor concentration (OC) prediction based on odor activity values (OAVs) during composting of solid wastes and digestates. *Atmos. Environ.* **2019**, *201*, 1–12.
5. Papurello, D.; Soukoulis, C.; Schuhfried, E.; Cappellin, L.; Gasperi, F.; Silvestri, S.; Santarelli, M.; Biasioli, F. Monitoring of volatile compound emissions during dry anaerobic digestion of the Organic Fraction of Municipal Solid Waste by Proton Transfer Reaction Time-of-Flight Mass Spectrometry. *Bioresour. Technol.* **2012**, *126*, 254–265. [CrossRef] [PubMed]
6. Wasajja, H.; Lindeboom, R.E.F.; van Lier, J.B.; Aravind, P.V. Techno-economic review of biogas cleaning technologies for small scale off-grid solid oxide fuel cell applications. *Fuel Process. Technol.* **2020**, *197*, 106215. [CrossRef]
7. Paolini, V.; Petracchini, F.; Segreto, M.; Tomassetti, L.; Naja, N.; Cecinato, A. Environmental impact of biogas: A short review of current knowledge. *J. Environ. Sci. Health Part A Toxic Hazard. Subst. Environ. Eng.* **2018**, *53*, 899–906. [CrossRef] [PubMed]
8. Rasi, S.; Veijanen, A.; Rintala, J. Trace compounds of biogas from different biogas production plants. *Energy* **2007**, *32*, 1375–1380. [CrossRef]
9. Santos-Clotas, E.; Cabrera-Codony, A.; Martín, M.J. Coupling adsorption with biotechnologies for siloxane abatement from biogas. *Renew. Energy* **2020**, *153*, 314–323. [CrossRef]
10. Santos-Clotas, E.; Cabrera-Codony, A.; Boada, E.; Gich, F.; Muñoz, R.; Martín, M.J. Efficient removal of siloxanes and volatile organic compounds from sewage biogas by an anoxic biotrickling filter supplemented with activated carbon. *Bioresour. Technol.* **2019**, *294*, 122136. [CrossRef]
11. Liu, Y.H.; Meng, Z.Y.; Wang, J.Y.; Dong, Y.F.; Ma, Z.C. Removal of siloxanes from biogas using acetylated silica gel as adsorbent. *Pet. Sci.* **2019**, *16*, 920–928. [CrossRef]
12. Ruiling, G.; Shikun, C.; Zifu, L. Research progress of siloxane removal from biogas. *Int. J. Agric. Biol. Eng.* **2017**, *10*, 30–39.
13. Shen, M.; Zhang, Y.; Hu, D.; Fan, J.; Zeng, G. A review on removal of siloxanes from biogas: With a special focus on volatile methylsiloxanes. *Environ. Sci. Pollut. Res.* **2018**, *25*, 30847–30862. [CrossRef] [PubMed]
14. Xiao, Y.; Yuan, H.; Pang, Y.; Chen, S.; Zhu, B.; Zou, D.; Ma, J.; Yu, L.; Li, X. CO2 Removal from Biogas by Water Washing System. *Chin. J. Chem. Eng.* **2014**, *22*, 950–953. [CrossRef]
15. Mandal, B.P.; Biswas, A.K.; Bandyopadhyay, S.S. Selective absorption of H2S from gas streams containing H2S and CO2 into aqueous solutions of N-methyldiethanolamine and 2-amino-2-methyl-1-propanol. *Sep. Purif. Technol.* **2004**, *35*, 191–202. [CrossRef]
16. Jassim, M.S.; Rochelle, G.; Eimer, D.; Ramshaw, C. Carbon Dioxide Absorption and Desorption in Aqueous Monoethanolamine Solutions in a Rotating Packed Bed. *Ind. Eng. Chem. Res.* **2007**, *46*, 2823–2833. [CrossRef]
17. Xu, H.J.; Zhang, C.F.; Zheng, Z.S. Solubility of hydrogen sulfide and carbon dioxide in a solution of methyldiethanolamine mixed with ethylene glycol. *Ind. Eng. Chem. Res.* **2002**, *41*, 6175–6180. [CrossRef]
18. Privalova, E.; Nurmi, M.; Marañón, M.S.; Murzina, E.V.; Mäki-arvela, P.; Eränen, K.; Murzin, D.Y.; Mikkola, J. CO_2 removal with 'switchable' versus 'classical' ionic liquids. *Sep. Purif. Technol.* **2012**, *97*, 42–50. [CrossRef]
19. Gao, J.; Cao, L.; Dong, H.; Zhang, X.; Zhang, S. Ionic liquids tailored amine aqueous solution for pre-combustion CO_2 capture: Role of imidazolium-based ionic liquids. *Appl. Energy* **2015**, *154*, 771–780. [CrossRef]
20. Shiflett, M.B.; Yokozeki, A. Solubilities and Diffusivities of Carbon Dioxide in Ionic Liquids: [bmim] [PF 6] and [bmim] [BF 4]. *Ind. Eng. Chem. Res.* **2005**, 4453–4464. [CrossRef]
21. Jou, F.Y.; Mather, A.E. Solubility of hydrogen sulfide in [bmim][PF 6]. *Int. J. Thermophys.* **2007**, *28*, 490–495. [CrossRef]

22. Pomelli, C.S.; Chiappe, C.; Vidis, A.; Laurenczy, G.; Dyson, P.J. Influence of the interaction between hydrogen sulfide and ionic liquids on solubility: Experimental and theoretical investigation. *J. Phys. Chem. B* **2007**, *111*, 13014–13019. [CrossRef]
23. Rahmati-Rostami, M.; Ghotbi, C.; Hosseini-Jenab, M.; Ahmadi, A.N.; Jalili, A.H. Solubility of H2S in ionic liquids [hmim][PF6], [hmim][BF4], and [hmim][Tf2N]. *J. Chem. Thermodyn.* **2009**, *41*, 1052–1055. [CrossRef]
24. Zhang, X.; Zhang, X.; Dong, H.; Zhao, Z.; Zhang, S.; Huang, Y. Carbon capture with ionic liquids: Overview and progress. *Energy Environ. Sci.* **2012**, *5*, 6668–6681. [CrossRef]
25. Xie, Y.; Zhang, Y.; Lu, X.; Ji, X. Energy consumption analysis for CO2 separation using imidazolium-based ionic liquids. *Appl. Energy* **2014**, *136*, 325–335. [CrossRef]
26. Thuy Pham, T.P.; Cho, C.W.; Yun, Y.S. Environmental fate and toxicity of ionic liquids: A review. *Water Res.* **2010**, *44*, 352–372. [CrossRef]
27. Abbott, A.P.; Capper, G.; Davies, D.L.; Rasheed, R.K.; Tambyrajah, V. Novel Solvent Properties of Choline Chloride/Urea Mixtures. *Chem. Commun.* **2003**, *0*, 70–71. [CrossRef]
28. Azizi, N.; Dezfooli, S.; Khajeh, M.; Hashemi, M.M. Efficient deep eutectic solvents catalyzed synthesis of pyran and benzopyran derivatives. *J. Mol. Liq.* **2013**, *186*, 76–80. [CrossRef]
29. Abbott, A.P.; Barron, J.C.; Ryder, K.S.; Wilson, D. Eutectic-Based Ionic Liquids with Metal-Containing Anions and Cations. *Chem. Eur. J.* **2007**, *13*, 6495–6501. [CrossRef]
30. Mainar, A.R.; Iruin, E.; Colmenares, L.C.; Kvasha, A.; de Meatza, I.; Bengoechea, M.; Leonet, O.; Boyano, I.; Zhang, Z.; Blazquez, J.A. An overview of progress in electrolytes for secondary zinc-air batteries and other storage systems based on zinc. *J. Energy Storage* **2018**, *15*, 304–328. [CrossRef]
31. Werner, S.; Haumann, M.; Wasserscheid, P. Ionic Liquids in Chemical Engineering. *Annu. Rev. Chem. Biomol. Eng.* **2010**, *1*, 203–230. [CrossRef]
32. Sowmiah, S.; Srinivasadesikan, V.; Tseng, M.C.; Chu, Y.H. On the Chemical Stabilities of Ionic Liquids. *Molecules* **2009**, *14*, 3780–3813. [CrossRef]
33. Dai, Y.; Van Spronsen, J.; Witkamp, G.J.; Verpoorte, R.; Choi, Y.H. Ionic liquids and deep eutectic solvents in natural products research: Mixtures of solids as extraction solvents. *J. Nat. Prod.* **2013**, *76*, 2162–2173. [CrossRef] [PubMed]
34. Turosung, S.N.; Ghosh, B. Application of Ionic Liquids in the Upstream oil Industry-A Review. *Int. J. Petrochem. Res.* **2017**, *1*, 50–60. [CrossRef]
35. Smith, E.L.; Abbott, A.P.; Ryder, K.S. Deep Eutectic Solvents (DESs) and Their Applications. *Chem. Rev.* **2014**, *114*, 11060–11082. [CrossRef] [PubMed]
36. Vanda, H.; Dai, Y.; Wilson, E.G.; Verpoorte, R.; Choi, Y.H. Green solvents from ionic liquids and deep eutectic solvents to natural deep eutectic solvents. *Comptes Rendus Chim.* **2018**, *21*, 628–638. [CrossRef]
37. Marcus, Y. Estimation of the Critical Temperatures of Some More Deep Eutectic Solvents from Their Surface Tensions. *Adv. Mater. Sci. Eng.* **2018**, *2018*, 2–5. [CrossRef]
38. Makoś, P.; Słupek, E.; Gębicki, J. Extractive detoxification of feedstocks for the production of biofuels using new hydrophobic deep eutectic solvents – Experimental and theoretical studies. *J. Mol. Liq.* **2020**, in press.
39. Makoś, P.; Przyjazny, A.; Boczkaj, G. Hydrophobic deep eutectic solvents as "green" extraction media for polycyclic aromatic hydrocarbons in aqueous samples. *J. Chromatogr. A* **2018**, *1570*, 28–37. [CrossRef] [PubMed]
40. Makoś, P.; Boczkaj, G. Deep eutectic solvents based highly efficient extractive desulfurization of fuels—Eco-friendly approach. *J. Mol. Liq.* **2019**, 111916. [CrossRef]
41. Makoś, P.; Fernandes, A.; Przyjazny, A.; Boczkaj, G. Sample preparation procedure using extraction and derivatization of carboxylic acids from aqueous samples by means of deep eutectic solvents for gas chromatographic-mass spectrometric analysis. *J. Chromatogr. A* **2018**, *1555*, 10–19. [CrossRef] [PubMed]
42. Zhong, F.-Y.; Huang, K.; Peng, H.-L. Solubilities of ammonia in choline chloride plus urea at (298.2–353.2) K and (0–300) kPa. *J. Chem. Thermodyn.* **2019**, *129*, 5–11. [CrossRef]
43. Zhang, K.; Hou, Y.; Wang, Y.; Wang, K.; Ren, S.; Wu, W. Efficient and Reversible Absorption of CO2 by Functional Deep Eutectic Solvents. *Energy Fuels* **2018**, *32*, 7727–7733. [CrossRef]
44. Moura, L.; Moufawad, T.; Ferreira, M.; Bricout, H.; Tilloy, S.; Monflier, E.; Costa Gomes, M.F.; Landy, D.; Fourmentin, S. Deep eutectic solvents as green absorbents of volatile organic pollutants. *Environ. Chem. Lett.* **2017**, *15*, 747–753. [CrossRef]

45. Słupek, E.; Makoś, P.; Gębicki, J. Deodorization of model biogas by means of novel non-ionic deep eutectic solvent. *Arch. Environ. Prot.* **2020**, *46*, 41–46.
46. Słupek, E.; Makoś, P. Absorptive Desulfurization of Model Biogas Stream Using Choline Chloride-Based Deep Eutectic Solvents. *Sustainability* **2020**, *12*, 1619. [CrossRef]
47. Zhang, Y.; Ji, X.; Xie, Y.; Lu, X. Thermodynamic analysis of CO2 separation from biogas with conventional ionic liquids. *Appl. Energy* **2018**, *217*, 75–87. [CrossRef]
48. Ali, E.; Hadj-Kali, M.K.; Mulyono, S.; Alnashef, I. Analysis of operating conditions for CO2 capturing process using deep eutectic solvents. *Int. J. Greenh. Gas Control* **2016**, *47*, 342–350. [CrossRef]
49. Makoś, P.; Słupek, E.; Małachowska, A. Silica Gel Impregnated by Deep Eutectic Solvents for Adsorptive Removal of BTEX from Gas Streams. *Materials* **2020**, *13*, 1894. [CrossRef]
50. Häckl, K.; Kunz, W. Some aspects of green solvents. *Comptes Rendus Chim.* **2018**, *21*, 572–580. [CrossRef]
51. Del Olmo, L.; Lõpez, R.; García De La Vega, J.M. Effect of the molecular structure in the prediction of thermodynamic properties for 1-butyl-3-methylimidazolium chloride ionic liquid. *Int. J. Quantum Chem.* **2013**, *113*, 852–858. [CrossRef]
52. Pye, C.C.; Ziegler, T. An implementation of the conductor-like screening model of solvation within the Amsterdam density functional package. *Theor. Chem. Acc.* **1999**, *101*, 396–408. [CrossRef]
53. AMS 2019.3 COSMO-RS, SCM, Theoretical Chemistry. Vrije Universiteit: Amsterdam, The Netherlands; Available online: http://www.scm.com (accessed on 4 May 2020).
54. Rasi, S.; Läntelä, J.; Rintala, J. Upgrading landfill gas using a high pressure water absorption process. *Fuel* **2014**, *115*, 539–543. [CrossRef]
55. Ma, C.; Liu, C.; Lu, X.; Ji, X. Techno-economic analysis and performance comparison of aqueous deep eutectic solvent and other physical absorbents for biogas upgrading. *Appl. Energy* **2018**, *225*, 437–447. [CrossRef]
56. Xie, Y.; Björkmalm, J.; Ma, C.; Willquist, K.; Yngvesson, J.; Wallberg, O.; Ji, X. Techno-economic evaluation of biogas upgrading using ionic liquids in comparison with industrially used technology in Scandinavian anaerobic digestion plants. *Appl. Energy* **2018**, *227*, 742–750. [CrossRef]
57. Scholz, M.; Frank, B.; Stockmeier, F.; Falß, S.; Wessling, M. Techno-economic analysis of hybrid processes for biogas upgrading. *Ind. Eng. Chem. Res.* **2013**, *52*, 16929–16938. [CrossRef]
58. Guthrie, H.M. Capital cost estimation for the chemical and process industries. *Chem. Eng.* **1969**, *32*, 114–142.
59. Euro Statistic Explained Electricity Prices First Semester of 2017_2019 (EUR Per kWh). Available online: https://ec.europa.eu/eurostat/statistics-explained/index.php?title=File:Electricity_prices,_first_semester_of_2017-2019_(EUR_per_kWh).png (accessed on 4 May 2020).
60. Alibaba.com Choline Chloride. Available online: https://www.alibaba.com/product-detail/choline-chloride-CAS-67-48-1_62448780792.html?spm=a2700.7724857.normalList (accessed on 4 May 2020).
61. Alibaba.com Urea. Available online: https://www.alibaba.com/product-detail/price-per-ton-urea-2019-high_60456580510.html?spm=a2700.7724857.normalList.188.c2e9746du64Jbd (accessed on 4 May 2020).
62. Alibaba.com Oxalic Acid. Available online: https://www.alibaba.com/product-detail/99-6-min-Oxalic-Acid-h2c2o4_60767148271.html?spm=a2700.7724857.normalList.11.27d05ea2G45DWC&s=p&bypass=true (accessed on 4 May 2020).
63. De Hullu, J.; Massen, J.L.W.; van Meel, P.A.; Shazad, S.; Vaessen, J.M.P. *Comparing Different Biogas Upgrading Techniques-Report*; Eindhoven University of Technology: Eindhoven, The Netherlands, 2008.
64. Chu, Y.; He, X. MoDoop: An Automated Computational Approach for COSMO-RS Prediction of Biopolymer Solubilities in Ionic Liquids. *ACS Omega* **2019**, *4*, 2337–2343. [CrossRef]
65. Makoś, P.; Słupek, E.; Gębicki, J. Hydrophobic deep eutectic solvents in microextraction techniques–A review. *Microchem. J.* **2020**, *152*. [CrossRef]
66. Han, J.; Dai, C.; Yu, G.; Lei, Z. Parameterization of COSMO-RS model for ionic liquids. *Green Energy Environ.* **2018**, *3*, 247–265. [CrossRef]
67. Mu, T.; Rarey, J.; Gmehling, J. Performance of COSMO-RS with sigma profiles from different model chemistries. *Ind. Eng. Chem. Res.* **2007**, *46*, 6612–6629. [CrossRef]
68. Mullins, E.; Oldland, R.; Liu, Y.A.; Wang, S.; Sandler, S.I.; Chen, C.C.; Zwolak, M.; Seavey, K.C. Sigma-profile database for using COSMO-based thermodynamic methods. *Ind. Eng. Chem. Res.* **2006**, *45*, 4389–4415. [CrossRef]

69. Salleh, Z.; Wazeer, I.; Mulyono, S.; El-blidi, L.; Hashim, M.A.; Hadj-Kali, M.K. Efficient removal of benzene from cyclohexane-benzene mixtures using deep eutectic solvents—COSMO-RS screening and experimental validation. *J. Chem. Thermodyn.* **2017**, *104*, 33–44. [CrossRef]
70. Liu, Y.-R.; Thomsen, K.; Nie, Y.; Zhang, S.-J.; Meyer, A.S. Predictive screening of ionic liquids for dissolving cellulose and experimental verification. *Green Chem.* **2016**, *18*, 6246–6254. [CrossRef]
71. Hosseinipour, S.A.; Mehrpooya, M. Comparison of the biogas upgrading methods as a transportation fuel. *Renew. Energy* **2019**, *130*, 641–655. [CrossRef]
72. Ryckebosch, E.; Drouillon, M.; Vervaeren, H. Techniques for transformation of biogas to biomethane. *Biomass Bioenergy* **2011**, *35*, 1633–1645. [CrossRef]
73. Petersson, A.; Holm-nielsen, J.B.; Baxter, D. Biogas upgrading technologies—Developments and innovations. *IEA Bioenergy Task* **2009**, *37*.
74. Warren, W.K.E.H. A Techno-economic Comparison of Biogas Upgrading Technologies in Europe. Master's Thesis, Jyväskylän yliopisto—University of Jyväskylä, Jyväskylä, Finland, 2012; p. 44.
75. Trading Economics. Available online: https://tradingeconomics.com/commodity/steel?fbclid=IwAR12n6kfQegnqpqrIPqJZ47gjDUh2Q7OIt9BChQosYiJnZSHskCDwczPqJ8 (accessed on 21 May 2020).
76. Stürmer, B.; Kirchmeyr, F.; Kovacs, K.; Hofmann, F.; Collins, D.; Ingremeau, C.; Stambasky, J. *Technical-Economic Analysis for Determining the Feasibility Threshold for Tradable Biomethane Certificates-Report*; European Renewable Gas Registry: Brussels, Belgium, 2016; pp. 1–24.
77. Chen, W.; Xue, Z.; Wang, J.; Jiang, J.; Zhao, X.; Mu, T. Investigation on the thermal stability of deep eutectic solvents. *Wuli Huaxue Xuebao/Acta Phys. Chim. Sin.* **2018**, *34*, 904–911. [CrossRef]
78. Haz, A.; Strizincova, P.; Majova, V.; Skulcova, A.; Jablonsky, M. Thermal stability of selected deep eutectic solvents. *Int. J. Recent Sci. Res.* **2016**, *7*, 14441–14444.
79. Brown, R.C.; Brown, T.R. Biorenewable Resources: Engineering New Products from Agriculture: Second Edition. *Biorenew. Resour. Eng. New Prod. Agric. Second Ed.* **2014**, *9781118524*, 1–375.

© 2020 by the authors. Licensee MDPI, Basel, Switzerland. This article is an open access article distributed under the terms and conditions of the Creative Commons Attribution (CC BY) license (http://creativecommons.org/licenses/by/4.0/).

energies

MDPI

Article

Lightweight Equipment Using Multiple Torches for Fast Speed Asphalt Roofing

Alberto Barragán-García [1,2], Miguel Fernández-Muñoz [1] and Efrén Díez-Jiménez [1,*]

1 Mechanical Engineering Area, Universidad de Alcalá, 28801 Alcalá de Henares, Spain; a.barragan@uah.es (A.B.-G.); miguel.fernandezmuno@uah.es (M.F.-M.)
2 Industrial Engineering Department, Ingeniería Industrial Ibérica S. A., 28034 Madrid, Spain
* Correspondence: efren.diez@uah.es; Tel.: +34-91-885-6771

Received: 7 April 2020; Accepted: 20 April 2020; Published: 2 May 2020

Abstract: In this work, we describe the design and test of a new piece of equipment, developed in order to enhance speed, gas consumption and safety during the manual asphalt roofing process. The novelty of the equipment is based on the use of a set of five parallel gas burners located in front of the roll to maximize heat transfer. The equipment is light and can be used by any worker on any type of roof. It also includes a thermal insulation cover to significantly reduce gas consumption and, thus, to reduce CO_2, SO_2, and other non-eco-friendly emissions. In this paper, we present the mechanical and thermal design and analysis of the equipment, Computer Fluid Dynamics (CFD) simulations for heat transfer calculation, a description of the manufacturing and assembly, a preliminary thermal test, and an operational test. The results demonstrate an installation speed of 1.75 m^2/min, for 3 kg/m^2 rolls, which translates to around 700–735 m^2 per person per day, more than twice the usual manual roofing rate. Nevertheless, some issues need to be resolved, such as the nonuniform heat distribution and the low heat transfer at the end of the roll installation.

Keywords: fuel burners; asphalt roofing; heating equipment

1. Introduction

Nowadays, roofing methods play an essential role in the energy efficiency and weather resistance of buildings and houses [1]. One common roof protection method is to install asphalt coatings. Specialized workers generally do asphalt roofing manually. They typically carry a single Liquefied Petroleum Gas (LPG) burner to apply the heat to the rolls. Workers perform several steps when roofing: locating the rolls, aligning them with respect to the previous rolls, unrolling them, applying heat, and finishing them off. A worker can install around 20–25 rolls per day (200–250 m^2) using this manual procedure. Such a reduced capacity for manual installation has motivated patents and developments of new systems and equipment to facilitate and enhance roll installation speed.

Patent EP0466249 [2]—a method and apparatus for applying a bituminous sheet to a substrate—presents a roller system with a previously heated additional drum wherein the asphalt roll is heated. Another patent US4725328 [3] was claimed for a mechanism that includes two heating torches perpendicular to the roll. Heating torches were located at both ends to increase adherence where the rolls overlap. Patent US 2013228287 [4] has been commercialized using the name Unify-ER by the company RES Automatisation Contrôle [5]. This device applies heat to the roll through a longitudinally distributed set of LPG burners. This equipment shows a roll installation speed of 1 m/min, claiming a rate of 18 rolls per hour, but requires two workers; thus, this device provides an installation of 480 m^2 per day per person. Although it is faster and much more efficient than the manual technique, the large number of frames, structure and auxiliary mechanisms leads to a total weight of 180 kg. Such a high weight considerably hinders its use on weak roofs. In addition, the excessive size of the equipment makes it difficult to be transported in small vans. Other commercial equipment are the Seal-Master

1030 from Schäfer Technic GmbH and the Bitumenbrenner from Bamert Spenglerei GmbH. From a practical point of view, the main disadvantages of all the previous systems are that they are heavy, too large for operation on small roofs and difficult to transport. On the other hand, manual installation only requires a single torch and propane bottle. The lightweight equipment that we present in this article provides a fast, high quality application capacity, while being light and small enough to be transported and used in most cases.

All previous patents and products enhance asphalt roofing speed but they all require gas burners to heat the asphalt up for its installation. Gas burners generate CO_2 and SO_2 emissions, which must be reduced as much as possible. Roofing has attracted the attention of environmental administrations in order to reduce their emissions. For example, the Asphalt Roofing Manufacturing Association (ARMA) developed emissions factors for asphalt-related air emissions for all the relevant processes in the manufacture of roofing asphalt [6]. Moreover, a measurement of 75.2 kg CO_2eq per roll was estimated in [7] and the installation stage in buildings was found to be responsible for the majority of those emissions. Infrared heating could be an alternative option to gas burners [8], but its applicability is limited to places where a high electrical power source is available. Therefore, it is necessary to reduce the gas consumption as much as possible by optimizing the heat transfer from torch burners to asphalt rolls.

Besides speed and gas consumption, roofing workers' health must also be guaranteed and protected. When doing manual installation, workers often do not use safety prevention equipment in order to improve their speed, at the expense of increasing the risk of accidents. A proper way to prevent occupational illnesses, injuries and fatalities is to design systems and equipment that allow workers to operate efficiently [9]. Moreover, inadequate posture during manual roofing may cause a rise in the number of illnesses and injuries. The equipment presented in this article allows workers to operate in safer conditions and to prevent work injuries. We hope that the equipment presented in this article will help workers during manual installation, reduce installation risks, increase installation speed, optimize gas consumption and will be light and practical for use on any type of roof. Furthermore, the shorter the installation time of the asphalt rolls, the lower the CO_2 emissions; thus, the presented equipment even provides an improvement in the human factor because, during the installation of asphalt rolls, humans are also a source of emissions (eating, creating waste, using electricity, etc.), and the impact of their emissions decreases if the working time is reduced.

The presented equipment consists of a lightweight trolley mechanism that allows the installation of asphalt rolls in a quicker, cleaner, and safer way. This trolley includes a set of five parallel torches located in front of the roll to heat the asphalt up. Asphalt rolls are directly placed on the floor, so the worker does not have to lift them. As the worker pushes the trolley, auxiliary wheels unroll the roll, while the worker can control the heat application. The equipment also includes an insulation cover to maintain and increase the heat transfer from the torches to the roll. This allows for the uniform and continuous heating of the roll for its installation on the ground, which is synchronized with the unrolling. Moreover, the trolley has two small compaction drum rollers in both lateral sides, in order to assure the adherence of the asphalt where the rolls overlap.

In this article, we present the mechanical and thermal design and an analysis of the equipment, CFD simulations for the calculation of heat transfer, a description of the manufacturing and assembly, a preliminary thermal test, and an operational test of this new equipment, demonstrating the claimed advantages. To conclude, we include a list of lessons learned after the tests that may help to improve the equipment in further developments.

2. Design and Analysis

The described lightweight equipment is basically a lightweight trolley design that facilitates the installation of asphalt rolls. This equipment is composed of the following elements: a light trolley structure, four wheels for the motion of the trolley, an auxiliary wheel for unrolling the asphalt roll, a set of five torches aligned parallel to the roll, a top thermal insulation cover to avoid heat loss, small drum rollers for the compaction of the asphalt in the overlaps and general gas flow regulators and switches.

2.1. Mechanical Design and Analysis

Figure 1 compiles three views of the trolley. The isometric view shows the trolley (1), which can be pushed using a handle placed at a comfortable height for the worker. A total of four wheels (3) allow for the smooth motion of the equipment. The rear wheels are blocked and the front wheels are free to allow steering, so that the worker can precisely drive and guide the trolley. The asphalt roll (4) is located inside the main frame. The trolley has two locating stops on both laterals to keep the roll in the right position and allow for the re-direction of the roll while it is being unrolled. In addition, the trolley has two auxiliary wheels (5) located behind the roll, and at a height lower than the radius of the roll. These auxiliary wheels transmit the pushing force tangentially to the roll. A thermal insulation cover made by rock wool (6), which maximizes the heat transfer, covers the whole roll. Additionally, the trolley includes two small drum rollers (12) for the compaction of the asphalt where rolls overlap.

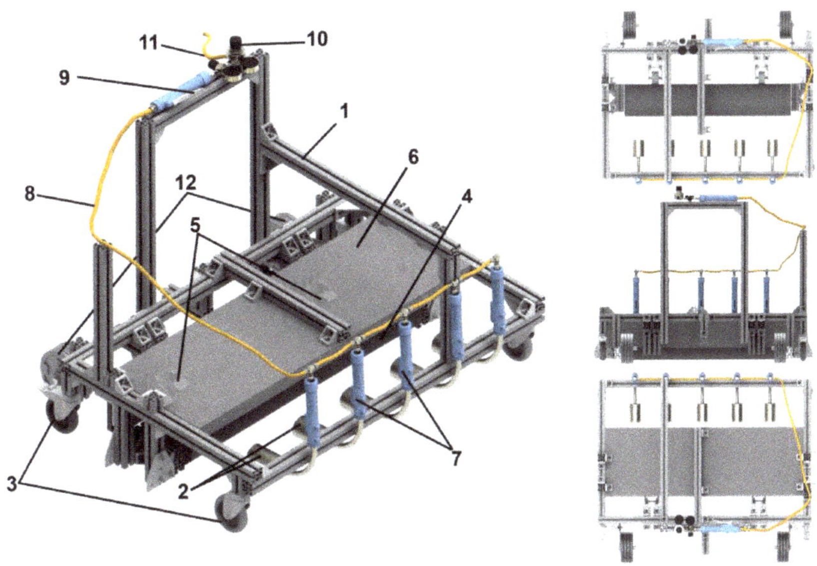

Figure 1. Isometric and orthogonal views of the design: 1—trolley, 2—torches, 3—wheels, 4—asphalt roll, 5—auxiliary wheels, 6—thermal insulation cover, 7—individual flow regulator, 8—gas pipeline, 9—on/off valve, 10—main flow regulator, 11—main pipeline from propane bottle, 12—drum rollers.

Figure 1 also shows the set of five torches (2), which are parallel and aligned in front of the asphalt roll. A flexible pipeline (8) connects all the torches with the main gas pipeline. Each torch flow can be individually regulated (7), but is also regulated by a general flow regulator (10) placed close to the handle. The activation of the torches is set and controlled by an on/off manual control valve (9). The gas is provided from a propane pressurized bottle through a flexible ten-meter-long pipeline (11).

The equipment is specially adjusted for ten-meter-long, one-meter-wide 3 kg/m^2 asphalt rolls. The main dimensions are depicted in Figure 2. Smaller roll sizes could be compatible with the equipment, but the positions of the locating stops may be modified, placing larger or shorter limits on the rolls. For wider rolls, modifications to the whole structure would be necessary, as well as the repositioning of the torches. The total trolley mass is 16 kg, excluding the compaction drum rollers.

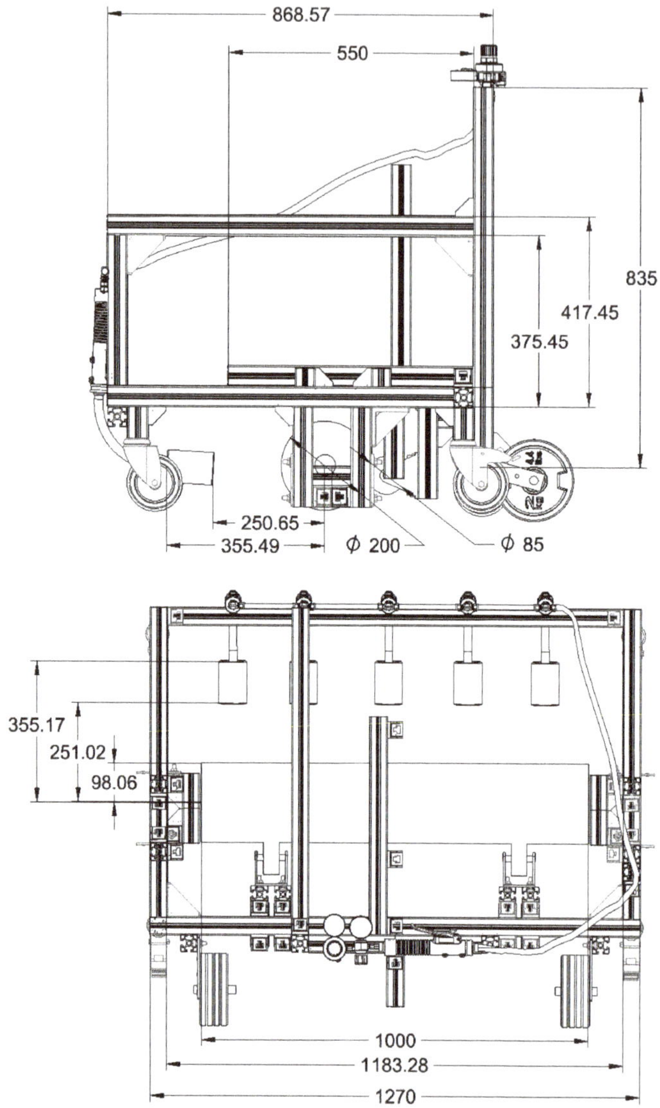

Figure 2. General dimensions of the equipment in millimeters.

During installation, it is not necessary to lift the roll from the ground, since the trolley directly moves and unrolls it. Hence, there is no need to add heavy auxiliary rolls for unrolling, which makes the entire system much lighter and saves workers' energy. The unrolling mechanism and structure of the trolley follows the design shown in [10] and the patent in [11]. The mechanism is based on a pair of auxiliary wheels.

Standard 3 kg/m^2 asphalt rolls weigh 30 kg and are 0.25 m in diameter. The friction coefficient between concrete and asphalt is estimated as μ = 0.4. Thus, with a maximum friction force of F = $\mu \cdot m \cdot g$ = 117.6 N, almost 12 kg of horizontal pushing force is needed. Such a value is easily achievable for a construction worker. This implies that there could be an issue—if the trolley directly pushes the roll, the worker might be able to slide the roll without unrolling it. This could be severely problematic at the start of the installation of the roll because, at this point, there is no adherence between the asphalt and ground. One solution to prevent sliding is to include auxiliary wheels in order to facilitate unrolling. These auxiliary wheels permit the conversion of the horizontal thrust into a tangent force, making the unrolling much easier.

2.2. Thermal Design

One of the principal issues during manual installation with a single torch is that the heat flux is only applied in a highly localized area. With just one torch, heat has to be applied progressively along a single area on the roll. This leads to nonuniform adhesion and problems in the overlapping areas. The uniformity can be improved by using a longitudinal five-torch layout, meaning that heat is uniformly distributed along the roll line. The heat power required for this fast installation method is calculated in Section 2.2.1.

Additionally, the localized heat transfer is dependent on the desired length of application, which is variable, as it is controlled by the worker. When performing manual installation, the heating time is usually much longer than required for a proper asphalt installation. Workers do not have any way to measure or control of the temperature beyond their own eyesight and experience. By fixing the distance between the roll and the torches and by adjusting the gas flow of the torches individually and generally, workers can perfectly control the amount of heat that is applied. The gas pressure and torch selection were calculated through an iterative process, with transient CFD simulation results. CFD analysis determined the percentage of heat transferred to the roll, as shown in Section 2.2.2.

Moreover, in the manual method, the torches' efficiency is reduced because most of the heat is lost due to the large environmental air convection that occurs far from the roll. In our model, we include a thermal insulation cover between the roll and the air flow, which forces the hot air to remain closer to the roll, increasing efficiency and reducing thermal losses. The effect of including this thermal insulation cover is quantified by a transient CFD analysis, which is also described in Section 2.2.2.

2.2.1. Requirement of Heat Power into the Roll

In order to select adequate torches and to adjust the gas flow consumption, it is necessary to first determine the necessary heat power that needs to be transferred to the roll for fast and adequate asphalt adhesion with the ground. However, before total heat, we have to determine the temperature the asphalt needs to reach for correct adhesion. This temperature can be set by analysing the viscosity behaviour of the asphalt with respect to temperature. The asphalt viscosity decreases asymptotically when the temperature increases [12]. With a value of 140 °C, the viscosity is lower than 125 mPa·s, which is very close to the final asymptotic value. As the asphalt's bonding with the ground is based on its viscosity, 140 °C could be considered a hot enough temperature point for asphalt–ground bonding. The physical properties of asphalt and rock wool are shown in Table 1 [13].

Table 1. Physical properties of asphalt and rock wool.

Property	Asphalt	Rock Wool
Heat capacity (Cp)	1014 J/kg·°C	840 J/kg·°C
Thermal conductivity (k)	0.5 W/m·K	0.035 W/m·K
Density (ρ)	1700 kg/m^3	40 kg/m^3

Thus, a rough first attempt was carried out, assuming no heat loss. The total power needed to heat up 1.75 m^2/min of a 3 kg/m^2 asphalt roll can be calculated as the necessary power to bring the complete mass from an initial temperature of 22 °C to a final temperature of 140 °C:

$$P = \frac{m \cdot c_p \cdot \Delta T}{t} = \frac{3 \cdot 1.75 \cdot 1000 \cdot (140 - 22)}{60} = 10.33\ \text{kW} \quad (1)$$

The environmental temperatures that can be found at workplaces can vary from 4 °C to 40 °C. For these conditions, the calculation returns the total power as 11.9 kW in the extreme case of an environmental temperature of 4 °C. Therefore, we consider the value of 11.9 kW to be the maximum total heat power needed.

During the second heat power calculation iteration, we have to take into consideration that asphalt rolls are made by consecutive asphalt layers. Hence, the heat applied to the roll does not only heat the first layer up, but it is transferred internally to the rest of the layers. This fact must be carefully analysed, since adhesion between layers must be avoided. Therefore, the temperature increase has to be sufficiently fast to maintain the installation speed, but slow enough not to make the layers adhere to one another. Accordingly, it is more accurate to determine an adequate temperature profile for the layers that is higher than the temperature of the asphalt. In this analysis, we will follow the same criterion used in the previous article [10]. This criterion is that the top surface temperature of the layer must be above 140 °C, the middle section of the layer must be at least 110 °C, and the interlayer point must be lower than 70 °C. In our previous work [10], we reached this point after 35 s when applying 11.73 kW/m^2 for a total installation speed of 1 m^2/min; if we target 1.75 m^2/min, then we should transfer at least 20.52 kW/m^2 to the roll (directly proportional to the calculation from [10], finite elements models results) in order to move the trolley at the targeted speed.

2.2.2. Heat Transferred and Thermal Insulation Cover

The next step in the thermal design is to estimate the heat percentage transfer from the hot combustion products coming from the torches to the actual roll. Two-dimensional CFD analysis has been used in order to determine the heat transfer efficiency for the exact geometry of the system, including the thermal insulating cover. Simulations have been set-up and solved by using CFD software ANSYS FLUENT software 2019 R2 (ANSYS, Inc, Canonsburg, PA, USA).

The geometry for the simulations is shown in Figure 3. The dimensions of each geometry correspond with the real prototype dimensions. This 2D geometry includes several boundaries:

- The background boundary considers a zero static pressure and an infinite ambient temperature of 25 °C;
- The inlet boundary supplies air a certain input speed, v_i, and input temperature T_i;
- The thermal insulation cover boundary is a solid–fluid interface line that links the thermal properties of adjacent fluids with the solid thermal conduction inside the rock wool (the fluid speed in this boundary shall be zero). This boundary is removed in the second simulation in order to quantify the improvement effect of the thermal insulation cover;
- The roll boundary is a solid–fluid interface line that links the thermal properties of adjacent fluids with the solid thermal conduction inside the asphalt (the fluid speed in this boundary shall be zero);
- The inner diameter of the roll is a thermal convective boundary with a free convective coefficient of 5 W/m^{2}K, which evacuates the heat to 25 °C ambient temperature;
- The floor is a wall-type boundary, which means that the fluid speed in this boundary shall be zero.

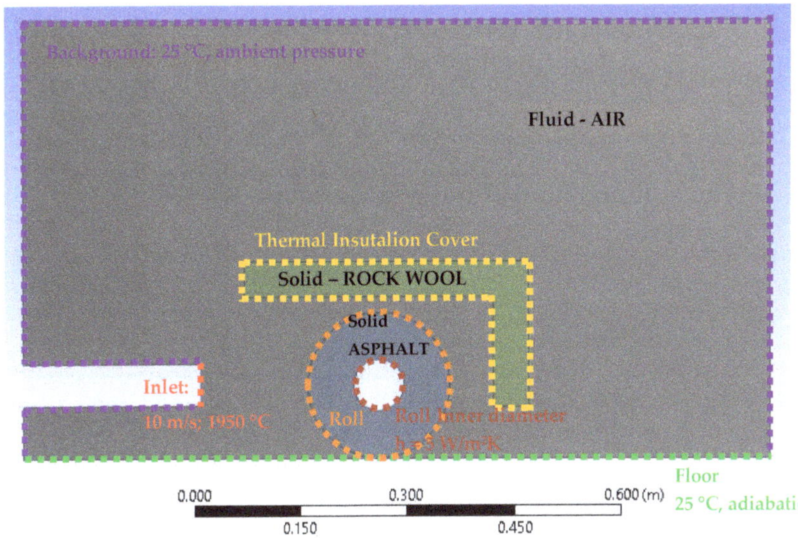

Figure 3. CFD simulation geometry and boundaries.

The inlet boundary length is 57 mm, which corresponds with the output geometrical diameter of the selected torches. The selected blowtorches are 70 mm in diameter and 58 kW of thermal power from company ECD-Germany (Neunkirchen-Seelscheid, Germany). This torch, shown in Figure 4, can provide up to 58 kW of thermal power with a propane consumption of 4000 g/h, at 4 bar of propane pressure in the gas pipeline orifice.

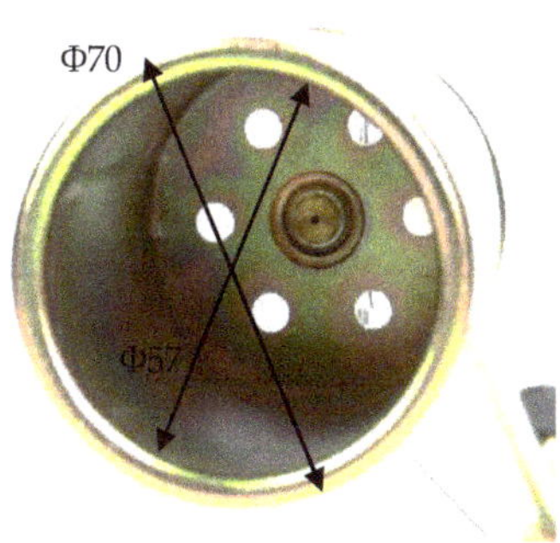

Figure 4. Torch (70 mm in diameter) from Neunkirchen-Seelscheid, Germany.

The inlet boundary fluid conditions are the inlet speed v_i = 10 m/s and the input temperature T_i = 1950 °C. The temperature of the flame at the ignition point is 1950 °C, as demonstrated for full propane combustion [14]. The inlet speed was selected after the preliminary laboratory tests. These kinds of assumptions must always be validated during the test phase [15], as we did in this study.

In the 2D CFD model, the extrapolated inlet area is the length of the inlet boundary line multiplied by a 1 m width, which makes a rectangle. This does not correspond exactly with the real model, which uses five circular torches. However, the objective of the 2D CFD analysis is to determine the relative percentage of transferred heat. In any case, both the thermal input power and the thermal transferred power are calculated per unit length. The thermal input power can be calculated as:

$$Q_i = \rho_i \cdot v_i \cdot \phi_t \cdot c_v \cdot (T_i - T_{amb}) = 0.148 \cdot 10 \cdot 0.057 \cdot 1260 \cdot (1950 - 25) = 205 \text{ kW/m} \quad (2)$$

where ρ_i is the air density at the case temperature T_i, v_i is the inlet speed, ϕ_t is the effective output diameter of the torch 57 mm and c_v is the isochoric specific heat capacity of air when the temperature is T_i. All values are expressed in international units. The thermal input power remains constant at each simulation time step.

The mesh element size is 0.004 m for most of the model, excluding the roll surface wherein the mesh is two orders smaller. The transient model was set up to solve the fluid–solid interaction in transient conditions. The air properties (density, cp, thermal conductivity and viscosity) were defined using a table as a function of the temperature, and the properties of the asphalt roll and the rock wool sheet as the constant values. For the model solver, the method and boundaries were: pressure-based, y-axis gravity, energy equation, standard k-epsilon, initializations of atmospheric pressure and 25 °C. The simulation time was set to 0–60 s. A constant time step of 0.1 s as set. Each time step simulation typically converged after 120 iterations in an Intel Core i5-9600K computer with 16 Gb of RAM, taking around 30 s for each step simulation. The complete transient simulation was done with and without thermal insulation cover.

The results of the air velocities for a time of 15 s with and without the thermal insulation cover are shown in Figure 5. For the result with the cover, it can be observed that most of the air flow goes between the middle height and the top cover, lifted up because lower density of the gas. Lower air speeds appears in the front-bottom part of the roll. The flame tends to rise at any moment. It can also be seen that the design with the cover increases the contact area between the hot air flow and the back of the roll. In the design without the cover, the air speed on the back of the roll is almost zero, and therefore the heat transmission in this area is very low.

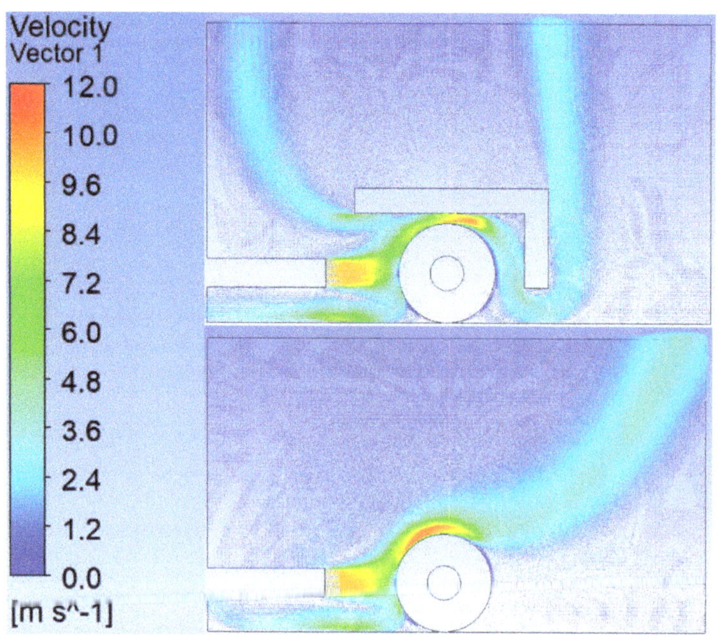

Figure 5. CFD results: velocity distribution with cover (**top**) and without (**bottom**), time = 15 s.

The results of the temperature distribution for a time of 15 s with and without the thermal insulation cover are shown in Figure 6. For the result with the cover, it can be observed that the cover acts as an oven, making the hot air remain close to the roll. It is also relevant that a column of hot air rises from the bottom, which could affect the worker. Because of the lower density, the flame appears to rise at any moment. It can also be seen that the design with the cover increases the contact area between the hot air flow and the back of the roll. In the design without the cover, the temperature on the back of the roll is almost ambient and, again, we can expect that the heat transmission in this area will be very low. The area affected by hot temperatures can be estimated to be around 75% of the total roll area when using the cover and around 55% of the total roll area when not using the cover.

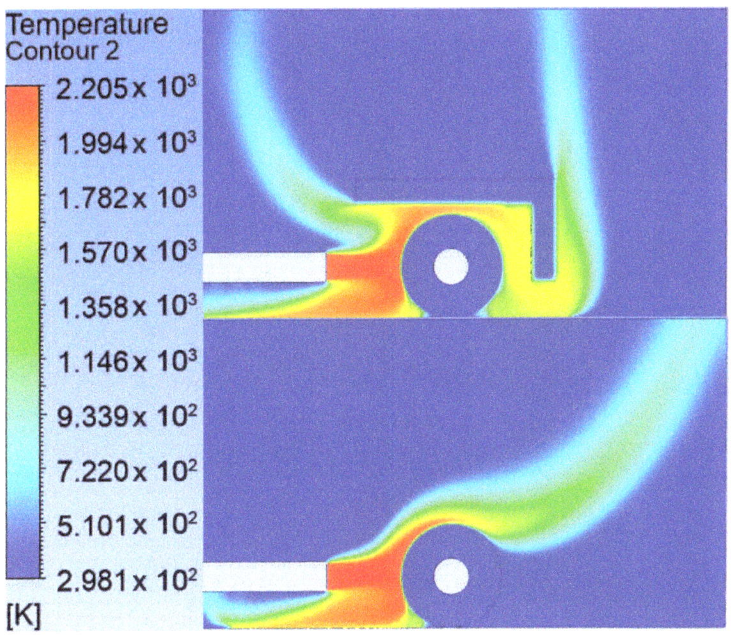

Figure 6. CFD results: temperature distribution with cover (**top**) and without (**bottom**), $t = 15$ s.

The details of the temperature evolution inside the roll with and without the cover are shown in Figure 7. The figure presents the temperatures at three different points, $t = 5$, 10 and 15 s. Again, it is clear that the back of the roll was heated differently depending on whether the cover was present. It is important to note that the colour scale corresponds to the temperature profile criteria described in the previous subsection. The roll should, at least, be above 70 °C (yellow contour) to guarantee a successful adhesion. Moreover, we included a black circumference, indicating the position of the second bitumen layer of the roll. In this way, we can determine whether the first and second layer interfaces are hot enough to be installed.

It can be observed that, with or without the cover, the increase in the temperature on the front of the roll is almost the same. If $t = 10$ s, the first layer would be within the yellow contour and, therefore, it would be ready for installation. However, the back of the roll behaves differently with and without the cover. Without the cover, the back of the roll does not get hot enough, even if $t = 15$ s. Therefore, we conclude that using the cover will surely speed up the installation process. By using the cover, we can determine that around 75% of the exposed roll area is ready to be installed after 15 s; this means that a total area of 0.031 m^2/s or 1.88 m^2/min was aimed for in the prototype.

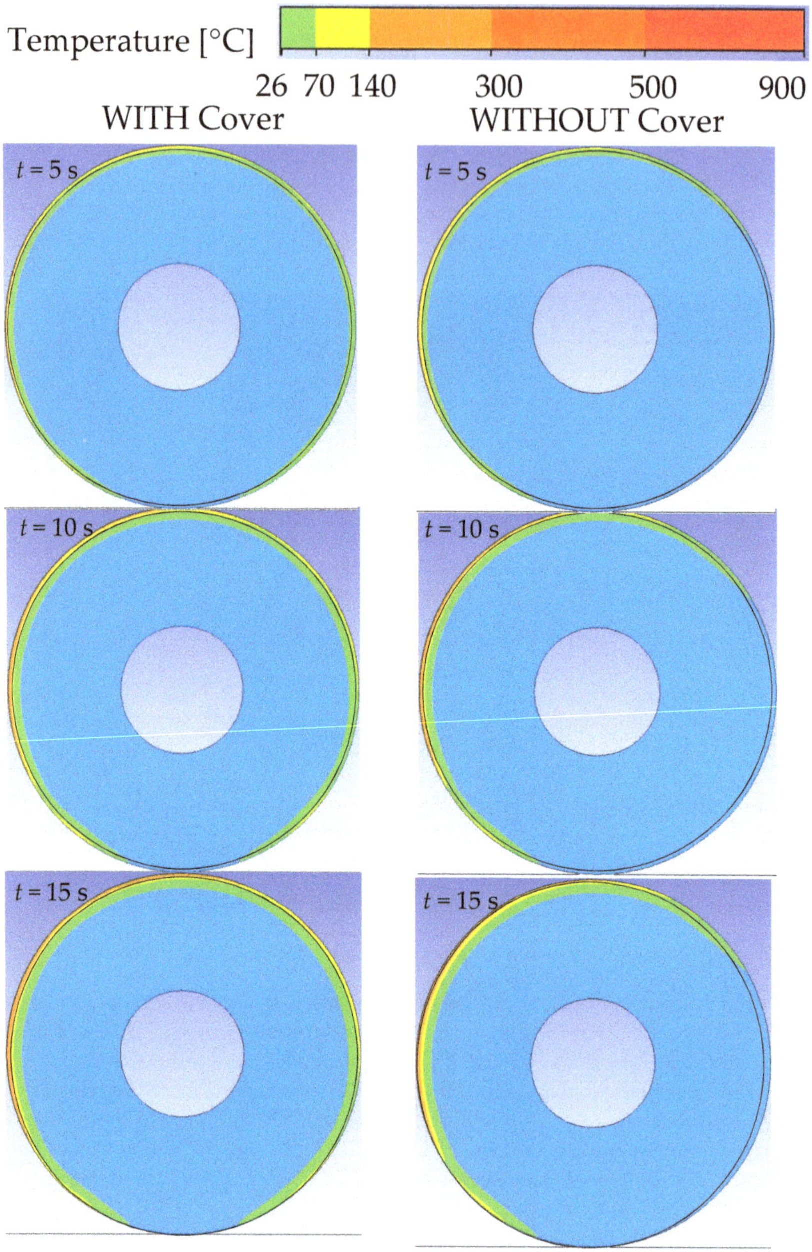

Figure 7. CFD results: temperature inside the roll, with cover (**left**) and without (**right**).

The heat transferred inside the roll, with and without the cover, as a function of time is shown in Figure 8. With the cover, the heat transferred was calculated to be around 20 kW. This coincides with the estimated requirement of the previous subsection, but the objective of the 2D model is to estimate the percentage of heat transfer and not the total transferred heat. Notably, the heat transfer rate decreases with time. This can be explained because, as the roll gets hotter, the total heat transfer is reduced until it reaches a certain stationary asymptotic value. The heat transferred without the cover is around 75% of the heat transferred with the cover, which was expected based on the temperature distribution analysis.

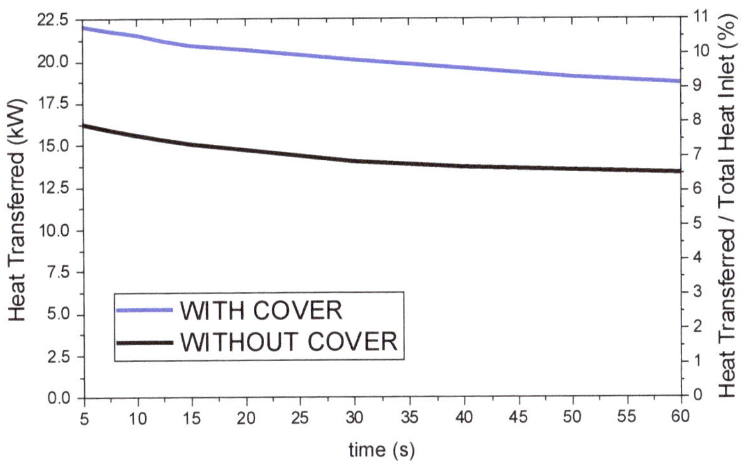

Figure 8. CFD results: heat transferred to the roll and relative percentage with respect to total heat inlet, with cover (**left**) and without (**right**).

However, when comparing the total heat transferred to the roll with the total heat generated, the values are low. It has been determined that only around 10% of the total generated heat will be transferred to the roll. This implies that the system shall generate a total heat of around 205 kW. Each torch shall generate up to 41 kW. If one torch has the capacity to generate 58 kW of thermal power with a consumption of 4000 g/h, at 4 bar of propane pressure, then it means that the system should operate with a pressure of 2.8 bar and a consumption of 2800 g/h per torch.

3. Manufacturing and Assembly of the Prototype

The prototype was assembled following the design described above. Normalized 45 × 45 aluminium profiles make up the main structure. Although they are not the optimal solution in weight terms, their use allowed for the realization of quick and small modifications and/or adjustments to correctly fit the rest of the parts. The main frame and auxiliary wheel mechanisms of the prototype are assembled together, as shown in Figure 9. The insulation cover was manufactured at Universidad de Alcalá workshop using 1-mm laminated stainless steel for the coating and 40-mm-thick rock wool as internal insulation material. The thermal isolation cover is mounted above the frame, and the lateral stops are adjusted in order to keep the roll within the trolley and to limit the degrees of freedom when unrolling. Five torches are mounted and equally spaced in the front, pointing toward the auxiliary wheel, as shown in Figure 9. The gas flow pipelines, flow and pressure regulators are connected and assembled on the trolley.

Figure 9. Real prototype assembled.

The overlap compaction drums are connected to the main frame through a bi-articulated connecting rod, as shown in detail in Figure 10. Drums are 10 cm wide and they are aligned in order to press the 10-cm-wide area required for the overlaps. Each compaction drum is constructed using four 2-kg weights jointed by a connecting axle. This axle can rotate around the linking rod, which can also rotate around the main frame. This allows for the continuous vertical pressing of the weights against the ground, compacting the asphalt overlaps with the subsequent asphalt roll.

Figure 10. Compaction drum at the bottom of the trolley.

The gas circuit scheme is depicted in Figure 11. The gas circuit starts from the propane bottle, which is linked through a free output regulator to a manual pressure regulator. The propane bottle provides gas at a pressure between 2 and 6 bar. During tests, the propane bottles provided 5 bar to the manual pressure regulator, as shown in Figure 12. The manual pressure regulator allows us to control and regulate the pressure reaching the torches. In the diagram shown in Figure 12, the output pressure was limited to 2.8 bar. Once the pressure is regulated, the flow can be controlled and modulated through an on/off valve which also has a second flow regulator. In the case of the prototype, this second flow regulator was fully open and so the activation or deactivation of the torches could be achieved using this on/off valve, which is also shown in Figure 12.

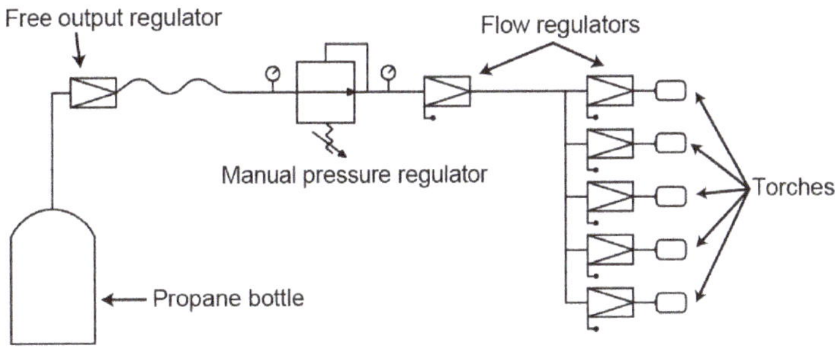

Figure 11. Schematics of gas circuit.

Figure 12. Manual pressure regulator and on/off valve.

The main pipeline is connected to the manual pressure regulator by a 1/4″ Left Handed (LH) nut connection. The pressure regulator is linked to the on/off valve with a shortcut using a 3/8″ LH nut attached to a 3/8″ LH screw. The on/off valve is linked to the torch pipeline using a 3/8″ LH nut connection.

The pipeline distributes the gas flow to the five torches through a T-type 3/8″ LH connector. Each torch has its own flow regulator, as shown in Figure 13. This allows for individualized heat flow adjustment. This can be useful for increasing heat flow in the two torches that heat the overlaps, since these areas are critical to ensure the impermeability of the roof. Torches are mounted in parallel at the distance and height depicted in the dimensional drawings in Figure 2.

Figure 13. Individual flow regulators and parallel torches.

In order to initiate the installation, the roll must be placed inside the trolley, in contact with the auxiliary wheels and between the lateral stops. The steps for placing the roll are to leave the roll on the ground and to front-lift the trolley using the rear wheels. The trolley has a little aluminium profile-made pedal, shown in Figure 12 next to the feet, which helps to lift the trolley when the worker steps on it. Then, the trolley is pushed against the roll to ensure contact against the auxiliary wheels, as shown in Figure 14. The asphalt roll has to be prepared and located in the proper position in order to overlap any previously installed asphalt. Furthermore, the surface has to be prepared before the installation. For this, we applied ChovASTAR®waterproofing surfaces preparer on the surface, which is a liquid product formed by resins and bitumen in aqueous solution. This product drastically improves the adherence between the ground and the asphalt roll. Once the preparer is applied and the roll is in position, installation can start.

Figure 14. Prototype and surface ready for asphalt roofing.

4. Test Results and Discussion

The test campaign is planned and executed in order to determine the real capacity, speed and equipment performance after installation. Preliminary laboratory tests are used to validate some considered assumptions during the design phase, such as inlet input speed.

4.1. Preliminary Laboratory Test

A preliminary torch flow analysis test was conducted in order to estimate the hot air flow temperature and speed. In addition, the propane consumption was measured during this test. We supplied propane to the torches by using two propane bottles, delivering the gas at a pressure of 5 bar. The pressure that reached the five torches was limited to 2.8 bar, with a consumption of 2800 g/h per torch. We activated the five torches at their maximum flow capacity; flow regulators were fully open.

The air-fed propane flame is shown in Figure 15. Air enters to the torch from the bottom, where there are five holes 8 mm in diameter. Air is mixed with propane inside the torch. Torches must be manually ignited. Once the ignition occurs, the flow can be controlled individually at each torch. Figure 15 (top) shows the different zones of the flame shape. A blue flame means the complete combustion of the gas. Pure hydrocarbons like methane (refined natural gas), propane, butane and ethane gases also burn with a blue flame. The flame turns blue at the output of the torch, which means that air reaches a temperature between 1800–1980 °C for propane [14]. In the second zone of the flame, which is yellow, temperatures are lower, at 500–750 °C. This is the temperature of the air when it encounters the roll. There is a zone where the combustion becomes incomplete due to the absence of gas.

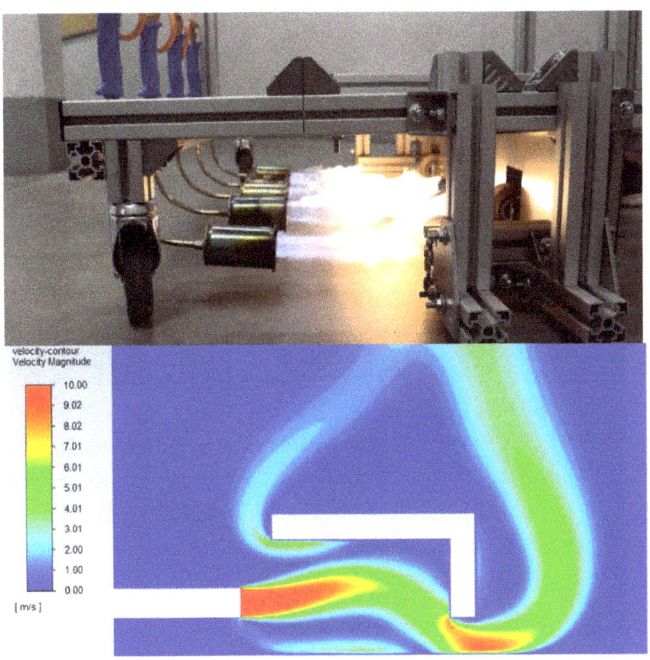

Figure 15. Flow and consumption test of the five torches, photograph and validation CFD analysis—speed results.

Air speed has been estimated at around 10 m/s by using the CFD 2D model. In this case, the CFD model did not include the roll. CFD stationary simulations return flame shapes similar to the one we observed in the test. It is interesting to notice that the cross section of the flow is slightly smaller than the torch diameter and it is oriented to the upper part. This is due to the lower density of the outlet air already present in the torch. Thus, the CFD inlet temperature and air speed conditions can be considered reasonable, although it is important to note that CFD considers air as fluid, which is not exactly the case in reality, because combustion products do not have the same properties as air.

4.2. Outdoor Asphalt Temperature and Adhesion Test

Once the temperature and speed of the hot air flow have been determined in the laboratory, the outdoor installation tests must begin. Tests were carried out on the rough roof of an individual parking structure located in Torres de la Alameda (Spain). Several rolls and the prototype were transported to the location. The preparer was applied and the roll was put in position. Then, the torches were activated using a manual lighter. One person was in charge of moving the trolley and activating the on/off valve in order to apply the heat. Thermal images of the roll were taken using an infrared Fluke Ti32 thermal camera.

Figure 16 shows the torches in operation at a pressure of 2.8 bar and the temperature achieved on the roll surface after 10 s of application. It can be observed that almost half of the roll surface was above 200 °C, which is high enough for adhesion. The temperature was even higher on most of the surface (white areas), while it was slightly lower in the fringes created between the flows of consecutive torches. The total asphalt distance ready for adhesion after 10 s of heating was approximately 0.314 m, which equates to a 0.0314 m/s or 1.8 m/min targeted speed for the equipment.

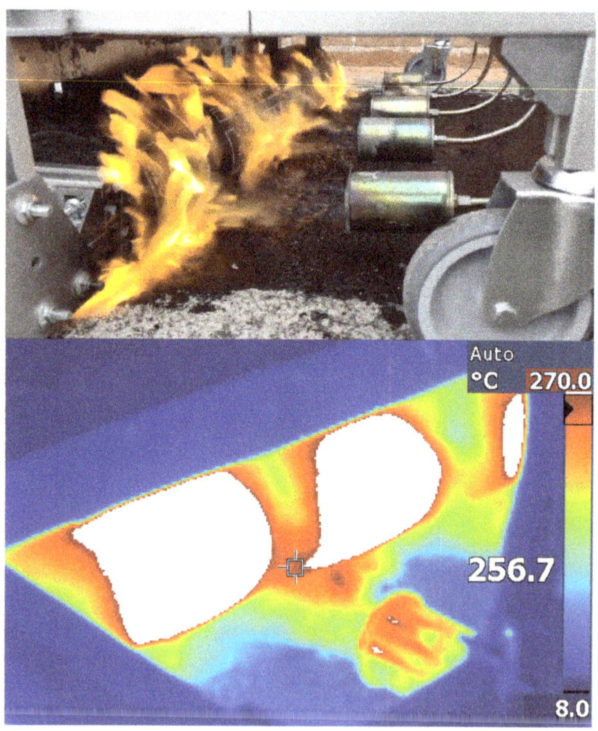

Figure 16. Torches applying heat to the roll and corresponding infrared thermal image after ten seconds of application.

It is also necessary to analyse the temperature of the layer when it goes out the heating and how is its adherence with the ground. The trolley has to be moved in a step-by-step way, taking care that all the areas are heated.

The adhesion results, after the heating phase, are shown in Figure 17. The asphalt is clearly softer, which indicates a good adhesion at the bottom. The exerted pressure by the compaction drums is shown in the overlaps, where the grooves from the weights can be clearly seen. The thermal picture shows the temperature of the layer when it is adhered, in the range from 90 to 40 °C. We must consider that the surface shown in the figure is not the surface that receives the heat, but is the bottom face of the layer. Temperature differences between along the roll area appear because the heat application is not entirely uniform. There are spaces between torches that affect the uniformity of the heat distribution. This is something that could be improved by adding more torches or using wider diffusors on each torch. In any case, we believe that the layer is correctly installed.

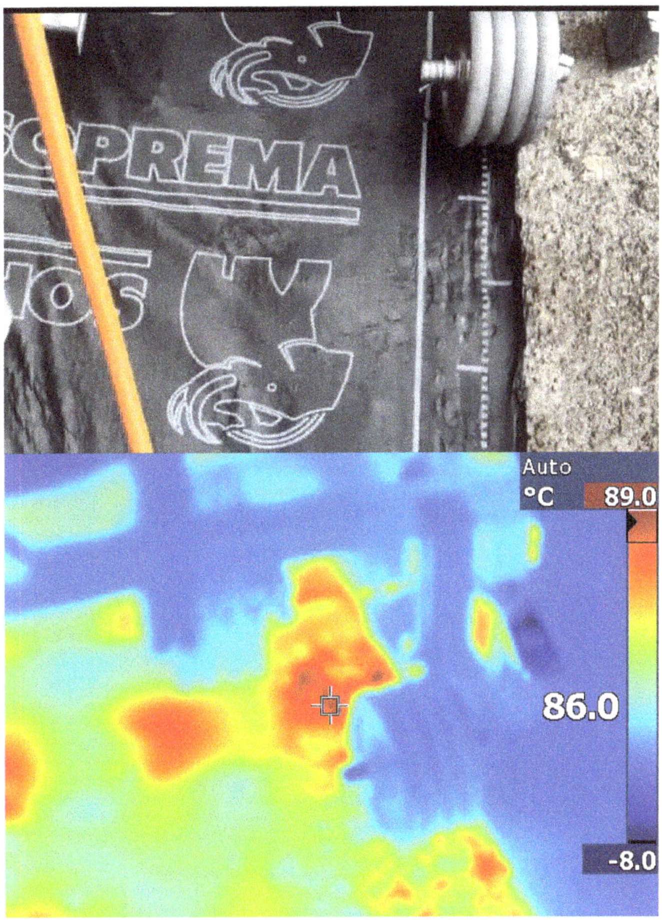

Figure 17. Detail of the adhered layer in the overlaps areas and corresponding thermal image.

From this experience, we determined that a proper method for the installation of the roll consists on, first, to sustain heating at maximum capacity, and then to move the trolley. More specifically, the operation is to activate the on/off valve for 10 s, and then move the trolley forward by approximately 0.3 m.

4.3. Outdoor Installation Speed Test

Following the defined installation method, we performed the installation of several adjacent rolls, as shown in Figure 18. During the continuous installation, we realized several problems that may limit the actual installation speed. One important issue is the fact that, as the roll gets installed and unrolled, the distance between the torches and the roll increases, negatively affecting the amount of heat transferred. This implied that we must move slower towards the end of the installation of the roll. Another significant issue appeared when wind started to blow. This significantly reoriented the torches' hot air flow, limiting the heat transfer. Adding some lateral walls may reduce the effect of external wind on the flows. Nevertheless, we reached an installation speed of approximately 1.75 m^2/min with a torch pressure of 2.8 bar and a corresponding consumption of 2800 g/h of propane per torch.

Figure 18. Asphalt rolls installation results.

4.4. Outdoor Installation Quality Test

The rolls obtained good adhesion on most of the surface. In addition, the overlaps, which must be completely adhered to each other in order to ensure isolation, were adhered perfectly, one over the other, assuring the waterproof properties of the roofing, as shown in Figure 19. We observed that the compression by the compaction drums was really effective and, in these areas, the adhesion was better than in the middle, where no pressing besides the worker's steps occurred. A potential improvement could be to add extra compaction drums all along the trolley. This will surely improve the adhesion in the middle, although the preparation of the trolley would be more difficult. In any case, the driving of the trolley would not be much more difficult, since the weight of the drums is not supported by the trolley.

Figure 19. Quality test at overlaps.

5. Lessons Learned and Suggested Improvements

Although the main objective of the development—installation speed—was achieved, we detected several elements that require improvement. These improvements would lead to faster installation speeds and/or lower consumption. Next, we itemized all weak points and provided suggestion for their improvement:

- The differences in temperature between the adjacent areas show that heat was not entirely uniformly distributed, as the spaces between torches affects the heat distribution. This is something that could be improved by adding more torches or using wider diffusors on each torch;
- As the roll gets installed and unrolled, the distance between the torches and the roll increases, negatively affecting the amount of heat transferred A simple cam-follower mechanism could be integrated between the torches and the roll in order to maintain the distance;
- Heat is lost through the lateral sides of the trolley. A special metallic lateral wall could prevent this loss in the same way that the thermal insulation cover prevents vertical heat losses;
- The lateral stops for the roll are very close to the ground level, and may collide with imperfectly flat areas of ground or surfaces. By elevating the lateral stops by 30–40 mm, these collisions could be avoided;
- If the roll is not perfectly aligned in the lateral sides it is hard to place, since the lateral margins are very small. The lateral stops should have a spring follower mechanism to allow the roll to be positioned in an easier way;
- The hot air flow becomes turbulent around the roll. Future CFD analysis should consider this;

- It is sometimes hard to slightly change the roll direction. By enabling the rotation of the rear wheels, the guidance of the trolley could be easier;
- Increasing the flow in the lateral torches and reducing it in the middle may reduce total consumption;
- Adding extra compaction drums all along the trolley will surely improve the adhesion in the middle without affecting the driving of the trolley;
- We detected slight differences in the flow, and thus on the applied heat, of the torch that was closer to the main pipeline in relation to the one at the other side. This may be due to differences in pressure decay along the pipeline. This difference could be avoided by adjusting the lateral torches and/or by connecting the main pipeline at the middle rather than at one side.

6. Conclusions

In this work, we have described the design and testing of a new piece of equipment, which was developed to enhance speed, gas consumption and safety during the manual asphalt roofing process. The novelty of the equipment is sustained by the use of a set of five parallel gas burners located in front of the roll to maximize the heat transfer. The equipment is light and practical for use by any worker on any type of roof. It also includes a thermal insulation cover to significantly reduce the gas consumption and, thus, to reduce CO_2, SO_2, and other non-eco-friendly emissions. We have shown the mechanical and thermal design and analysis of the equipment, CFD simulations for the heat transfer calculations, a description of the manufacturing and assembly, a preliminary thermal test, and an operational test.

The equipment demonstrates an installation speed of 1.75 m^2/min for 3 kg/m^2 rolls, which leads to around 700–735 m^2 per person per day, more than twice the usual manual roofing rate. Nevertheless, some issues need to be resolved, such as the nonuniform heat distribution or the low heat transfer at the end of the roll installation. Finally, a complete list of weak points and suggestions for improvement was given.

Author Contributions: Conceptualization, E.D.-J. and A.B.-G.; methodology, E.D.-J. and A.B.-G.; mechanical design, A.B.-G. and M.F.-M.; manufacturing and assembly, A.B.-G. and E.D.-J.; simulation, A.B.-G.; validation test, A.B.-G., E.D.-J., and M.F.-M.; formal analysis, A.B.-G.; literature review, E.D.-J.; data curation, E.D.-J. and A.B.-G.; writing—original draft, A.B.-G. and E.D.-J.; writing—review and editing, E.D.-J. and M.F.-M.; supervision, E.D.-J.; project administration, E.D.-J.; funding acquisition, E.D.-J. and A.B.-G. All authors have read and agreed to the published version of the manuscript.

Funding: The research leading to these results was partially funded by Ingeniería Industrial Ibérica S.A. and Cubiertas ALMADI 2017 S.L.

Acknowledgments: The authors wish to recognize the work of Alba Martínez Pérez during preparation of the figures and 3D pictures.

Conflicts of Interest: The authors declare no conflict of interest.

References

1. Abuseif, M.; Gou, Z. A review of roofing methods: Construction features, heat reduction, payback period and climatic responsiveness. *Energies* **2018**, *11*, 3196. [CrossRef]
2. Van Toor, A.C.; Bax, N.X.C. Method and Apparatus for Applying a Bituminous Sheet to a Substrate. Patent EP 0466249, 15 January 1993.
3. Warren, A. Single Ply Roofing Applicator. U.S. Patent 4725328, 16 February 1988.
4. Bessette, R. Membrane Applying Apparatus. U.S. Patent 2013228287, 5 September 2013.
5. Unify-er-Produits. Available online: www.modbitapplicator.com/produits.php (accessed on 4 November 2019).
6. Trumbore, D.; Jankousky, A.; Hockman, E.L.; Sanders, R.; Calkin, J.; Szczepanik, S.; Owens, R. Emission factors for asphalt-related emissions in roofing manufacturing. *Environ. Prog.* **2005**, *24*, 268–278. [CrossRef]
7. Vaz, W.; Sheffield, J. Preliminary assessment of greenhouse gas emissions for atactic polypropylene (APP) modified asphalt membrane roofs. *Build. Environ.* **2014**, *78*, 95–102. [CrossRef]

8. EL-Mesery, H.S.; Abomohra, A.E.-F.; Kang, C.-U.; Cheon, J.-K.; Basak, B.; Jeon, B.-H. Evaluation of Infrared Radiation Combined with Hot Air Convection for Energy-Efficient Drying of Biomass. *Energies* **2019**, *12*, 2818. [CrossRef]
9. Young-Corbett, D.E. Prevention through Design: Health Hazards in Asphalt Roofing. *J. Constr. Eng. Manag.* **2014**, *140*, 06014007. [CrossRef]
10. Díez-Jiménez, E.; Vidal-Sánchez, A.; Barragán-García, A.; Fernández-Muñoz, M.; Mallol-Poyato, R. Lightweight equipment for the fast installation of asphalt roofing based on infrared heaters. *Energies* **2019**, *12*, 4253. [CrossRef]
11. Díez-Jiménez, E.; Vidal-Sánchez, A.; Corral-Abad, E.; Gómez-García, M.J. Mecanismo Ligero para la Puesta Rápida de Láminas Bituminosas en Impermeabilizaciones de Cubiertas Planas. Patent P 201830702, 13 July 2018.
12. Alade, O.; Al Shehri, D.; Mahmoud, M.; Sasaki, K. Viscosity-temperature-pressure relationship of extra-heavy oil (bitumen): Empirical modelling versus artificial neural network (ANN). *Energies* **2019**, *12*, 2390. [CrossRef]
13. Ministerio de Industria, Gobierno de España. *Código Técnico de la Edificación*, 1st ed.; Garceta Grupo Editorial: Madrid, Spain, 2009; ISBN 9788493720896.
14. Haynes, W.M. *CRC Handbook of Chemistry and Physics;* Apple Academic Press: Palm Bay, FL, USA, 2014; ISBN 9781482208672.
15. Musavi, Z.; Kusar, H.; Andersson, R.; Engvall, K. Modelling and Optimization of a Small Diesel Burner for Mobile Applications. *Energies* **2018**, *11*, 2904. [CrossRef]

© 2020 by the authors. Licensee MDPI, Basel, Switzerland. This article is an open access article distributed under the terms and conditions of the Creative Commons Attribution (CC BY) license (http://creativecommons.org/licenses/by/4.0/).

energies

MDPI

Review

Modern Small and Microcogeneration Systems—A Review

Marcin Wołowicz [1,*], Piotr Kolasiński [2] and Krzysztof Badyda [1]

1 Faculty of Power and Aeronautical Engineering, Institute of Heat Engineering, Warsaw University of Technology, 00-661 Warsaw, Poland; krzysztof.badyda@pw.edu.pl
2 Department of Thermodynamics and Renewable Energy Sources, Faculty of Mechanical and Power Engineering, Wrocław University of Science and Technology, 50-370 Wrocław, Poland; piotr.kolasinski@pwr.edu.pl
* Correspondence: marcin.wolowicz@pw.edu.pl

Abstract: Small and micro energy sources are becoming increasingly important in the current environmental conditions. Especially, the production of electricity and heat in so-called cogeneration systems allows for significant primary energy savings thanks to their high generation efficiency (up to 90%). This article provides an overview of the currently used and developed technologies applied in small and micro cogeneration systems i.e., Stirling engines, gas and steam microturbines, various types of volumetric expanders (vane, lobe, screw, piston, Wankel, gerotor) and fuel cells. Their basic features, power ranges and examples of implemented installations based on these technologies are presented in this paper.

Keywords: microturbine; stirling engine; fuel cell; expander; vane; lobe; screw; piston; Wankel; gerotor; microcogeneration; CHP

check for updates

Citation: Wołowicz, M.; Kolasiński, P.; Badyda, K. Modern Small and Microcogeneration Systems—A Review. *Energies* **2021**, *14*, 785. https://doi.org/10.3390/en14030785

Received: 20 December 2020
Accepted: 20 January 2021
Published: 2 February 2021

Publisher's Note: MDPI stays neutral with regard to jurisdictional claims in published maps and institutional affiliations.

Copyright: © 2021 by the authors. Licensee MDPI, Basel, Switzerland. This article is an open access article distributed under the terms and conditions of the Creative Commons Attribution (CC BY) license (https://creativecommons.org/licenses/by/4.0/).

1. Introduction

For several decades the continuous energy consumption growth has been observed globally. This trend is mainly caused by the growing energy needs of an increasing world population, life quality improvement and technological development. The social access to different energy receivers (e.g., household appliances, cars, electronic devices, etc.) and their availability is nowadays easier than before, which directly translates into increasing consumption of fuels, heat, electricity and chill. Research on the energy consumption forecasting and modelling its variations is proceeding by different agencies and researchers around the world [1]. Reported data show [1] that the global primary energy consumption (i.e., the energy contained in fuels and renewable energy sources which is then converted into electricity, heating and sanitary heat and chill) in 2020 was ca. 1.58×10^8 GWh. The forecast for 2050 [1] indicates that the global primary energy consumption will probably increase nearly by 50% up to ca. 2.64×10^8 GWh. It is expected that ca. 28% of this energy demand will be covered by renewable energy sources, 27% by petroleum products and other liquid fuels (including biofuels), 22% by natural gas, 29% by coal and 4% by nuclear energy. Reliable and highly efficient energy conversion devices and systems (additionally characterized by low emission of harmful substances into the environment) have to be applied to meet this constantly growing energy demand and at the same time fulfil the strict regulations related to the natural environment protection. Therefore, the research, design and optimization activities related to modern energy conversion systems should be focused on limiting the fossil fuels consumption and increasing the use of alternative energy sources or clean fuels (e.g., natural gas). Energy conversion systems can be classified by different criteria (e.g., by the operating principle, design, cost, etc.). One of the commonly applied classification criteria is the system power output. By this criterion, systems can be classified into large-power (1.5 MW and more), medium-power (500 kW–1.5 MW), small-power (15 kW–500 kW) and micro power (up to 15 kW). Large- and medium-power energy conversion systems (such as e.g., large steam power plants) are usually highly efficient and used for industrial energy generation (they are supplying cities, regions or countries). Small- and micro-power systems are mainly used in distributed energy systems or by individual prosumers. Currently, different research works are carried out on the development of modern energy conversion technologies. In the field of large-power energy conversion technologies new solutions (such as the International Thermonuclear Experimental Reactor, i.e., ITER [2]) are investigated. In addition to these emerging technologies, research is still proceeding on the design optimization and improving the conversion efficiency of traditionally used units (e.g., steam power plants [3] and combined cycle gas turbine, i.e., CCGT units [4]). In addition to activities aimed at large-power systems, important research works are proceeded on small- and micro-power units dedicated for application in distributed energy systems. For many years, one of the visible development trends in modern power sector has been pursuing the diversification of the energy systems [5–7] by supporting activities related to the implementation of energy clusters [8] and small energy conversion units. Therefore, much attention is now paid to research and development works on small and micro-power systems which can be used by individual energy recipients (e.g., apartments, houses, shops or small industrial plants) for covering their own energy needs. Nowadays special focus is paid on cogeneration systems. Cogeneration, i.e., Combined Heat and Power (CHP), is an important tool for achieving significant fuel savings and reduction of CO_2 emissions. Compared to separate fossil fuel-based electricity and heat generation, fuel savings and corresponding CO_2 reductions will often be in the order of 30%. The CO_2 reductions can sometimes be higher as a fuel conversion, e.g., from coal or oil to natural gas or biomass, can be taken into account. These benefits have been recognized by the European Commission and several initiatives have been taken to increase the use of cogeneration in the generation of heat and electricity.

Cogeneration is commonly used in large generating units-combined heat and power plants. However, there is a noticeable trend towards the use of cogeneration in smaller systems, especially those designed for local and distributed applications. Currently, there are several technologies used in micro-cogeneration such as small gas turbines, small steam turbines, Stirling engines, organic Rankine cycle systems (ORC systems) and fuel cells. Technological progress [9–12], as well as the general trend towards smaller generating units, resulted in an increased interest in small cogeneration units, hoping that they would be able to efficiently supply electricity and heat to individual facilities [13,14]. Small heat and power plants are usually based on internal combustion gas engines and are fully adapted to the needs of customers. They are characterized by high efficiency, thus providing cheaper energy, and they also allow for long-term cost planning. It should also be noted that such systems meet the strictest European environmental standards. The EU Cogeneration Directive defines micro-cogeneration as a unit featuring a maximum power of less than 50 kWe, while in Germany micro-cogeneration systems are treated as those that feature a power below 15 kWe. This is due to the following reasons: these systems are clearly intended for use in single family homes, apartment buildings, small businesses or hotels [15,16]. In the following part of this paper modern small and microcogeneration systems are reviewed, with special focus on the technologies utilizing energy sources of high-quality. The features and operating parameters of these systems are discussed together with their possible applications.

2. Stirling Engines

There has recently been an interest in using Stirling engines as electricity generators in domestic CHP systems. Many households, especially in Europe, use natural gas for heating. The natural gas burner is easily customizable to provide a heat source for a small 1 to 10 kW Stirling engine, sized to meet the electricity needs of a typical household or a small business unit. The engine then forms a part of a system in which the waste heat that first heats the Stirling cylinder head is then directed to domestic hot water distribution system and/or central heating system. Several commercial systems based on this type of engine are already commercially available.

2.1. Historical Background

The Stirling engine was invented in 1816 by Robert Stirling in Scotland, some 80 years before the invention of the diesel engine, and enjoyed considerable commercial success until the early 20th century.

Before the Stirling engine was created, however, attempts were made to develop a hot air engine. The first hot air engine was the atmospheric fire-mill. It was an air engine in which heat was transformed in mechanical power. Its inventor was Guillaume Amontons [17–19], a Frenchman, who invented it in 1669. The engine proposed by Guillaume was a pistonless engine, and the air was heated and cooled in a closed circuit. Heat was supplied from external source (i.e., fossil fuel combustion chamber). The drawing of Guillaume engine is presented in Figure 1.

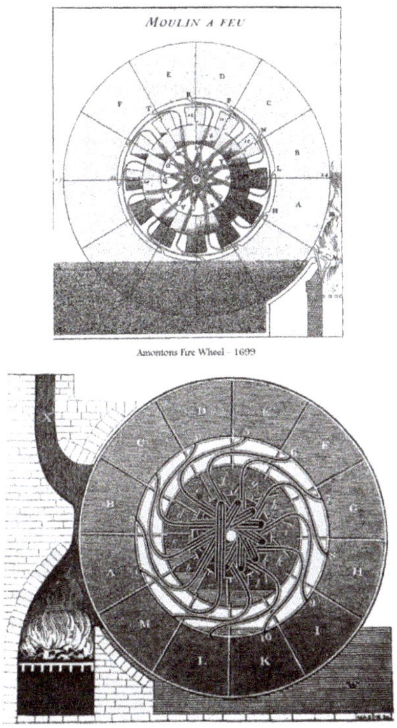

Figure 1. View on the Guillaume Amontons hot air engine [20].

Another example of a hot air engine is the engine proposed by Sir George Cayley [21,22]. It is considered to be the first hot air engine that worked successfully. Cayley has shown great ingenuity in overcoming the practical difficulties of high operating temperatures. This type of engine was one in which the fire is shielded and fed with air forced under the grate in an amount sufficient to maintain combustion, while by far the largest part of the air goes above the fire to be heated and expanded; the air with the combustion products then acts on the piston and passes through the operating cylinder, no metal heating surface is required, the heated air is brought into direct contact with the fire. One of these engines worked for many months for testing. It was better than any design of steam engine known at the time in terms of fuel economy compared to the power output. However, the joints were very troublesome, and the cylinder and piston seal were quickly destroyed by dust and gravel particles from the fuel, which acted as abrasive and prevented lubrication. An attempt was made to filter the air before entering the cylinder with sheets of wire mesh, but these either subsided or were soon choked and rendered useless [20]. The drawing of the Cayley engine is presented in Figure 2.

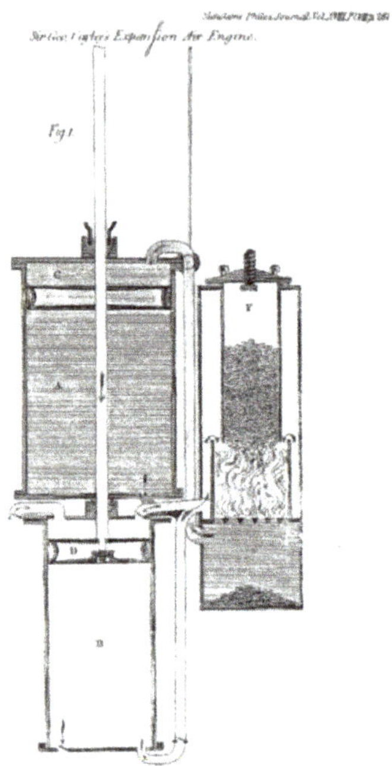

Figure 2. View on the Cayley hot air engine [20,23].

The Stirling engine was invented and patented in 1816 by Robert Stirling. It was originally used in Scotland, Ayrshire, in 1818 to drive a water pump. Unfortunately, due to the shortcomings of the materials at that time, the engine only worked for two years and was then replaced by a steam engine. In later years, Robert Stirling and his brother James improved the design by, among other changes, adding a second piston. The end result of these works was that the engine obtained a higher efficiency than steam engine, but unfortunately there were still problems with the materials, which caused users to return to steam engines. A view of Stirling's engine patent is presented in Figure 3.

Figure 3. View of Stirling's engine patent drawing, 1816 [24].

Until the First World War, many other interesting concepts of using a hot air engine, e.g., for powering aircrafts, were created, but they were not widely used. These engines were mainly used in small workshops and for driving water pumps because, unlike a steam engine, they did not require a qualified engineer to operate them. Unfortunately, their power and efficiency remained low compared to their size [25–27].

The renaissance of interest in the Stirling engine took place in the interwar period thanks to the concern of the Dutch Philips, who was looking for a simple, light engine for powering a radio. Thanks to the invention of steel resistant to high temperatures after the First World War and its application in the Stirling engine, it was possible to reduce the failure rate. At that time, Stirling engines sometimes had an efficiency of less than 1% (while theoretically they could achieve an efficiency of 60%). Professor Holst believed in this possibility when he started his research on the Stirling engine. Unfortunately, the German occupation during the Second World War significantly inhibited this research. However, in 1947, after ten years of development, a small 30-horsepower engine was presented, featuring a rotational speed of 3000 rpm and efficiency similar to internal combustion engine. Another result of this work was a small engine that ran for over 2000 h without any visible damage. The important direction of research on Stirling machine design was also the Stirling cooler based on reversed engine cycle, which application gave the opportunity of obtaining of very low temperatures. Initially, the obtained temperature was around −190 °C, but in later years even the temperature of −260 °C was achieved. However, the main goal of Philips, who was creating a small power source, has not been achieved despite the use of solutions such as the diamond-shaped mechanism [28].

In late 1950s, Philips engines achieved an efficiency of about 38%, which was higher than the efficiency achieved by the gasoline and diesel engines. In 1958, the automotive concern General Motors was interested in these works, looking for a new type of propulsion in the automotive industry, as well as for powering generators and submarines. The result was a developed prototype of 150 hp Rinia engine [29]. However General Motors senior management decided to abandon the program before the engine was put into production. The Stirling engine concept was revisited at the turn of the 1960s and 1970s, mainly due to the rising oil prices. The first prototypes of buses powered by Stirling engines were then created, but putting this type of bus into production after the end of the oil crisis was found unprofitable. As a curiosity, it is worthy to mention that Stirling engines have had a significant impact on the design of today's conventional submarines. Thanks to application of the Stirling engine, it was possible to extend the time they can be submerged. The first class of submarines in which the Stirling engine is applied for propulsion is the Swedish project A-19 Gotland, equipped with two Stirling engines. Thanks to their application, the ship can stay fully submerged for 2 weeks and travel at a speed of 5 knots. The efficiency of these engines is ca. 40%. Nowadays, in addition to the above-mentioned applications, the possibility of using Stirling engines in thermal energy and renewable energy for electricity production is also considered [30].

Some engine companies, such as MAN-MWM, United Stirling of Sweden, and Ford Motor Company of Detroit, have started research programs to develop Stirling engines for automotive applications. To date, few multi-cylinder Stirling engines featuring different power ranges have been prototyped by these companies. The thermal efficiency of Stirling engines designed for automotive applications is higher than 40%. The main design and application problem that needs to be solved in the future is the high weight and large size of such an engine [31–33].

2.2. Functional Description

A Stirling engine [27,34,35] is an external combustion reciprocating engine that uses one or more pistons to achieve useful operation by supplying heat from an external source. They differ significantly from the internal combustion engines found in most vehicles. Stirling engines use the same gas during operation, as opposed to internal combustion engines which constantly take in and discharge gas (they take the air-fuel mixture and discharge the exhaust gases). In addition, Stirling engines do not use the combustion effect inside the chamber as is the case of conventional internal combustion engines, which makes their operation very quiet.

A key unique feature of Stirling engines is that there is a constant amount of gas inside the cylinders. The gas pressure can be regulated by supplying or receiving heat, i.e., thanks to the changes in gas volume. Thanks to the supplied heat pressure and increasing temperature of the gas contained in the cylinders, on the contrary, removing heat form cylinders reduces pressure and temperature. By changing the way these two processes are performed engine deliver useful work. The engine operates according to the "Stirling cycle" described below [28,35]. The operation cycle is organized as follows:

- heating and expansion-heat is supplied from an external source, raising the gas temperature and thus the pressure. This causes the piston to slide and provides useful work;
- flow and cooling-the piston moves to force gas into another cylinder where it is cooled. Cooling the gas allows for easier compression, meaning less work is required to proceed this process than in step 1;
- compression-the gas is compressed and the excess heat resulting from compression is removed via the cooling source;
- reverse flow and heating-the pressurized gas returns to the starting cylinder and the cycle repeats.

The principle of the Stirling engine operation and its basic components are shown in Figure 4.

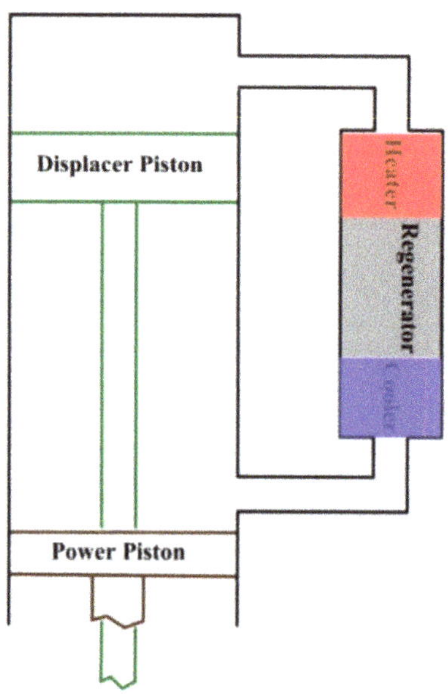

Figure 4. Example schematic diagram of a free-piston Stirling cycle engine with a linear alternator for energy extraction [36].

Heat transport in the Stirling engine is realized by a displacer piston. There is more gas in the working chamber on the hot side of the engine when the displacer piston is closer to the bottom dead center. When the heat is supplied to cylinder from the heat source the gas pressure inside the engine also increases. When the displacer is closer to the top dead center, the gas pressure drops and most of the gas is cooled. An additional effect is the movement of the power piston which interacts directly with the displacement element. By balancing the area and masses of the pistons, the dynamics of the pistons movement, and the restriction of mass flow from one side of piston to the other, a self-sustained cycle can be achieved to convert the heat absorbed by the engine into useful work [36].

To increase efficiency, most Stirling engines use a regenerative heat exchanger, simply referred as to "regenerator". A regenerator works like a thermal condenser in which heat is absorbed and released from the gas as it passes from one cylinder to another. This heat transfer takes place cyclically. This cyclicality corresponds to the operating frequency of the engine. The regenerator is clearly visible in Figure 5, which shows a block diagram of the operation of a Stirling engine [36–38].

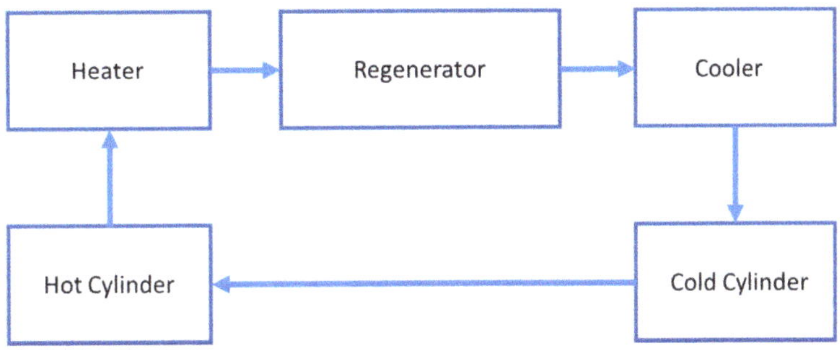

Figure 5. Block diagram showing working fluid flow in the Stirling engine and its main components.

2.3. Stirling Engine Designs

Stirling engines can have many different designs. Three classic designs are described in the following. An alpha-type engine has two cylinders in which two pistons move. Pistons are loaded on the one hand with a variable pressure of the working gas, and on the other hand with a constant gas pressure prevailing in the so-called buffer space. The phase shift (between 85° and 120°) of the pistons is required. It is important that when assembling this type of engine, the structural and kinematic elements of the working mechanism and engine block are so arranged that the piston working in the hot cylinder is shifted in phase ahead to the piston moving in the cold cylinder. More details about the alpha-type Stirling engine can be found in [37,39–46].

The beta-type design [47–54] has one cylinder in which two pistons move coaxially with the required phase shift. The upper piston moves in the cylinder, forcing the gas twice in circulation between the compression and expansion chamber through a set of heat exchangers. As a result, the engine is only loaded by the pressure difference resulting from the gas flow through the heat exchangers and through the resistances. The group of heat exchangers is connected to the compression and expansion chamber, and the buffer space is located under the piston. The movement of the lower piston, which is phase-lagged, compresses and decompresses the gas.

The gamma-type design is the simplest and easiest to manufacture of the Stirling engines. Similar to the beta design, the gamma design has two cylinders (sometimes the diameter of one cylinder is larger and diameter of the other is smaller). Cylinders are connected by channels to the built-in set of heat exchangers. A more detailed description of the gamma-type Stirling engine's application, as well as results of its operation modeling is presented in [55–64]. Starting this type of engine can be proceeded with smaller heat input, compared to the alpha and beta designs. Schemes of alfa, beta and gamma Stirling engines are presented in Figure 6.

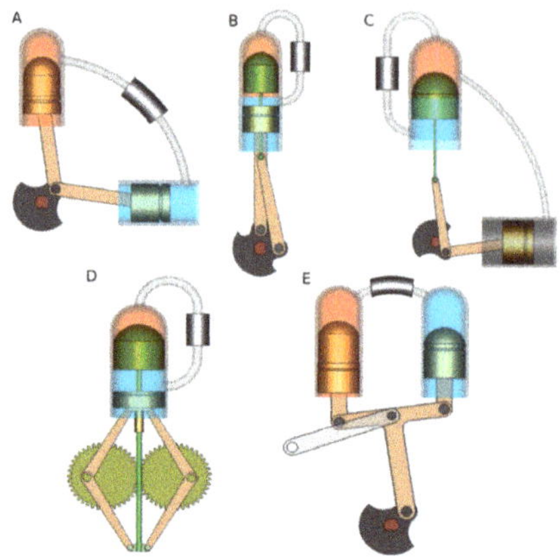

Figure 6. Alpha, beta and gamma configuration of a Stirling engine. (**A**) alpha configuration with crank drive; (**B**) beta configuration with crank drive; (**C**) gamma configuration with crank drive; (**D**) beta configuration with a rhombic drive replacing the crank drive; (**E**) alpha configuration with Ross yoke instead of a crank drive [65].

2.4. Stirling Engine Applications

At the beginning of the 19th century, as a result of the rapid development of internal combustion engines and electrical machines, the further development of Stirling engines was severely impeded. However, due to the high thermal efficiency, quiet operation and the ability of Stirling engines to use multiple fuels, it meets today's requirements related to energy efficiency and environmental protection. Cogeneration units based on a Stirling engine are considered to be one of the best among the low power range electricity generation units.

Stirling engines are built in a fairly large power range. These units can have a power of a few watts to over 1 MW, but the most popular are those for single kW. Stirling engines are a viable alternative to currently used heat engines. The main barrier to the development of this technology is the presence on the market of competitive solutions in the form of well-developed classic technologies.

Microgen is a leader in the production of small Stirling engines. It offers engines with a capacity of several kilowatts. An example of a Stirling engine manufactured by Microgen is shown in Figure 7.

Figure 7. View on the Stirling engine made by Microgen [66,67].

Stirling engines are used in micro-cogeneration mainly due to the nature of their construction, i.e., their operation requires an upper and lower heat source [67–75]. However, the source from which this heat comes is not significant. This means that these engines can operate with virtually any fuel. The most popular microcogeneration systems found today are those based on gas fuel.

An example of such systems based on gas fuel are the systems of the German company Viessmann. These systems are known under trade names Vitotwin 350-F and Vitotwin 300-W. Their view is shown in Figure 8.

Figure 8. Combined Heat and Power (CHP) system based on a Stirling engine made by Viessmann [76,77]. 1—peak gas burner, 2—stainless steel heat exchanger, 3—valve dosing air supplied to the burner, 4—ring gas burner to feed the Stirling engine, 5—Stirling engine, 6—control panel, 7—hot water buffer tank, 8—vessel for pressure equalization.

Viessmann cogeneration modules can be used in almost any facility, e.g., in residential houses, office buildings, industrial plants and local district heating networks.

The Vitotwin 300-W micro cogeneration system with an electrical output of 1 kW and a thermal output of 26 kW is a real alternative to conventional heating systems in single- and two-family houses. It works particularly efficiently with an annual gas consumption of at least 20,000 kWh and an electricity consumption of over 3000 kWh. As heat is constantly generated during operation, a combination with a heating water buffer cylinder is required. The storage tank can be installed in devices that only need 0.36 m^2 of floor space to be installed, such as the Vitotwin 350-F.

Compact micro-cogeneration systems with an integrated condensing boiler can be a self-sufficient source of heat and electricity for a household. The Stirling engine in the Vitotwin can work with power modulation in the range of 0.3 to 1 kW of electrical power and requires virtually no maintenance. The electricity generated this way covers the basic demand of the building, correspondingly reducing the consumption of electricity from the power grid and contributing to savings.

Increased demand for electricity at home can be signaled to the micro-cogeneration system using the function of producing electricity on demand. This function is activated by the timer or by a button on the maintenance-free remote control or by means of a wireless socket. This way, electricity can be generated during greater demand for it, e.g., during washing and cooking. Lower the consumption of electricity from the grid will be, the more electricity will be produced by the micro-cogeneration device. An exemplary installation diagram is shown in Figure 9.

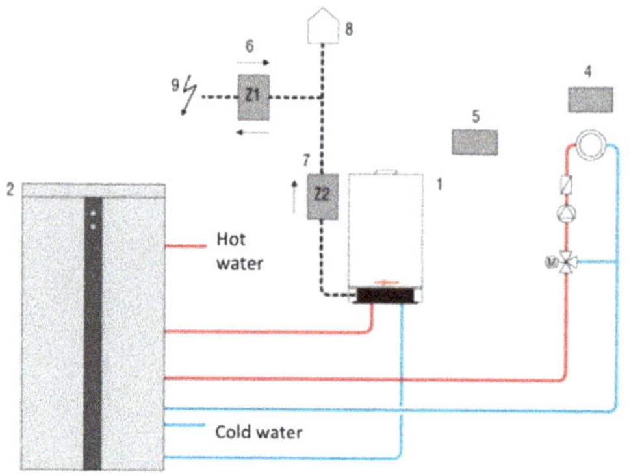

Figure 9. An exemplary installation diagram with CHP system Vitotwin [77]. 1—Vitotwnin 300-W microcogeneration unit, 2—heating water buffer cylinder with built-in domestic hot water heater-Vitocall 340-M with a capacity of 400 L, 3—heating installation, 4—wireless remote control module, 5—monitoring device-remote control, 6—bidirectional meter, 7—meter of produced electricity (installed in the Vitotwin system), 8—home electrical network, 9—external power grid.

Thanks to the aforementioned advantage of the Stirling engine, i.e., the possibility of cooperation with any heat source, microcogeneration systems powered by solid fuel were created. Such a system was implemented by Okofen, which offered a Pellematic biomass boiler with an integrated Stirling engine on the market. The view of such a set is shown in Figure 10. The Stirling engine is located partially outside, while the engine head is located inside the biomass boiler.

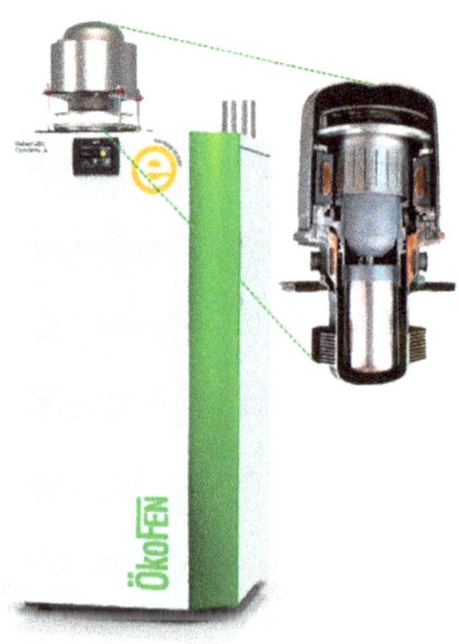

Figure 10. View of the Okofen biomass boiler coupled with a Microgen Stirling engine [77,78].

A major problem in microcogeneration systems with a Stirling engine is the transfer of heat generated in the combustion process to the engine head. Manufacturers of engines as well as boilers and burners are still carrying out optimization work in order to better receive heat by the Stirling engine head. Various methods are used, such as ribbing. This problem does not arise in the case of gaseous fuel systems, since the configuration of the burner can be adapted to the shape of the head of the engine. Stirling engine heads adapted to receive heat from solid fuel systems are intensively developed by means of advanced numerical analyses. The example of the concept Stirling engine with an additional heat exchanger located at an angle of 45°, designed specifically for the system analyzed by the authors [79,80] is shown in Figure 11. The manufactured engine based on this design is shown in Figure 12.

Figure 11. Visualization of a Stirling engine with an additional heat exchanger attached to work with a solid fuel boiler [79,80].

Figure 12. Manufactured Stirling engine with an additional heat exchanger attached for operation with a solid fuel boiler [79,80].

The views on the heat exchanger which was designed specifically for this system, are presented in Figures 13 and 14.

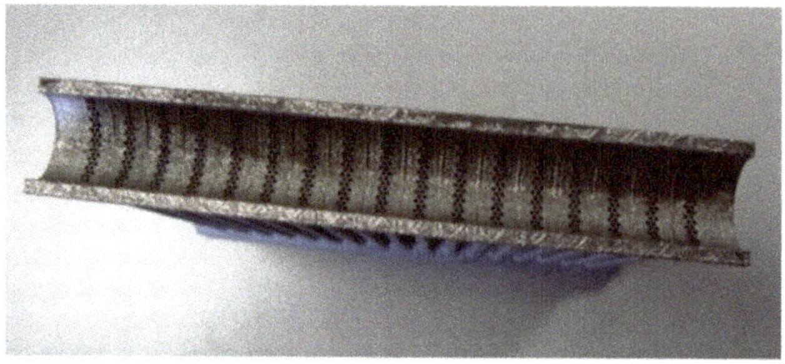

Figure 13. Side view of a heat exchanger adapted to operate with a Stirling engine and a solid fuel system [80].

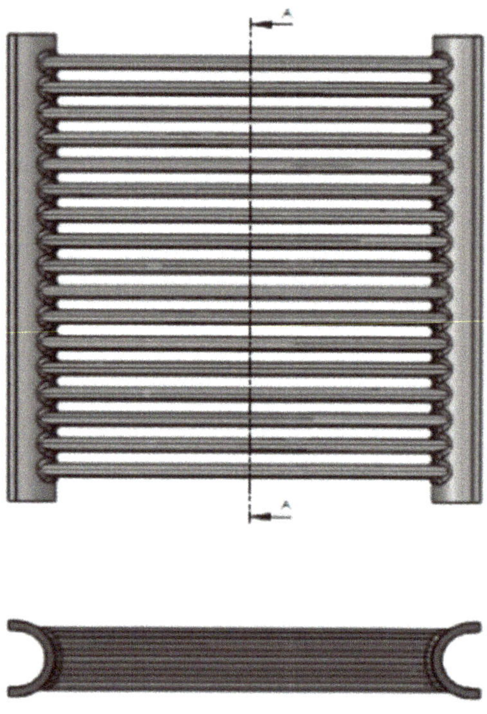

Figure 14. A view of a specially designed exchanger for the operation of a Stirling engine with a solid fuel system [80].

A scheme of the micro-cogeneration system with the previously presented heat exchanger collecting heat from the exhaust gases and transmitting it to the Stirling engine head is shown in Figure 15. Sample temperatures at characteristic points of the system are plotted. The heat exchanger of the Stirling engine located at an angle of 45° in the combustion chamber of the boiler can be also noticed.

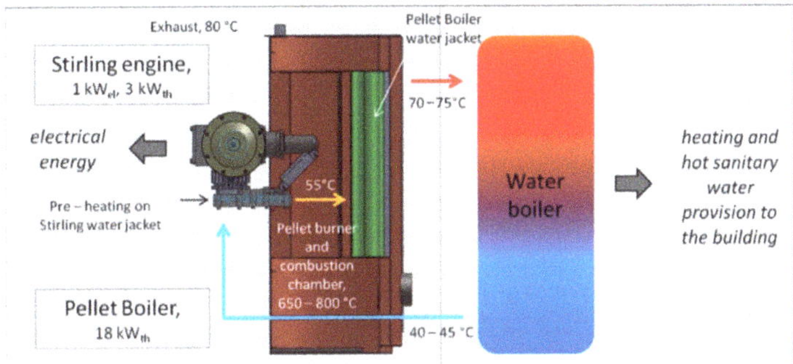

Figure 15. View of the cogeneration system with a Stirling engine with a visible heat exchanger [80].

Another example of a heat exchanger cooperating with a Stirling engine head is presented in Figure 16. This heat exchanger was designed at West Virginia University, Morgantown, USA [81,82]. The concept of this exchanger differs from the previous one mainly in shape. It features toroidal design, which is the most common design of Stirling engine heads.

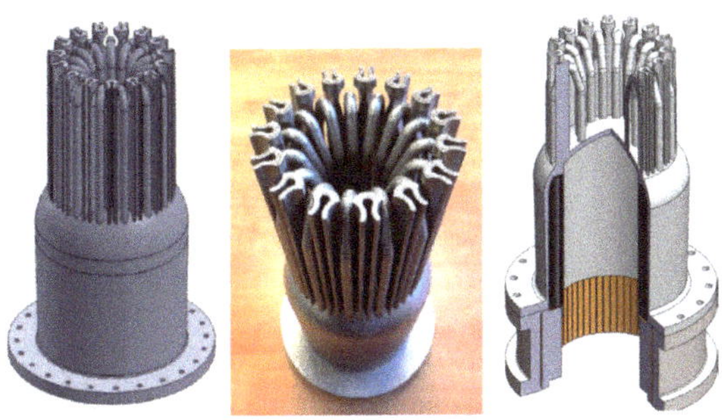

Figure 16. View of the Stirling engine head designed for operation with a solid fuel boiler [77,81].

The aforementioned Microgen company, as a leading manufacturer of small Stirling engines in the world, has developed different types of heads dedicated to Stirling engines. However, their design is kept confidential and not widely published. The example of 1 kW engine head implemented by Microgen company is presented in Figure 17.

Figure 17. View of the Stirling engine head designed by Microgen [66,77].

Stirling engines can be used in many industries and domestic applications. First of all, they can be used as small cogeneration systems for the needs of domestic [83–85] customers and for industrial purposes [86–91]. Due to the development of renewable energy, especially wind and solar, Stirling engines have found application in solar power plants [92–99]. Another sector where they are used is the refrigeration sector [100–108]. They are an alternative to the commonly used compressor systems. The helium used as a working medium is safe in terms of toxicity and flammability.

2.5. Summary

The Stirling engine has the unquestionable advantage of being able to work with any heat source. It can be powered by heat obtained, for example, from the combustion of biomass or other solid fuel source. However, as mentioned in this article, there are technological problems, primarily in the process of efficient heat transfer from its source to the engine head. For this reason, among microcogeneration systems utilizing Stirling engines those fed by natural gas are much more popular. The challenge of better designing of solid fuel-based systems may be perspective. There are many areas of world (such as the countries of far Asia) where there is a shortage of gaseous fuel, while solid fuels are widely and cheaply available. Today's Stirling engines are refined and virtually maintenance-free units. They do not generate vibrations or noise.

The application of a Stirling engine as part of a small CHP system also has some disadvantages. The production of electricity is closely related to the production of heat, while the demand for electricity and the demand for heat often do not coincide. Therefore, there may be an overproduction of electricity or an insufficient amount of it. It is therefore necessary to verify contracts with electricity suppliers in terms of contracted power as well as to enable the opportunity for sale of excess, not consumed, electricity generated by the Stirling engine. This can be done through two-way electricity meters and appropriate contracts with the operator of the distribution grid. For example, Polish regulations allow electricity to be fed into the grid and treated as an energy storage. Unfortunately, for such a service there is a fee in the form of 20% of the delivered energy. This means that the excess of produced electricity can be "pumped" to the grid at any time and 80% of it can be taken back at any time. It should be noted that this regulation applies to installations featuring a maximum power of 10 kW (for larger installations the percentage increases up to 40%) and utilizing renewable energy sources. Thus, in order to take advantage of this possibility, the Stirling engine would have to be powered by heat obtained from biomass combustion.

Other disadvantages include, for example, long start-up times in the case of solid fuel systems and low flexibility in the case of load changes.

3. Microturbines

Research and implementation activities on microturbines are currently proceeding in different scientific units and companies. Microturbines are investigated to be applied as expansion machines in many technical applications, such as power generating systems and CHPs [109], unmanned airplanes [110] and hybrid cars [111]. The microturbine implementation was possible thanks to recent progress in different fields of science (especially materials science, thermodynamics, fluid mechanics and computer aided design). Advanced computer aided design (CAD) techniques and numerical flow modeling are widely used to design microturbines, and computer numerical control (CNC) machines are used for manufacturing microturbine parts. By the direction of the working fluid flow through the microturbine, they can be classified into radial and axial machines and by the type of the applied working fluid into gas microturbines and steam microturbines [112]. A characteristic feature of microturbines is their high rotational speed, ranging from several dozen to several hundred thousand revolutions per minute [113]. Therefore, very precise tools must be applied to manufacture microturbines. The microturbine shaft is coupled with the shaft of the generator in a different way than in the case of standard large-power turbines (i.e., small-scale, specially designed high-speed generators are applied and electronic systems are used to convert the generated current into a current of frequency and voltage that can be transferred directly to the grid). What is more, due to high rotational speeds of the microturbine shaft and large heat load, magnetic or foil bearings are often applied instead of classically used slide bearings [113]. Compared to volumetric expanders, microturbines are featuring smaller dimensions, a smaller number of moving parts, lower friction losses and higher efficiency [114]. They are also lighter. However, working fluid flow through microturbine and machine cost are much higher.

3.1. Gas Microturbines

In recent years, gas microturbines have gained a reputation as a refined technology and are boldly entering a variety of municipal and industrial facilities where reliable, independent electricity and/or heat generation at competitive prices is required. The leaders of this technology are mainly British [115,116], Italian [117] and American [118–120] companies. Currently, their offer includes devices with a power of 50-several hundred kW, and in the future, it is planned to gradually expand this range. Recently, microturbines have had a number of original applications. One of the largest sports and recreation centers in London used a micro-turbine to generate 80 kW of electricity and 150 kW of heat for its own facilities and equipment, including a swimming pool, sports hall and other rooms using the Bowman Power TG80CG gas microturbine [115]. This original 80% efficient power plant has been supplemented with a conventional boiler to cover peak heat loads. The microturbine itself is only a slightly more complex design than a typical low power turbo generator. What distinguishes this device from classic machines of this type is a high-speed, four-pole self-excited generator and dedicated software controlling the operation of the unit. The alternator was made of rare earth metals of extremely high density, which allowed for such a significant reduction of elements that the turbine and generator rotors were placed on one shaft, thus eliminating the troublesome mechanical transmission. This single shaft assembly rotates at over 100,000 rpm producing an output voltage with a frequency in the range of 1000 to 3000 Hz. A special power electronic converter converts them into voltage with a mains frequency of 50 or 60 Hz and an ideal sinusoidal shape and value. Thus, the unit becomes a reliable power source with a quality that meets the most stringent requirements [121]. Manufacturers produce microturbines in two main types: with and without exhaust gas heat recovery. In microturbines without heat recovery, a compressed mixture of natural gas and air is burnt at constant pressure, and the resulting hot exhaust gas stream expands in the gas turbine, driving the generator. These systems follow a simple thermodynamic cycle and are cheaper and more reliable than microturbines with heat recovery. The latter devices have an exchanger in which part of the heat contained in the turbine exhaust gas stream is transferred to the inlet air. As a result, microturbines with heat recovery are characterized by higher efficiency (fuel savings up to 30–40%), comparable to diesel-based combined heat and power plants. In some implementations the exhaust/air heat exchanger has been replaced with a hot water boiler. In other applications, the microturbine exhaust stream without heat recovery is routed to the furnace, eliminating traditional gas burners. Currently, microturbines with a capacity of 25–250 kW and an electricity generation efficiency of 30% are offered in the world. With combined production of electricity and heat, this ratio can reach 80%. Gas microturbines have many advantages and offer a number of advantages, especially when used in small- and micro-power distributed energy. A small number of rotating and moving parts, compact design, small dimensions and weight-facilitate assembly and maintenance. At the same time, very low emission of pollutants and noise level allow their use in virtually every facility. Microturbines can be supplied by different types of fuel, e.g., dairy cattle biogas [122], syngas [123] and biofuels [124].

In [125] authors presented very small model of gas microturbine, featuring only 500 W of power. The turbine is called the Ultra Micro Gas Turbine (UMGT). The test bench of this turbine is presented in Figure 18.

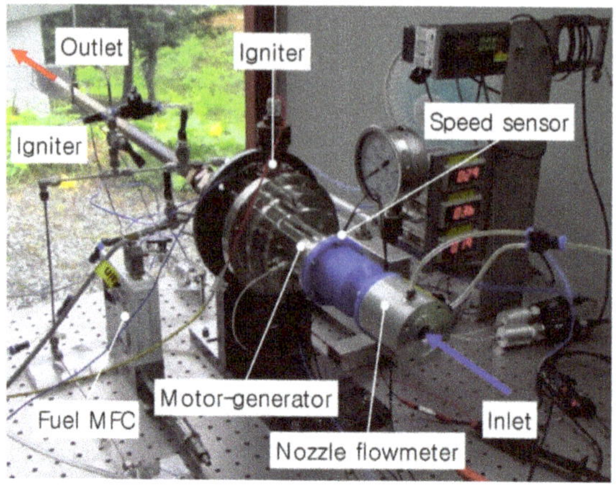

Figure 18. The test bench of the Ultra Micro Gas Turbine (UMGT) power generator [125].

The elements of the tested gas turbine are shown in the Figure 19.

Figure 19. Elements of the first integrated test rig: (**a**) compressor, (**b**) turbine, (**c**) rotor shaft with compressor and turbine, (**d**) radial-thrust integral static air bearing, (**e**) graphite hot-bulb igniter and (**f**) annular-type combustor with 12 nozzles [125].

Furthermore, the technology of ultra-small gas turbines is presented in [126–130].

Ansaldo Energia [131–133] and Capstone [134] are one of the world's leading manufacturers of micro gas turbines. A view of the Capstone 30 kW gas turbine is shown in Figure 20.

Figure 20. View of the Capstone 30 kW gas turbine [134].

Ansaldo Energia is a global manufacturer in a distributed generation market with its AE-T100 Gas Microturbine, available in three different versions: natural gas-fired AE-T100NG, biogas-fired AE-T100B, and fed by heat from external combustion: An example ofEFGT technology is shown in Figure 21. The AE-T100 is producing 100 kW_e of electrical power and about 200 kW_{th} of thermal power. The efficiency of this system is up to 90%. A broad power range is achieved by good modularity of this systems and can spread by adding additional units.

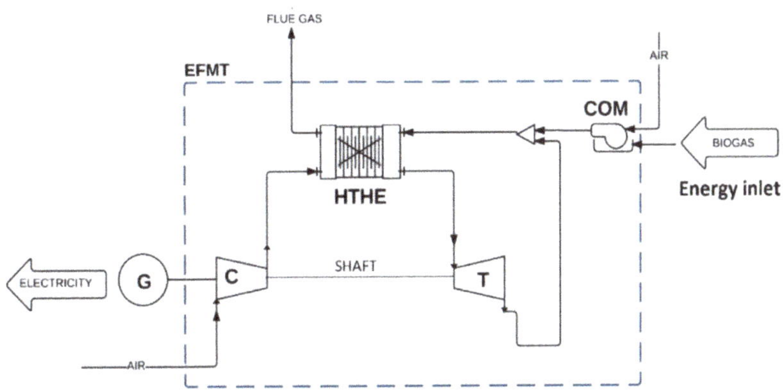

Figure 21. View of the external combustion gas turbine schematic layout [135].

3.2. Steam Microturbines

A characteristic feature of microturbines is their high durability and reliability, which results from their relatively simple design, as they have only one rotating element in the form of a shaft assembly with rotors and a generator [136–139]. High rotational speeds are also typical of this type of turbine machine, thanks to which, with small overall dimensions, the microturbines enable a high power output. However, high rotational speeds lead to complications in the design of the clutch connecting the turbine to the generator and bearings [140–144].

Many research centers around the world conduct research and development work on the continuous improvement of steam microturbine technology. Works on small steam turbines are carried out by, among others, The Institute of Fluid Flow Machinery of the Polish Academy of Sciences in Gdańsk and the Institute of Turbomachinery of the Łódź University of Technology.

As part of the research tasks proceeded at Institute of Fluid Flow Machinery of the Polish Academy of Sciences, several alternative solutions for devices enabling the conversion of thermal energy into electricity were developed and tested. With the assumed power level and limitations resulting from the target place of operation of domestic conditions, steam microturbines turned out to be the optimal solution. Among the examined expansion devices, the most promising results were obtained for the variant of the four-stage radial microturbine and the single-stage radial microturbine. It was decided to couple the steam microturbine with the ORC system [145–147]. The working fluid in the ORC system is a low-boiling fluid. To drive the microturbine, it is firstly heated in a heat exchanger to the temperature at which the state changes from liquid to gas (evaporation takes place).

The gaseous medium at the appropriate pressure is fed to the microturbine blade system, causing its acceleration and then maintaining a constant rotational speed. The mechanical energy of the shaft rotation is then converted into electricity (by means of a generator) which, after appropriate preparation, can be used e.g., in a household. The low-boiling fluid used to drive the microturbine circulates in a closed system; after condensation, it flows through the pump and then it is reheated [148].

The four-stage microturbine developed at Institute of Fluid Flow Machinery of the Polish Academy of Sciences has two centripetal and two centrifugal stages. The shaft is supported by two radial-thrust gas bearings in which a low-boiling fluid is used as the lubricant. At the nominal rotational speed of approx. 24,000 rpm, the microturbine allows to obtain ca. 2 kW of electric power [148]. A cross-section through the turbine and the rotor disk is shown in Figure 22.

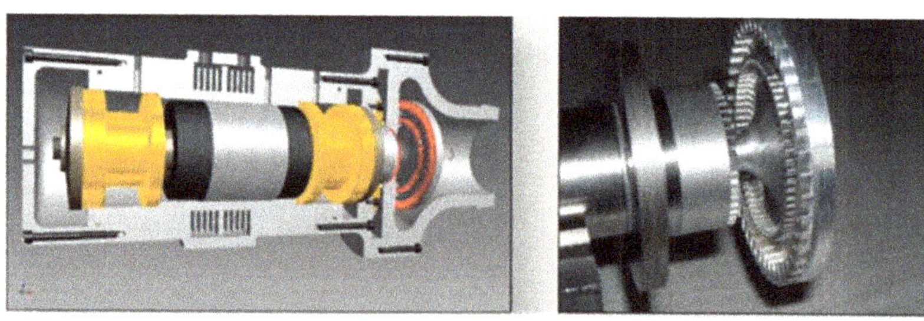

Figure 22. Cross-section and manufactured rotor disc of the four-stage radial microturbine made by Institute of Fluid Flow Machinery [148].

A single-stage microturbine has also been developed at the Institute of Fluid Flow Machinery (see, Figure 23). It is characterized by a centrifugal stage where the flow velocity is more than twice larger than the velocity of sound. It was built on the basis of the experience gained in the implementation of a four-stage microturbine. It uses the previously proven radial-thrust gas bearings, lubricated with a low-boiling fluid. The nominal rotational speed of this micro-turbine is ca. 30,000 rpm, which allows it to generate ca. 2.5 kW of electricity.

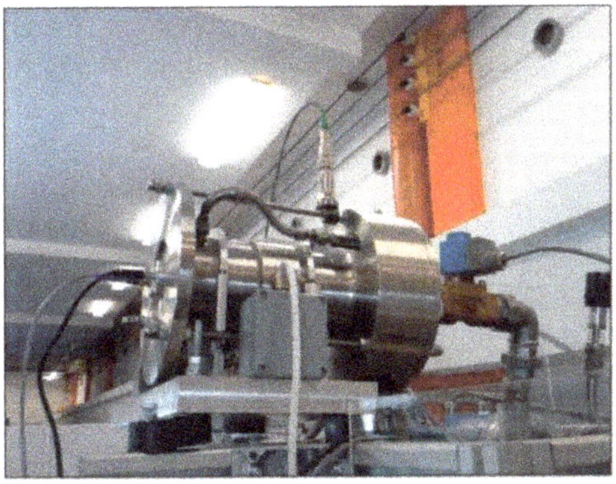

Figure 23. Radial, single-stage steam microturbine installed on the Institute of Fluid Flow Machinery test stand [148].

The developed steam microturbines were tested in the laboratory of Fluid Flow Machinery in conditions corresponding to their operation in the ORC system, using their own control system. Research has shown that prototype microturbine solutions have many advantages. Thanks to the use of high rotational speeds, machines with a compact structure and small dimensions were developed. Unlike other devices that enable the conversion of heat into electricity, steam microturbines are characterized by high durability and reliability because they do not have wear parts and parts that require periodic replacement or repair. Since the same low-boiling fluid is used in the flow system and the bearings, the risk of mixing the working medium, e.g., with oil, has been eliminated. The test results also confirmed the very low vibration level and quiet operation of the developed turbines. The mentioned advantages of microturbines mean that they can be successfully used in domestic ORC micro-cogeneration installations, as well as in other installations requiring a small and reliable device that enables electricity generation [148].

As mentioned earlier, work on small steam turbines is also carried out at the Institute of Turbomachinery of the Łódź University of Technology [136].

The experimental steam turbine with a nominal power of 50 kW was built on the basis of the Institute of Turbomachinery project. The turbine is powered by a steam generator that uses waste heat from a biogas combustion engine. Firstly, the following live steam parameters were considered: temperature equal to 613 K, pressure equal to 12 bar, while the mass flow rate was 0.075 kg/s. The turbine works in condensing mode. For these parameters, the isentropic drop in enthalpy is as high as 732 kJ/kg, which excludes a single-stage design, except for the Curtis two-ring (or three-ring) stage, operating in the range of very high Mach numbers. Therefore, after discussion, the live steam temperature was limited. Ultimately, the following turbine design parameters were established [136]:

- live steam: pressure 12 bar, temperature 573 K;
- steam mass flow 0.075 kg/s;
- pressure in the condenser 0.25 bar;
- rotational speed 6276 rpm.

The rotational speed was imposed due to the unavailability of a suitable gear.

A small backpressure turbine was also designed at Institute of Turbomachinery of the Łódź University of Technology. It generates 165 kW of power. The turbine was designed to cooperate with the existing technological installation; therefore, its operating parameters were strictly defined and were not subject to any discussion. These parameters were established as follows:

- live steam: pressure 5.8 bar, temperature 553 K (280 °C);
- steam mass flow 2 kg/s;
- 3.5 bar back pressure.

In this case, the isentropic enthalpy drop is only 119.7 kJ/kg, which greatly facilitates the adoption of the advantageous design solution. After preliminary calculations, it was assumed that the turbine would be implemented as a single action stage, powered on the entire circuit.

Experimental research on the application of a microturbine (featuring a maximum power of 1.9 kW) in domestic ORC CHP systems using ethanol as a working fluid was also proceeded at the Gdańsk University of Technology. The results of these experiments were reported in [149–152]. The experimental tests were proceeded for varied thermodynamic parameters of the working fluid at the inlet and at the outlet of the microturbine. The pressure at the inlet to the machine was varied between 0.36 and 0.6 MPa, while the working fluid flow was varied between 15 and 20 g/s. The maximum temperature of the working medium at the inlet to the machine was equal to 143 °C. For these experimental conditions, the obtained electric power of the ORC system was ranging between 0.66 and 0.76 kW, electrical efficiency was ranging between 6.40 and 6.65% and the total efficiency of the ORC CHP system was ranging between 22.53 and 23.54% [150].

There are also suppliers on the world market that offer microturbine technology. One of them is Spirax Sarco. This technology is described in more detail in [153]. An example of a Spirax Sarco steam microturbine in a container version is presented in Figure 24.

Figure 24. View of the Spirax Sarco steam microturbine [153].

Other companies that have in their portfolio steam microturbines are e.g., Siemens Dressel-Rand and General Electric.

4. Volumetric Expanders (Vane, Lobe, Screw, Piston, Wankel, Gerotor)

Volumetric expanders can be applied in small CHP steam and ORC systems as an alternative to the earlier described microturbines. The principle of operation of volumetric expander differs from that of turbine. In the case of volumetric machine, the working fluid expansion proceeds in a working chamber which volume is limited by the cylinder and the displacer. The operation of volumetric expander is cyclical and working chamber volume changes during machine operation. For this reason, gas expansion processes are proceeding periodically. Compared to microturbines, volumetric expanders are featuring simpler design and lower investment costs. What is more, they are also featuring lower rotational speeds, higher pressure drops that can be obtained in one stage, lower mass flows of the working fluid and the possibility of wet-gas expansion. In selected cases, it is possible to design and implement oil-free volumetric expanders.

4.1. Vane Expanders

The design of the vane expanders is very simple, which translates into low production costs. This machine has a favorable ratio of the output power to its external dimensions. The use of special construction materials makes it possible to eliminate the need for lubrication. Compared to other types of volumetric machines and microturbines, vane expanders are characterized by a lower fluid flow rate and a lower pressure ratio. Moreover, the vane expanders can be hermetically sealed, which is one of the key issues in the cooperation of this type of expander with the ORC system [154]. This type of machine is insensitive to the negative impact of the expansion of the gas-liquid mixture, which is a great advantage when ORC systems are supplied by heat sources with variable thermal parameters.

Vane compressors and expanders are used in many industries, including mining, refrigeration and pneumatic systems [155]. Vane expanders have an output power from several dozen watts to about a dozen kilowatts. The maximum gas pressure at the inlet to the vane expander is approximately 30 bar. These machines are characterized by rotational speeds from several hundred to 10,000 rpm. Research on vane expanders is carried out, among others at the Wrocław University of Science and Technology. The implemented experimental CHP ORC test stand uses multi-vane expander featuring the maximum power output of 300 W and gas central heating boiler featuring a thermal power of 24 kW serves as a heat source. The view of the test stand and expander is shown in Figure 25.

Figure 25. General view of the test stand and expander. Left: 1—plate evaporator, 2—plate condenser, 3—pump, 4—tank 5—multi-vane expander; on the right-general view of the expander [154].

As already mentioned, the design of the vane expanders is simple. Figure 26 shows the components of a vane expander.

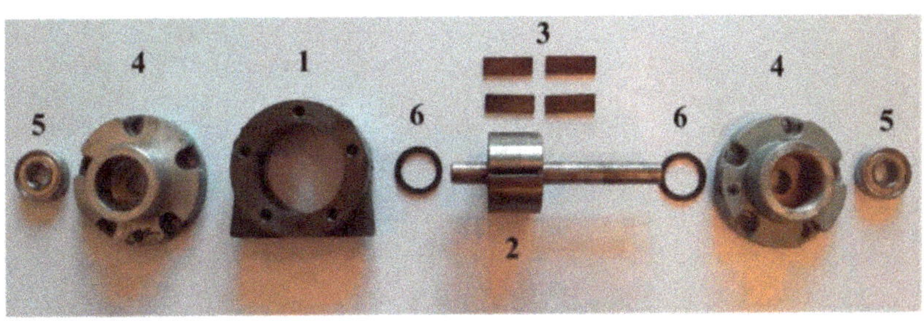

Figure 26. The individual elements of a vane expander: 1—body, 2—rotor, 3—blades, 4—housing, 5—bearings, 6—rings [154].

Based on the available literature [156–159], it can be stated that vane expanders are devices on a rather smaller scale. The maximum currently available units, when coupled with the ORC system, reach the power of 7.5 kW.

4.2. Lobe Expanders

Rotary lobe expanders are devices that are not currently in mass production. Work on them is still ongoing. Their design is derived from pneumatic motors [160]. The design of such an engine is shown in Figure 27.

Figure 27. View on the Armak pneumatic motor [160].

Rotary lobe expanders can be used as heat engines in energy systems powered by different heat sources (e.g., biomass boilers, waste heat recovery boilers, parabolic solar collectors, etc.). The experimental units work in small steam plants. The company that is currently conducting research on this type of expanders is the Polish company Termo2Power. The works are carried out under the research project "PBSE Power Sector Research Program" carried out under the contract with the National Center for Research and Development. Part of the substantive work is carried out by a team from the Faculty of Power and Aeronautical Engineering of the Warsaw University of Technology. The company also conducts tests of its own designs, including multi-stage expanders. These expanders can work with medium pressure from 8 to 40 bar. The inlet medium temperature should not exceed 350 °C due to sealing problems. In this type of devices, no labyrinth seals are used, but classic seals that must withstand high temperatures. The capacities of these devices range from single kilowatts up to ca. 150 kW for a single stage expander. It is also practiced to combine expanders into multi-stage systems. Then a system with a power of several hundred kilowatts can be configured. Figure 28 shows the Termo2Power rotary lobe expander and Figure 29 shows this expander connected to the generator and with flexible connectors for supplying and discharging steam.

Figure 28. View of the Termo2Power rotary lobe expander [161].

Figure 29. View of Termo2Power rotary lobe expander connected to the generator and pipings [161].

The figure below (Figure 30) shows a system with two rotary lobe expanders and interstage superheating. The boiler is visible on the left side. The expanders are coupled to the generators (blue). The power take-off (electric heaters) is visible under the expanders system.

Figure 30. View of a two-stage steam rotary lobe expander with inter-stage superheating [161].

In addition to the aforementioned Termo2Power company, Katrix from Australia is another manufacturer that produces lobe expanders. The expander of this company is of a slightly different design from the Termo2Power expander. A description of this expander and analysis of its operation can be found in [162,163]. Its view and principle of operation are shown in the Figure 31.

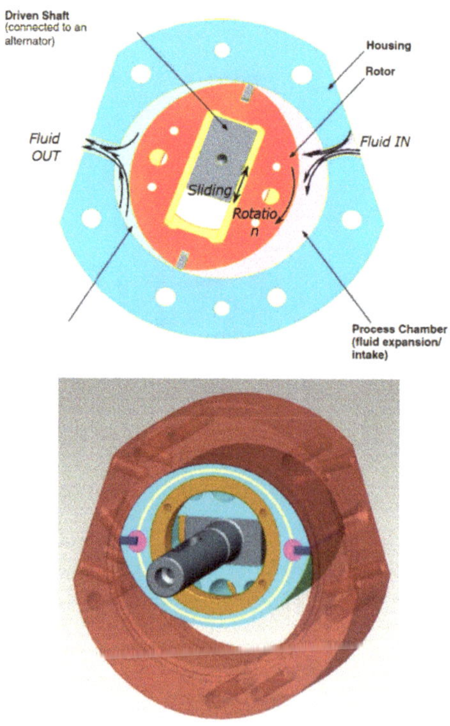

Figure 31. View of a Katrix expander construction [162,163].

Expander made by the Katrix company was under investigation as a part of the electricity generation system. The system based on solar collector to produce hot working fluid can be seen in the Figure 32. The expander connected to the pipings and generator is presented in Figure 33.

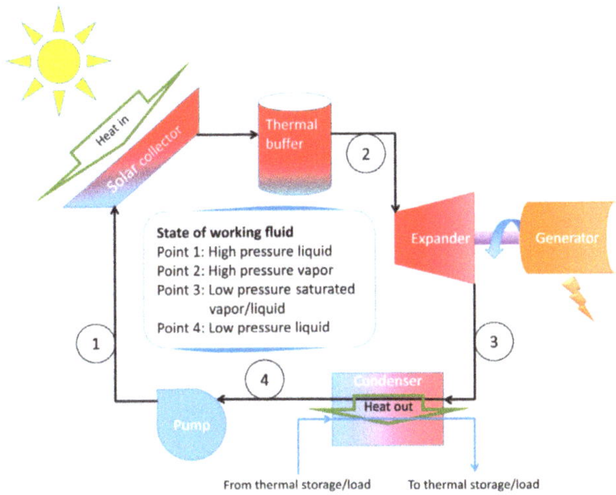

Figure 32. A solar Rankine microcogeneration system with Katrix expander [162,163].

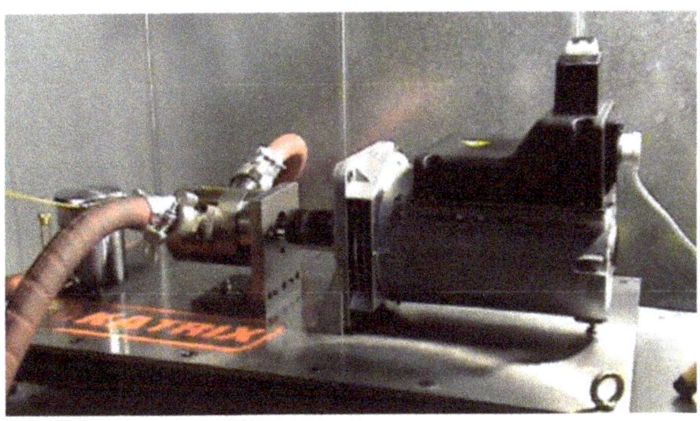

Figure 33. View of a Katrix expander connected to the source of working fluid and generator [162,163].

Roots expanders are the other lobe-type expanders that are currently investigated to be applied in ORCs. Roots expander is a two-shaft, rotary, positive displacement machine, featuring the transverse flow of the working fluid [164–167]. The results of experimental tests of a Roots expander operating in the ORC system are presented in [166]. The tests were carried out using a mixture of R245fa and oil to provide the expander lubrication. The working fluid pressure at the inlet to the machine was varied in the range between 3 and 10.8 bar. The working fluid pressure at the outlet of the machine was ranging between 1.35 and 2.25 bar. The rotational speed of the expander was varying between 1500 and 11,000 rpm. The highest isentropic efficiency of the expander (ca. 50%) was achieved for the rotational speed of ca. 4500 rpm and for small expansion ratio of ca. 1.5. However, for these operating parameters low power output was achieved (ca. 100 W). In order to increase the power output to ca. 3 kW it was necessary to increase the expansion ratio to ca. 3.5.

4.3. Screw Expanders

The screw expander is a displacement type device. It consists of an interlocking pair of spiral rotors placed in a housing, which together form the working chamber. In addition to the pair of screw rotor and cylinder, the structure consists of bearings, synchronous gears, a seal assembly, etc.

The expander has the opposite operating principle to the screw compressor, which has a similar basic structure. The production process and control system are much more complex for an expander than for a screw compressor.

A cross-sectional view of the screw expander is shown in Figure 34.

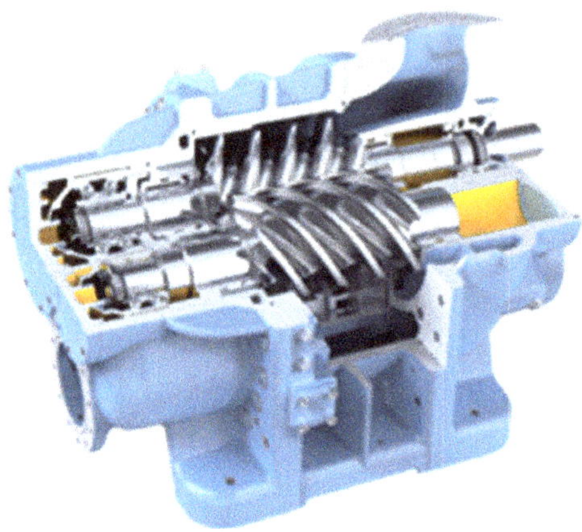

Figure 34. View of the screw expander made by Green Secure Power Systems [164].

Screw expanders can work with superheated steam and wet steam. Currently available units are powered by a working medium with a pressure of 3 to 25 bar. The power range of screw expanders ranges from a few kilowatts to 630 kW.

The world leader in the production of systems with a screw expander is the British company Heliex Power. One of the Heliex devices was installed in Poland in Kędzierzyn-Koźle at Grupa Azoty in 2016. The installation was performed together with the Polish partner of Heliex-the Zamkon company. The device installed in Poland is 160 kW. Figure 35 shows a container system with a screw expander installed at Zakłady Azotowe Kędzierzyn.

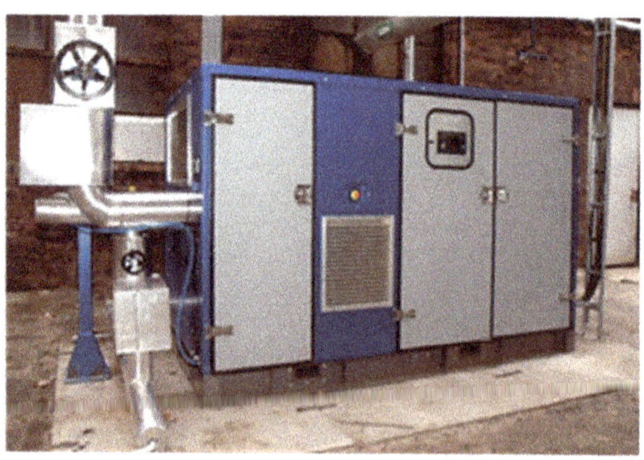

Figure 35. View of the screw expander made by Heliex Power installed at Zakłady Azotowe Kędzierzyn [164].

The layout of the example system based on screw is presented in Figure 36.

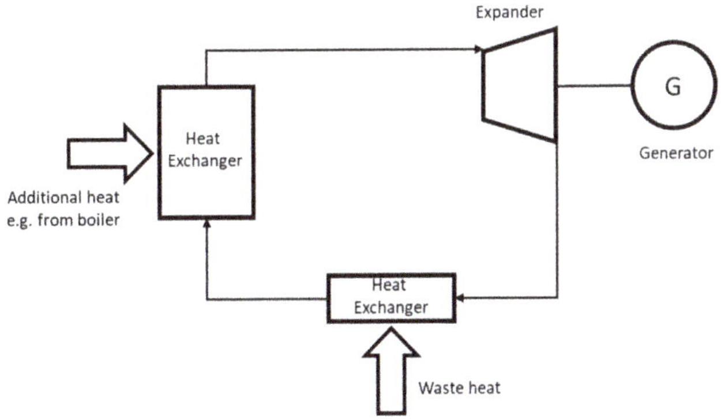

Figure 36. View of the system with screw expander.

The American company ELECTRA THERM [167] developed ORC system using screw expander that can be fed by different heat sources (i.e., geothermal water, biomass combustion products, waste steam or hot water from industrial processes. The minimum temperature of these sources should be 90 °C. The system can be also supplied with waste gases featuring different temperature (205–540 °C). The power output of this system ranges between 50 and 500 kW, depending on the thermal power of the heat source.

Screw expander is also applied in the ORC system developed by the German company KÖHLER-ZIEGLER [167]. This system can utilize waste heat carriers featuring a minimum temperature of 90 °C. The power output of this system ranges from 50 to 200 kW, depending on the heat source thermal parameters.The ORC systems with screw expanders are also manufactured by the Swedish company OPCON [167]. These ORC systems are featuring a modular design and a power output of 350, 500 and 800 kW. These systems can utilize waste heat sources featuring a temperature of 55–120 °C. Substances featuring a temperature of 0–30 °C can be used as the condenser coolant. The efficiency of these systems is ranging between 5 and 10%.

The other types of screw expanders that are applied in ORCs are single-screw expanders. The results of research on these expanders are presented in [168–177]. Single-screw expanders consist of three movable rotors, one of which is the main rotor, located centrally. The main rotor has helical teeth cut around the circumference. The other two rotors are having outer teeth of a rectangular cross-section. The teeth of the cooperating rotors mesh with the teeth of the main helical rotor. In this way, working chambers are formed on both sides of the central rotor.

Single-screw expanders can operate with different working fluids, including natural gas, superheated and saturated steam and liquid-gas mixtures. The power output of these expanders is in the range of 1–100 kW. Single-screw expanders feature a simple design, good performance at part load, high volumetric efficiency and low leakage rate. They are operating quietly and do not generate vibrations. Experimental research on application of single-screw expanders in ORC systems were proceeded in different research centers. In [173] studies on the ORC system utilizing a single-screw expander and using Solkatherm SES36 as a working fluid were described. The working fluid temperature at the inlet to the expander was varied in the range of 119.3–125 °C, the inlet pressure was varied in the range 4.5–10.2 bar, the expansion ratio in the range 3.12–10.97 and the expander rotational speed in the range 2000–3000 rpm. For these experimental conditions, the internal efficiency of the expander was ranging between 51% and 64.78% and expander rotational speed was ranging between 2000 rpm and 3000 rpm. The maximum power output of the expander of 7.8 kW and the efficiency of the ORC system of 9.8% was achieved during experiments.

Preliminary tests on single-screw expander prototype, which was designed for application in ORC systems, were also carried out with the use of air [171,176]. The total expander efficiency of 55%, the maximum torque of 100 Nm and the power output of 22 kW were achieved experimentally [176]. In [171] the results of research on a single-screw expander, which was supplied with compressed air (pre-heated to temperature of 80 °C) were reported. The experimental results showed that depending on the expander rotational speed (which was varied between 500 and 3500 rpm) the expander power output varied in the range of 1–5 kW. Expansion ratio was kept in the range 5–6.75, gas temperature drop was observed in the range 32.5–62.5 °C and the expander efficiency was varying in the range 10–60%. Single-screw expanders were also tested using R123. The results of these experiments are reported in [175]. The expander power output of 8.35 kW, internal efficiency of 56% and the maximum efficiency of the ORC system of 7.98 % were achieved. In [168] the results of experiments on a single-screw expander using R245fa as a working fluid were presented. In [173] the results of experimental studies on a single-screw expander applied in an ORC system utilizing SES36 as a working fluid were presented. The experimental tests were proceeded for varied thermodynamic parameters of the gas at the inlet to the expander. The temperature of gas at the inlet to the expander was varied in the range of 100–125 °C, the inlet pressure was varied in the range of 3–7.5 bar and the outlet pressure was kept in the range of 0.6–0.8 bar. The expander rotational speed was 3000 rpm. For these experimental conditions, the achieved expander output power was ranging between 1.5 and 6 kW and the achieved internal efficiency was ranging between 51 and 66%. In [170] the results of experimental studies which were proceeded on six prototypes of single-screw expanders featuring a power output between 5 and 172 kW were reported. The applied working fluid was R123. For varied experimental conditions the expanders efficiency varied between 52% and 73% and the efficiency of the ORC system reached 9.3%.

4.4. Piston Expanders

In a piston expander, the working chamber is formed by the inner surface of the cylinder and the surface of the moving piston. The piston reciprocates in the cylinder between top and bottom dead position and is driven by the crank mechanism. Studies on the possible application of piston expanders in micro steam and ORC systems have proceeded for many years.

The results of research on application of a small piston expander in the ORC system utilizing R134a as a working fluid are reported in [178–180]. The tested expander was manufactured by StarEngine company. The expander features a total displacement of 230 cm^3 and is hermetically coupled in one casing with a generator. The general view of this expander is presented in Figure 37.

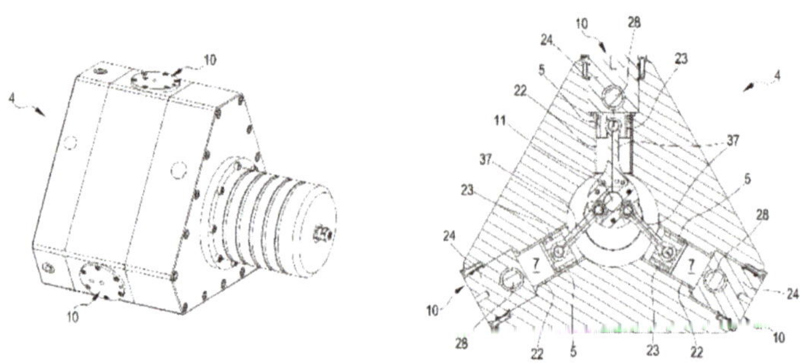

Figure 37. StarEngine piston expander [178–180].

The experimental research on this expander was carried out for varied heat source and heat sink temperature. Heat source temperature was varied in the range 65–85 °C and the temperature of the heat sink was varied in the range 28–27 °C. The mass flow rate of the working fluid was varied between 0.05 and 0.14 kg/s, pressure in the evaporator was varied between 11 and 19 bar and pressure in the condenser was varied between 6 and 7 bar. For these experimental parameters, an electric power output of 250–1200 W was achieved while expander rotational speed was ranging between 320 and 1100 rpm. The total efficiency of the expander-generator unit was varying between 38 and 42%. The achieved gross efficiency of ORC system was 4.5% while the achieved net efficiency was 2.2%.

In [181] the results of experimental tests carried out on a prototype oil-free piston expander designed for application in steam distributed generation systems are reported. The electric power output of the tested expander ranges between 740 and 2400 W depending on the parameters of the working medium. During the experiments the thermal parameters of the working fluid at the inlet to the expander were varied in the range 260–340 °C and 20–34 bar. The experimentally achieved internal efficiency of the expander was varying between 19 and 40%. This type of steam expander is also promising for application in ORC systems [181]. In [182,183] the results of research on a piston expander featuring a power of 3 kW and applied in the ORC system are presented. The expander is a modified reciprocating compressor with a specially designed control valve.

In [184] the authors presented the results of modeling of the operation of a piston expander designed for application in waste heat recovery system from the passenger car exhaust gases. Modeling results showed the possibility of obtaining an expander power output of 7 kW and an isentropic efficiency between 55 and 70%.

In [185] the results of experimental research on a swash-plate piston expander featuring a displacement of 195 cm^3, which was implemented in the ORC system using blowing agent R245fa as a working medium are presented. This type of expander uses a swash-plate to transmit torque from the pistons to the shaft. The cross-section of swash-plate expander is presented in Figure 38.

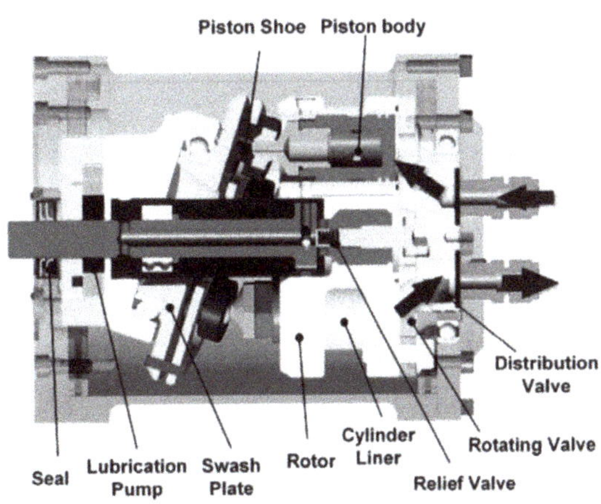

Figure 38. Cross-section of a swash-plate piston expander [185].

The tests of the expander were proceeded for gas inlet pressures ranging between 18 and 30 bar and the rotational speed of the expander ranging between 1000 and 4000 rpm. For these experimental conditions working fluid pressure at the outlet of the expander was ranging between 2.9 and 4.02 bar. The inlet pressure was regulated by changing the mass flow rate of the working fluid by means of a pump. Depending on the operational conditions working fluid was superheated in the range of 4–17 K and working fluid mass flow rate was varied between 29 and 105 g/s. The obtained mechanical power of the expander was ranging between 0.3 and 2 kW while the maximum achieved internal efficiency of the expander was 53%. The mechanical efficiency of the expander was ranging between 50 and 85%. The results of research on swash-plate expanders were also reported in [186,187].

In addition to the piston expanders discussed above, linear piston expanders are also investigated to be applied in ORC systems. Research results on this type of expansion machines were reported in [188,189]. In a linear piston expander, the linear arrangement of the cylinders is applied. This design is similar to the boxer arrangement of cylinders, but does not use a crankshaft. A linear generator is placed between the cylinders. The piston rods of the opposing pistons are connected to each other by a piston rod of a linear generator. The kinetic energy of the reciprocating movement of the piston rod is converted into electricity in a linear generator. The cross-section of linear piston expander is presented in Figure 39. In [190] the test results and design guidelines for a linear piston expander that can be applied in a micro-power ORC systems used for waste heat recovery from automotive engines were presented.

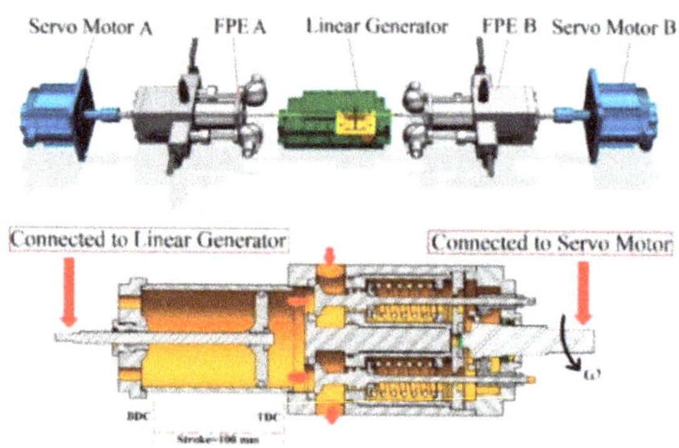

Figure 39. Cross-section of a linear piston expander [190].

It has been reported [190] that incomplete expansion (pressure of the working medium in the cylinder after expansion is higher than the pressure of the working medium in the outlet channel from the expander), heat transfer, flow losses during filling and evacuation of the working fluid from cylinder, friction and leakages are the main phenomena limiting the efficiency of linear piston expanders. It has also been shown that appropriate valve control has a significant impact on the linear piston expander operation and pressure losses occurring on the valves. The results of the tests carried out on the prototype of the linear piston expander showed that for gas inlet pressures ranging between 0.13 and 0.21 MPa, the internal efficiency of the expander varies in the range 66.2–93% and decreases with increasing inlet pressure. The highest power output of 22.7 W was achieved for the gas inlet pressure of 0.2 MPa and internal efficiency of 66.2%.

Compared to the other types of volumetric expanders (such as e.g., lobe, screw and Wankel expanders), piston expanders are characterized by a much simpler design; however, they require lubrication, valve timing, and generate vibrations during operation.

4.5. Wankel Expanders

Wankel expanders are positive displacement machines featuring rotational displacement movement. These machines are currently experimentally tested to be applied in small steam and ORC systems [191–196]. The principle of operation of the Wankel expander is similar to that of the Wankel engine, but unlike in the engine, the combustion of fuel-oxidant mixture is not proceeded in the expander. A Wankel expander use a triangular rotor which moves in an oval cross-section cylinder and the side edges of which are curve-shaped. The cylinder is closed on both sides with covers in which inlet and outlet ports are placed. The crank shaft, which drives the rotor, is mounted on bearings embedded in the side covers. The shaft is coupled to the rotor by means of a gear that synchronizes their mutual movement. The gear consists of a fixed rack, which is embedded in the side cover of the machine, and a ring rack assembled inside the rotor. Vane seals are placed in the tips of the triangular rotor. Seals limit gas leakage between the working chambers and separate the working chambers from each other. Seals are also placed on the rotor faces to limit gas leakage between the piston and side covers. The cross-section through the Wankel expander, with a description of the most important components of this expander, is presented in Figure 40a. The machine has two inlet and two outlet ports, thanks to which the gas can be expanded in two working chambers at the same time. The working fluid is supplied to the machine through two inlet ports (see Figure 40a) and the working fluid pressure exerted on the rotor causes its motion. Dosing of the working fluid to the working chambers is proceeded via the inlet and outlet valves which opening is controlled by timing belt driven by the rotating crank shaft (see, Figure 40b).

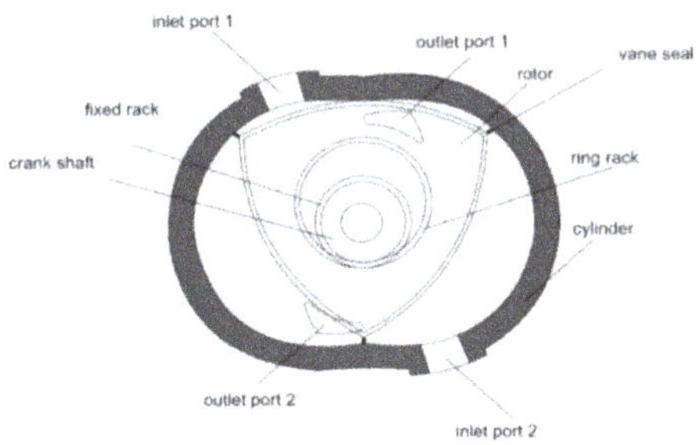

(a)

(b)

Figure 40. Design and assembly details of a Wankel expander: (**a**) cross section of Wankel expander; (**b**) a general view on Wankel expander [192].

Wankel expanders have a number of advantages when compared to the other positive displacement and turbine expanders. The main advantages of these machines are high power-to-weight ratio, compact design and small external dimensions, lack of reciprocating parts, high rotational speeds, lack of vibrations generated during operation and a small number of moving parts. The main disadvantages of Wankel expanders include piston face seals issues (piston face seals are stressed by temperature variation during the machine operation), piston apex seals issues (piston apex seals are receiving significant loads related to the difference in gas temperature and pressure in adjacent working chambers). At low rotational speeds or low expander load, it is possible that the seal does not fully adhere to the cylinder surface, which may result in an increase of internal gas leakages between adjacent working chambers.

Early works on the possibility of using a Wankel machine as a steam expander were started in 1970s [193–195]. In [195] the results of experimental tests carried out on a Wankel expander using steam as a working fluid were presented. During the tests the steam pressure at the inlet of the expander was varied between 2.76 and 6.5 MPa while the steam temperature was varied between 231 and 410 °C. For these experimental conditions, the obtained expander power was ranging between 12 and 17 kW and the rotational speed was ranging between 2196 and 2578 rpm. Further works on the application of these machines as steam expanders were carried out in the 1990s [193–195] and are continued currently. In [192] the results of experimental tests of a prototype of Wankel expander designed for steam expansion are reported. The authors developed an expander prototype using parts of a standard Wankel engine (i.e., bearings, shaft and seals were used) and a new specially designed cylinder that was adapted to the supply system consisting of control valves in order to increase expander compression. The view of this prototype is presented in Figure 40b. The valves are controlled by means of a mechanical system based on a timing belt driven from the main expander shaft.

4.6. Gerotor Expanders

The other type of volumetric expanders that are used in prototypes of small ORC systems are gerotor expanders [197–199]. The design of this type of volumetric expander is similar to the design of a gear pump. The basic components of gerotor expander are a cylinder and two rotors—internal and external. The internal rotor is assembled on the shaft. The cylinder is closed on both sides by side covers with inlet and outlet ports. The inner rotor is placed eccentrically to the outer rotor. Figure 41 shows the view of the components of the gerotor expander.

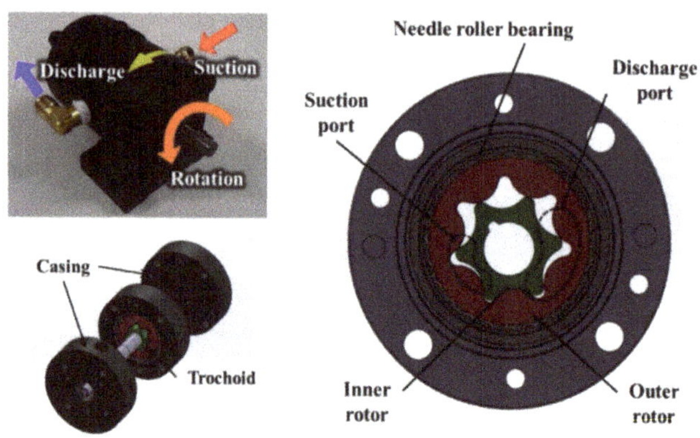

Figure 41. View of the components of the gerotor expander [199].

Gerotor expanders were experimentally tested for their applicability in micro-power ORC systems. In [197] the results of tests on an expander featuring a power output of 1 kW designed for application in the ORC system using the solar heat were presented. The tests were carried out with the use of the R134a. The pressure of the working fluid at the inlet to the expander was 3.28 MPa, while inlet temperatures were varied between 80 and 100 °C. The working fluid pressure at the outlet of the expander was 1.64 MPa and the rotational speed of the expander was 3000 rpm. For these experimental conditions, expander power output was ranging between 0.2 and 1 kW and efficiency was ranging between 35 and 75%. The optimal expansion ratio was found between 3.0 and 4.0. In [198] the results of experimental tests of three gerotor expanders, which were characterized by different geometrical parameters were presented. These tests were carried out using the ORC system utilizing R123 as a working fluid. The tests were carried out for various parameters of the working fluid at the inlet and outlet of the expander. The working fluid pressure at the inlet to the expanders was varied between 412 and 1878 kPa, the working fluid pressure at the outlet of the expanders was varied between 139 and 331 kPa, the working fluid temperature at the inlet to the expanders was varied between 84 and 160 °C, while the working fluid temperature at the outlet was varying between 61 and 129 °C. For these experimental conditions, the power output of these expanders was ranging between 0.28 and 2.07 kW and the achieved efficiency was ranging between 59 and 85%. It was also indicated that, compared to other positive displacement machines, gerotor expanders are characterized by a lower internal friction.

5. Fuel Cells

In 1839, British physicist William R. Grove demonstrated that an electrochemical reaction of combining hydrogen with oxygen produces an electric current [200–204]. Such a cell has no moving parts, works noiselessly, and its only waste substance is water. However, fuel cells based on this phenomenon were merely a laboratory curiosity for over a century. It was not until the sixties of the last century that NASA started to install light and compact (though expensive) versions in spacecraft to supply them with electricity. Today, this technology, which is promising, ecologically clean, efficient and silent, is being used in many new earthly applications, including powering mobile phones, notebook computers, homes and apartments and electric car engines. Chemical energy is directly converted into electricity in a fuel cell. It is a cell in which the fuel—hydrogen in a pure state or in a mixture with other gases—is fed continuously to the anode, and the oxidant—pure oxygen or a mixture (air)—is fed continuously to the cathode. Electrochemical processes are accompanied by the flow of an electron from the anode to the cathode. The closure of the circuit is carried out by ions that are transferred through the electrolyte. As a result of the electrochemical reaction of hydrogen and oxygen, electricity, water and heat are generated. Reagents are fed continuously to the fuel cell and theoretically it will not discharge; in fact, degradation or component failure will limit the life of any fuel cell.

Most fuel cells use hydrogen to produce electricity and heat [205–209]. Nevertheless, high-temperature fuel cells can run on natural gas due to the possibility of using the so-called internal reforming. The electrical efficiency of modern fuel cells is ca. 40–60 percent.

Different types of fuel cells are developed and are generally classified according to the type of electrolyte used, as it determines the operating temperature of the system and the type of fuel that can be used. The comparison of the different types of fuel cells is presented in Table 1.

Table 1. Comparison of different types of fuel cells [117].

Fuel Cell Type	Operating Temperature (°C)	Power Range (kW)	Efficiency (%)	Application
PEM–Polymer Electrolite Membrane	60–110	0.01–250	40–55	Portable, Mobile, Low power generation.
AFC–Alkaline Fuel Cell	70–130	0.1–50	50–70	Mobile, space, military.
PAFC–Phosforic Acid Fuel Cell	175–210	50–1000	40–45	Medium to large scale power generation and CHP
MCFC–Molten Carbonate Fuel Cell	550–650	200–100,000	50–60	Large scale power generation
SOFC–Solid Oxide Fuel Cell	500–1000	0.5–2000	40–72	Vehicle, medium to large scale power generation and CHP, and micro-CHP.
DMFC–Direct Methanol Fuel Cell	70–130	0.001–100	40	Mobile, portable.

Fuel cells are used both in small domestic power and heat generating units or auxiliary power sources with a capacity of several dozen kilowatts, as well as in large power plants with a capacity of several megawatts. Small systems with a power of 1–10 kW with fuel cells are able to provide electricity and heat to residential houses, offices and public buildings.

Another advantage is the design based on a modular system, which allows for relatively quick and easy construction of the installation and its possible expansion. Fuel cells are characterized by a high power yield per unit volume of fuel, and at the same time the process of direct conversion of fuel chemical energy into electricity takes place without the emission of toxic components and while maintaining high efficiency of fuel energy use. If the waste heat from the cell installation is used in combined CHP systems, the total energy efficiency may increase even up to 95% [210–212]. Waste heat can be used for heating, domestic hot water heating, cooling or air conditioning. Hybrid fuel cell installations connected to the gas turbine cycle achieve efficiency of 70% and more. The dynamic development of fuel cells in recent years means that they are more and more often alternative sources of electricity and heat.

There are many types of fuel cells. They differ mainly in the type of electrolyte, and thus in the operating temperature (see, Table 1). This indirectly also affects the power density that can be taken from the surface of the fuel cell. A Polymer Electrolyte Membrane

Fuel Cell (PEMFC) and Solid Oxide Fuel Cell (SOFC) are the most popular in small- and micro-generation. PEM fuel cells belong to the group of low temperature fuel cells. Their working temperature does not exceed 100 °C. Their power ranges from a few watts to hundreds of kilowatts. The unquestionable advantage of the operation of these fuel cells is a very quick response to load changes. The disadvantage, however, is the need to use clean fuel due to the platinum catalyst used in this type of fuel cell. An example of a small cogeneration source based on a PEM cell is shown in the Figure 42.

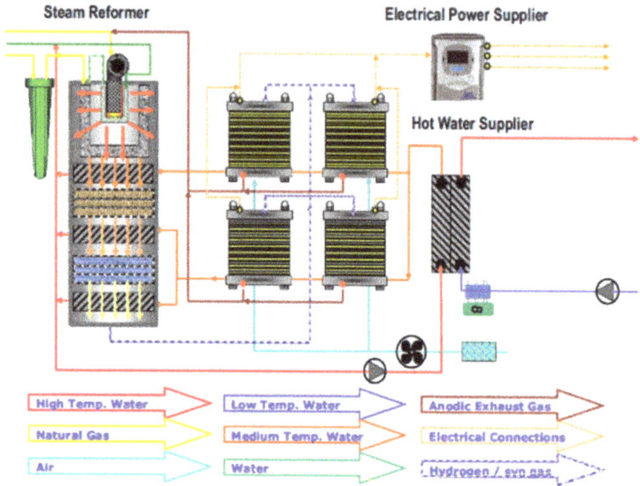

Figure 42. View of the small CHP system based on a Polymer Electrolyte Membrane Fuel Cell (PEM) fuel cell [213].

The system consists of a PEM fuel cell, a methane steam reformer, an inverter, a pump system and a heat exchanger. It produces 36 kW of electricity and 50 kW of heat. The system has been built in a container which is divided into mechanical and electrical parts. In the mechanical part there is a reformer and a fuel cell, while in the electrical part there is an inverter and automation and control systems. The view of the container can be seen in the Figure 43.

Figure 43. View of the container with 36 kW PEM fuel cell system [213,214].

The most important part of the system is the fuel cell. It is a cell produced by the Canadian company Ballard with a nominal power of 36 kW and a maximum power of 80 kW. The cell is connected with a fan in order to supply the appropriate amount of oxidant to the reaction process. The fan has high power in relation to the cell's power (3 kW), which is almost 10%. Figure 44 shows a fuel cell (four stacks at the top) and a fan (below the fuel cell at the bottom).

Figure 44. PEM fuel cell with the fan delivering air as an oxidant [213,214].

Another type of fuel cell used in small and micro cogeneration is an SOFC fuel cell. The systems based on this technology offer higher electrical efficiency than the systems based on PEM technology and are especially focused on the continuous operation mode. Contrary to what it was said in the case of PEM cells, which show great keeping up with the demand. In the case of SOFC fuel cells, the surplus electricity is sent to the grid or accumulated if the system is equipped with a battery.

As SOFC fuel cells are one of the most popular types of fuel cells, there have been many studies and publications on combining these fuel cells into systems for the simultaneous generation of electricity and heat. Basic information on such systems can be found in [215–220].

An example of micro cogeneration system based on an SOFC fuel cell is presented in Figure 45.

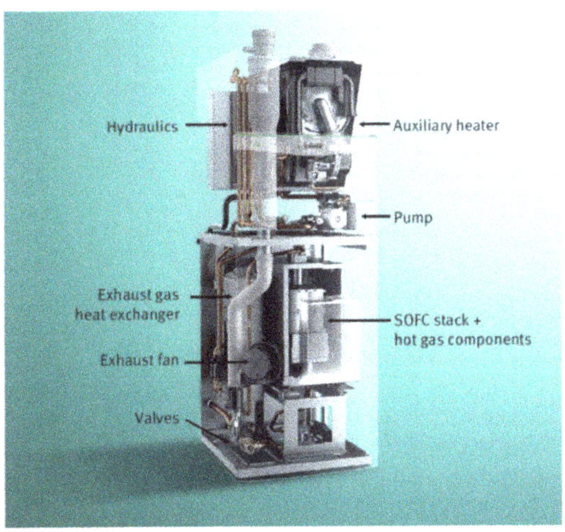

Figure 45. Micro cogeneration system based on a Solid Oxide Fuel Cell (SOFC) fuel cell made by Vaillant [164,221].

Due to the high operating temperature (800–1000 °C), SOFC fuel cells can also be combined into systems with other energy sources, such as gas turbines [222–229] and burners [230–235]. An example of an SOFC fuel cell coupled to a gas turbine can be seen in Figure 46, while an SOFC fuel cell coupled to an additional heat source in Figure 47.

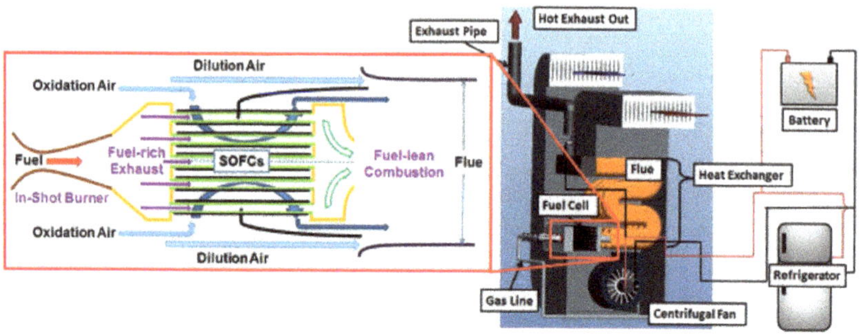

Figure 46. Micro cogeneration system with a flame-assisted SOFC fuel cell [234,235].

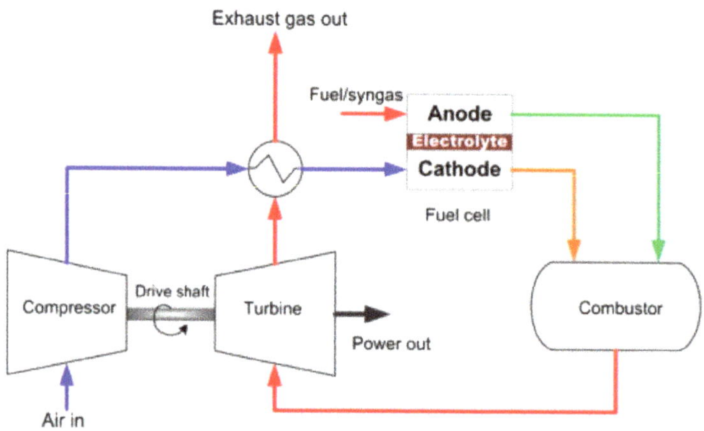

Figure 47. Cogeneration system with an SOFC fuel cell and gas turbine (the so-called hybrid cycle) [229].

The use of the small and microcogeneration systems based on fuel cells in countries where the energy sector is characterized by low CO_2 emissions or is largely based on renewable resources will not always bring the expected benefits. Sometimes it can even contribute to the deterioration of the current condition. In the above situation, it is necessary to carry out a detailed profitability analysis for various possible operating modes of the system:

- maximization of generated electricity;
- following the instantaneous demand for electricity;
- maximization of heat production;
- following the instantaneous demand for heat;
- maximization of prosumer profits.

For example, in Scandinavian countries, due to the significant share of renewable energy sources in the energy sector and thus the low emission of harmful compounds, the only economically viable solution is to follow the small and microcogeneration system with instantaneous heat demand. Electricity is a by-product of this operating mode of the system.

6. Conclusions

The article presents an overview of currently used and developed technologies for the production of electricity and heat in the so-called cogeneration (CHP) in small and micro scale. These technologies are becoming more and more popular and needed in relation to the development trend of distributed generation and the so-called virtual power plants. The technologies presented in the article are inter alia, Stirling engines, gas and steam microturbines, various types of volumetric expanders (vane, lobe, screw, piston, Wankel, gerotor), and fuel cells. There is no definite answer as to which of these technologies is the best. Each of them has its advantages and disadvantages and can be adapted to the specific conditions in which it has to operate. These conditions are, for example, the type and availability of fuel, the ability and speed of reaction to load changes, reliability, noise, generated power, environmental and social conditions, availability of service personnel and many others, which should be taken into account when selecting a given technology to meet needs.

Stirling engines are mature technology. Nevertheless, over the years, research centers have been working on their improvement. Microgen company, previously mentioned in the text, has developed an advanced technology that is commercially available. The company does not produce ready-made systems containing a Stirling engine, but supplies engines to system manufacturers. They are used in systems powered by both solid and gaseous fuels. Stirling engines are most often featuring the power output ranging from hundreds of watts to hundreds of kilowatts. In addition to applications in small and micro-cogeneration, Stirling engines can be used in solar energy systems and in the refrigeration industry. These engines are also applied in submarines. Works on their application in cars are ongoing.

The review of the volumetric expanders that are currently applied in small and micro cogeneration systems gave an outlook on their operating conditions and technical details. The following conclusions can be drawn on volumetric expanders basing on the reviewed literature.

- Multi-vane expanders are applied in experimental test-stands of micro CHP ORC systems. The experimental tests were proceeded for working fluid absolute pressure at the inlet to the multi-vane expander ranging between 1.5 and 6.39 bar. The power output of tested multi-vane expanders was ranging between 65 W and 8 kW, rotational speed was ranging between 1200 and 4100 rpm and isentropic efficiency was ranging between 17.2 and 55.8%. Efficiency of the ORC systems was ranging between 0.75 and 7.65%.
- Rotary lobe expanders are applied in experimental test-stands of small steam plants. The working fluid pressure at the inlet to the rotary lobe expander ranges between 8 and 40 bar. The working fluid pressure at the inlet should not exceed 350 °C due to sealing problems. The power output of these machines ranges from single kilowatts up to ca. 150 kW for a single stage expander. Lobe expanders were also applied in solar-powered ORCs.
- Root expanders are applied in experimental test-stands of micro CHP ORC systems. The experimental tests were proceeded for working fluid pressure at the inlet to the Roots expander ranging between 3 and 10.8 bar. The working fluid pressure at the outlet of the machine was ranging between 1.35 and 2.25 bar. The power output of tested Roots expanders was ranging between 100 W and 3 kW. The rotational speed of the expander was varying between 1500 and 11000 rpm. The highest expander isentropic efficiency of ca. 50% was obtained.
- Screw expanders are successfully applied in commercially available steam and ORC CHP systems. They can operate in different conditions. The absolute pressure of the working fluid at the inlet to the screw expander is ranging between 3 and 25 bar, while the temperature of the heat sources is ranging between 90 and 540 °C. The power output of these systems ranges between few kW and 630 kW while efficiency of the ORC systems adopting screw expanders ranges between 5 and 10%.

- Single-screw expanders are applied in experimental test-stands of small steam and ORC CHP plants. The experimental tests were proceeded for working fluid absolute pressure at the inlet to the screw expander ranging between 3 and 6.75 bar, and the temperature of the heat sources ranging between 80 and 120 °C. The power output of these expanders ranges between 5 kW and 172 kW, rotational speed ranges between 2000 and 3000 rpm, obtained efficiency of the expanders ranges between 10 and 66% while efficiency of the ORC systems adopting single-screw expanders ranges between 7.98 and 9.3%.
- Piston expanders are applied in experimental test-stands of small steam and ORC CHP plants. The experimental tests were proceeded for working fluid absolute pressure at the inlet to the piston expander ranging between 11 and 34 bar and the inlet temperature of the working fluid ranging between 65 and 340 °C. The power output of these expanders ranges between 0.25 kW and 3 kW, rotational speed ranges between 320 rpm and 4000 rpm and expanders efficiency ranges between 55 and 70%.
- Linear piston expanders are applied in experimental test-stands of ORC CHP plants. The experimental tests were proceeded for working fluid absolute pressure at the inlet to the linear piston expander ranging between 13 and 21 bar. The maximum power output of 22.7 W was obtained in case of these expanders, and internal efficiency was ranging between 66.2 and 93%.
- Wankel expanders are applied in experimental test-stands of steam plants. The experimental tests were proceeded for working fluid absolute pressure at the inlet to the Wankel expander ranging between 27.6 and 65 bar and the temperature of the working fluid ranging between 231–410 °C. The power output of Wankel expanders ranges between 12 and 17 kW and rotational speed ranges between 2196 and 2578 rpm.
- Gerotor expanders are applied in experimental test-stands of ORC plants. The experimental tests were proceeded for working fluid absolute pressure at the inlet to the gerotor expander ranging between 27.6 and 65 bar and the temperature of the working fluid ranging between 80 and 160 °C. The power output of Wankel expanders ranges between 0.2 and 2.07 kW, rotational speed was equal to 3000 rpm and obtained efficiency of the expanders ranges between 35 and 85%.

Both gas and steam microturbines are mature technology dating back several dozen years. Research on microturbines is carried out in scientific units around the world, but more often they result from attempts to use ready-made microturbines from well-known manufacturers for use in systems with their participation. There are few research centers that deal with the design of microturbines alone. As a mature technology, microturbines are used mainly in small and industrial microcogeneration. This technology is not used in households, mainly due to the costs and noise generated by microturbines. Due to the nature of their construction, they are high-speed machines. In addition to the noise generated, this has an impact on problems in the use of generators (a gear or a high-speed generator is necessary). The power range of microturbines is in the range of hundreds of watts to several hundred kilowatts. Their efficiency is 20–30% for gas microturbines (electrical efficiency) and 60–70% for steam microturbines (internal efficiency). It should be noted that the gas microturbine can work as an independent device for the supply of fuel, and in the case of a steam microturbine it must be part of the system (e.g., the Rankine cycle) and then the system's electrical efficiency should be taken as an output parameter of the system, which is ca. 20%.

Fuel cells are a commercialized technology. The most common types of cells used in microcogeneration are PEM fuel cells and SOFC fuel cells. The first one is a fully mature technology and commercially produced in the power range from a few watts to hundreds of kilowatts. They are characterized by a quick start-up and low operating temperature, and hence high flexibility in terms of load change. The problem is the fuel that should be clean (CO free) due to the platinum catalytic converter. It requires the use of pure hydrogen, e.g., from the electrolysis process, or, if it comes from other processes, such as methane steam reforming—purifying it. As for SOFC fuel cells, they are also a relatively mature technology. Unfortunately, the problem is the complicated production process that involves sintering components together. Scientific centers conduct research on the improvement of components (electrodes and electrolyte) in order to achieve better performance, but the finished devices are manufactured and sold, especially in Asian countries. The disadvantage is the high operating temperature, which is associated with a longer start-up and the need for more stable operation when it comes to changing loads. In practice, devices with a power of several hundred watts to hundreds of kilowatts are the most popular. The authors of this article believe that the long-observed trend towards decentralizing the production of electricity and heat will contribute to even more dynamic development of small and micro technologies in the CHP sector.

It is also important that most of the technologies presented here can be used to use waste heat from industrial plants. As a result of electricity and heat generation processes or other technological processes taking place in industrial plants, some energy is irretrievably lost to the environment. Especially in small and medium-sized industrial plants, energy and environmental awareness are at a low level. These plants were often built a long time ago, have old machinery and are located in energy-inefficient halls and buildings. The application of the systems presented here can contribute to the improvement of energy efficiency thanks to the recovery of waste heat. Thanks to their positive features and utilization of high-quality fuels, domestic micro CHP systems can possibly contribute to the significant reduction of the amount of pollutants emitted into the environment from standard heating systems. Standard furnaces that are often used for heating the houses during winter are often low-efficient and fed by low-quality fuels. Therefore, worse and worse air quality is being observed in many countries. Many of the currently applied furnaces can be successfully replaced with domestic CHP units based on the technologies described in this article. In this way, their application may have a positive influence on air quality.

Author Contributions: Conceptualization, M.W.; writing—original draft preparation, M.W. and P.K.; writing—review and editing, P.K., M.W. and K.B.; visualization, M.W. and P.K.; project administration, K.B.; funding acquisition, K.B. All authors have read and agreed to the published version of the manuscript.

Funding: This research received no external funding.

Acknowledgments: The authors would like to thank the Scientific Council of the Discipline of Environmental Engineering, Mining and Power Engineering and the Dean of the Faculty of Power and Aeronautical Engineering of the Warsaw University of Technology for their support. The publication is the result of an Internal Grant for employees of the Warsaw University of Technology in 2020.

Conflicts of Interest: The authors declare no conflict of interest.

References

1. Capuano, D.L. *Energy Information Administration*; Annual Energy Outlook: Washington, DC, USA, 2020.
2. Aymar, R.; Barabaschi, P.; Shimomura, Y. The ITER design. *Plasma Phys. Control. Fusion* **2002**, *44*, 519–565. [CrossRef]
3. Opriş, I.; Cenuşă, V.; Norişor, M.; Darie, G.; Alexe, F.-N.; Costinaş, S. Parametric optimization of the thermodynamic cycle design for supercritical steam power plants. *Energy Convers. Manag.* **2020**, *208*, 112587. [CrossRef]
4. Garievskii, M. Optimization of CCGT operating modes at variable loads taking into account equivalent operating hours. *J. Phys.* **2020**, *1683*, 042022.
5. Alanne, K.; Saari, A. Distributed energy generation and sustainable development. *Renew. Sustain. Energy Rev.* **2006**, *10*, 539–558. [CrossRef]
6. Wang, X.; Shu, G.; Yan, F.; Feng, W.; Wang, R.; Pan, J. Optimization of a distributed energy system with multiple waste heat sources and heat storage of different temperatures based on the energy quality. *Appl. Therm. Eng.* **2020**, *181*, 115975. [CrossRef]
7. Luo, Z.; Yang, S.; Xie, N.; Xie, W.; Liu, J.; Agbodjan, Y.S.; Liu, Z. Multi–objective capacity optimization of a distributed energy system considering economy, environment and energy. *Energy Convers. Manag.* **2019**, *200*, 112081. [CrossRef]
8. Lowitzsch, J.; Hoicka, C.E.; Van Tulder, F.J. Renewable energy communities under the 2019 European Clean Energy Package–Governance model for the energy clusters of the future? *Renew. Sustain. Energy Rev.* **2020**, *122*, 109489. [CrossRef]

9. Rosato, A.; Sibilio, S. Performance assessment of a micro-cogeneration system under realistic operating conditions. *Energy Convers. Manag.* **2013**, *70*, 149–162. [CrossRef]
10. Rosato, A.; Sibilio, S.; Ciampi, G. Energy, environmental and economic dynamic performance assessment of different micro-cogeneration systems in a residential application. *Appl. Therm. Eng.* **2013**, *59*, 599–617. [CrossRef]
11. Rosato, A.; Sibilio, S. Energy performance of a micro-cogeneration device during transient and steady-state operation: Experiments and simulations. *Appl. Therm. Eng.* **2013**, *52*, 478–491. [CrossRef]
12. Rosato, A.; Sibilio, S. Calibration and validation of a model for simulating thermal and electric performance of an internal combustion engine-based micro-cogeneration device. *Appl. Therm. Eng.* **2012**, *45*, 79–98. [CrossRef]
13. De Paepe, M.; Mertens, D. Combined heat and power in a liberalised energy market. *Energy Convers. Manag.* **2007**, *48*, 2542–2555. [CrossRef]
14. De Paepe, M.; D'Herdt, P.; Mertens, D. Micro-CHP systems for residential applications. *Energy Convers. Manag.* **2006**, *47*, 3435–3446. [CrossRef]
15. Pehnt, M. Environmental impacts of distributed energy systems—The case of micro cogeneration. *Environ. Sci. Policy* **2008**, *11*, 25–37. [CrossRef]
16. Pehnt, M.; Fischer, C. Environmental Impacts of Micro Cogeneration. In *Micro Cogeneration*; Springer Nature: Berlin/Heidelberg, Germany, 2006; pp. 87–116.
17. Dowson, D. Men of Tribology: Guillaume Amontons (1663–1705) and John Theophilus Desaguliers (1683–1744). *J. Lubr. Technol.* **1978**, *100*, 2–5. [CrossRef]
18. Talbot, G.R.; Pacey, A.J. Antecedents of Thermodynamics in the Work of Guillaume Amontons. *Centaurus* **1972**, *16*, 20–40. [CrossRef]
19. Guillaume, A. Ciencias Químicas. *Rev. CENIC* **2005**, *36*, 187–195.
20. Woerlen, I. Hot Air Engines. Available online: www.hotairengines.org (accessed on date 20 December 2020).
21. Cayley, S.G.; Gordon, A.; Gurney, G.; Brunel, I.K.; Stephenson, R. Discussion. Sir George Cayley's Hot–Air Engine. (Includes Plate). In *Minutes of the Proceedings of the Institution of Civil Engineers*; Thomas Telford Ltd.: London, UK, 1850; Volume 9, pp. 9197–9203.
22. Gibbs-Smith, C.H. Sir George Cayley* 'Father of aerial navigation' (1773–1857). *Notes Rec. R. Soc.* **1962**, *17*, 36–56. [CrossRef]
23. Cooke, C.W. On Wenham's Heated-Air Engine. *Proc. Inst. Mech. Eng.* **1873**, *24*, 63–86. [CrossRef]
24. Stirling International. Available online: www.stirlinginternational.org (accessed on 20 December 2020).
25. Harlow, H.F. Hot Air Engines. *Sci. Am.* **1852**, *7*, 293. [CrossRef]
26. Zanzig, J. *The Stirling Cycle Engine*; SAE Technical Paper Series; SAE International: Warrendale, PA, USA, 1963.
27. Otaka, T. Stirling Engines, Their History and Basic Theories. *J. Inst. Electr. Eng. Jpn.* **2016**, *136*, 596–600. [CrossRef]
28. Kazmierczak, J.; Zmudzki, S. Relative Analysis of free–Piston Stirling Engine Dynamics. *J. Propulsion Power* **1986**, *2*, 505–513.
29. Moscrip, W. The Moscrip–Stirling engine–A new departure. In Proceedings of the Intersociety Energy Conversion Engineering Conference, San Francisco, CA, USA, 19 August 1984.
30. Szałwiński, P. *Silnik Stirlinga i Możliwości Jego Zastosowania w Energetyce*; European Union funded Research Grant; Wrocław University of Science and Technolgy: Wrocław, Poland, 2010.
31. Walker, G. Elementary Design Guidelines for Stirling Engines. *Proc. Intersoc. Energy Convers. Eng. Conf.* **1979**, *1*, 1066–1068.
32. Walker, G.C.; Senft, J.R. *Free Piston Stirling Engines. Lecture Notes in Engineering*; Springer Verlag: Berlin/Heidelberg, Germany, 1985; ISBN 978-3-642-82526-2.
33. Karabulut, H.; Yücesu, H.S.; Koca, A. Manufacturing and testing of a V-type Stirling engine. Turkish. *J. Eng. Environ. Sci.* **2000**, *24*, 71–80.
34. Yoshihara, S.; Hoshino, T. Model free piston Stirling engine. In Proceedings of the Symposium on Stirlling Cycle, Japan Society of Mechanical Engineers, Tokyo, Japan, 17–18 October 2003; pp. 87–88.
35. Żmudzki, S. *Silniki Stirlinga*; WNT—Wydawnictwa Naukowo-Techniczne: Warsaw, Poland, 1993.
36. Rahmati, A.; Varedi-Koulaei, S.M.; Ahmadi, M.H.; Ahmadi, H. Dimensional synthesis of the Stirling engine based on optimizing the output work by evolutionary algorithms. *Energy Rep.* **2020**, *6*, 1468–1486. [CrossRef]
37. Thombare, D.G.; Umale, N. Theoretical Analysis of Effect of Regenerator Geometry and Material on Stirling Engine Performance. In Proceedings of the 1st National, P.G. Conference RIT NCon PG–2015, Sangli, Maharashtra, India, 1 June 2015.
38. Thombare, D.G.; Karmare, S.V. Theoretical and experimental investigation of Alfa type bio mass Stirling engine with effect of regenerator effectiveness, heat transfer, and properties of working fluid. *J. Renew. Sustain. Energy* **2012**, *4*, 43126. [CrossRef]
39. Karabulut, H.; Okur, M.; Halis, S.; Altin, M. Thermodynamic, dynamic and flow friction analysis of a Stirling engine with Scotch yoke piston driving mechanism. *Energy* **2019**, *168*, 169–181. [CrossRef]
40. Bulinski, Z.; Szczygieł, I.; Krysiński, T.; Stanek, W.; Czarnowska, L.; Gładysz, P.; Kabaj, A. Finite time thermodynamic analysis of small alpha-type Stirling engine in non-ideal polytropic conditions for recovery of LNG cryogenic exergy. *Energy* **2017**, *141*, 2559–2571. [CrossRef]
41 Kropiwnicki, J. Analysis of start energy of Stirling engine type alpha. *Arch. Thermodyn.* **2019**, *40*, 243–259. [CrossRef]
42. Bataineh, K. Mathematical formulation of alpha -type Stirling engine with Ross Yoke mechanism. *Energy* **2018**, *164*, 1178–1199. [CrossRef]
43. Almajri, A.K.; Mahmoud, S.; Al-Dadah, R. Modelling and parametric study of an efficient Alpha type Stirling engine performance based on 3D CFD analysis. *Energy Convers. Manag.* **2017**, *145*, 93–106. [CrossRef]

44. Altin, M.; Okur, M.; Ipci, D.; Halis, S.; Karabulut, H. Thermodynamic and dynamic analysis of an alpha type Stirling engine with Scotch Yoke mechanism. *Energy* **2018**, *148*, 855–865. [CrossRef]
45. Ipci, D.; Karabulut, H. Thermodynamic and dynamic analysis of an alpha type Stirling engine and numerical treatment. *Energy Convers. Manag.* **2018**, *169*, 34–44. [CrossRef]
46. Bataineh, K.M. Numerical thermodynamic model of alpha-type Stirling engine. *Case Stud. Therm. Eng.* **2018**, *12*, 104–116. [CrossRef]
47. Cinar, C.; Yücesu, H.S.; Topgül, T.; Okur, M. Beta-type Stirling engine operating at atmospheric pressure. *Appl. Energy* **2005**, *81*, 351–357. [CrossRef]
48. Solmaz, H.; Ardebili, S.M.S.; Aksoy, F.; Calam, A.; Yılmaz, E.; Arslan, M. Optimization of the operating conditions of a beta-type rhombic drive stirling engine by using response surface method. *Energy* **2020**, *198*, 117377. [CrossRef]
49. Chahartaghi, M.; Sheykhi, M. Thermal modeling of a trigeneration system based on beta-type Stirling engine for reductions of fuel consumption and pollutant emission. *J. Clean. Prod.* **2018**, *205*, 145–162. [CrossRef]
50. Uchman, W.; Remiorz, L.; Grzywnowicz, K.; Kotowicz, J. Parametric analysis of a beta Stirling engine–A prime mover for distributed generation. *Appl. Therm. Eng.* **2018**, *145*, 693–704. [CrossRef]
51. Shendage, D.; Kedare, S.B.; Bapat, S. An analysis of beta type Stirling engine with rhombic drive mechanism. *Renew. Energy* **2011**, *36*, 289–297. [CrossRef]
52. Chahartaghi, M.; Sheykhi, M. Energy and exergy analyses of beta-type Stirling engine at different working conditions. *Energy Convers. Manag.* **2018**, *169*, 279–290. [CrossRef]
53. Ahmed, F.; Huang, H.; Khan, A.M. Numerical modeling and optimization of beta-type Stirling engine. *Appl. Therm. Eng.* **2019**, *149*, 385–400. [CrossRef]
54. Caetano, B.C.; Lara, I.F.; Borges, M.U.; Sandoval, O.R.; Valle, R.M. A novel methodology on beta-type Stirling engine simulation using CFD. *Energy Convers. Manag.* **2019**, *184*, 510–520. [CrossRef]
55. Sowale, A.; Kolios, A.; Onabanjo, T.; Somorin, T.; Parker, A.; Williams, L.; Collins, M.; McAdam, E.; Tyrrel, S. Thermodynamic analysis of a gamma type Stirling engine in an energy recovery system. *Energy Convers. Manag.* **2018**, *165*, 528–540. [CrossRef]
56. Alfarawi, S.; Al-Dadah, R.; Mahmoud, S. Enhanced thermodynamic modelling of a gamma-type Stirling engine. *Appl. Therm. Eng.* **2016**, *106*, 1380–1390. [CrossRef]
57. Katooli, M.H.; Moghadam, R.A.; Hooshang, M. Investigation on effective operating variables in gamma-type Stirling engine performance: A simulation approach. *SN Appl. Sci.* **2020**, *2*, 1–7. [CrossRef]
58. Gheith, R.; Aloui, F.; Tazerout, M.; Ben Nasrallah, S. Experimental investigations of a gamma Stirling engine. *Int. J. Energy Res.* **2011**, *36*, 1175–1182. [CrossRef]
59. Araoz, J.A.; Cardozo, E.; Salomon, M.; Alejo, L.; Fransson, T.H. Development and validation of a thermodynamic model for the performance analysis of a gamma Stirling engine prototype. *Appl. Therm. Eng.* **2015**, *83*, 16–30. [CrossRef]
60. Hooshang, M.; Moghadam, R.A.; Alizadehnia, S. Dynamic response simulation and experiment for gamma-type Stirling engine. *Renew. Energy* **2016**, *86*, 192–205. [CrossRef]
61. Vahid, D.J.; Oskouei, H.D. Design and anylysis of gamma type Stirling engine. *Mech. Ind.* **2020**, *21*, 511. [CrossRef]
62. Parlak, N.; Wagner, A.; Elsner, M.; Soyhan, H.S. Thermodynamic analysis of a gamma type Stirling engine in non-ideal adiabatic conditions. *Renew. Energy* **2009**, *34*, 266–273. [CrossRef]
63. Damirchi, H.; Najafi, G.; Alizadehnia, S.; Mamat, R.; Azwadi, C.S.N.; Azmi, W.; Noor, M. Micro Combined Heat and Power to provide heat and electrical power using biomass and Gamma-type Stirling engine. *Appl. Therm. Eng.* **2016**, *103*, 1460–1469. [CrossRef]
64. Li, R.; Grosu, L.; Li, W. New polytropic model to predict the performance of beta and gamma type Stirling engine. *Energy* **2017**, *128*, 62–76. [CrossRef]
65. Egas, J.; Clucas, D. Stirling Engine Configuration Selection. *Energies* **2018**, *11*, 584. [CrossRef]
66. Microgen Engine Corporation. Available online: www.microgen-engine.com (accessed on 20 December 2020).
67. Li, T.; Tang, D.; Li, Z.; Du, J.; Zhou, T.; Jia, Y. Development and test of a Stirling engine driven by waste gases for the micro-CHP system. *Appl. Therm. Eng.* **2012**, *33–34*, 119–123. [CrossRef]
68. Zhu, S.; Yu, G.; Jongmin, O.; Xu, T.; Wu, Z.; Dai, W.; Luo, E. Modeling and experimental investigation of a free-piston Stirling engine-based micro-combined heat and power system. *Appl. Energy* **2018**, *226*, 522–533. [CrossRef]
69. Khoshbazan, M.; Ahmadi, M.H.; Ming, T.; Arjmand, J.T.; Ahmadi, M.H. Thermo-economic analysis and multi-objective optimization of micro-CHP Stirling system for different climates of Iran. *Int. J. Low-Carbon Technol.* **2018**, *13*, 388–403. [CrossRef]
70. Cardozo, E.; Malmquist, A. Performance comparison between the use of wood and sugarcane bagasse pellets in a Stirling engine micro-CHP system. *Appl. Therm. Eng.* **2019**, *159*, 113945. [CrossRef]
71. Napitupulu, F.H.; Ambarita, H. Manufacturing and testing prototype of a gamma type Stirling engine for micro-CHP application. *IOP Conf. Ser. Mater. Sci. Eng.* **2020**, *725*, 012016.
72. Stamford, L.; Greening, B.; Azapagic, A. Life cycle environmental and economic sustainability of Stirling engine micro-CHP systems. *Energy Technol.* **2018**, *6*, 1119–1138. [CrossRef]
73. Chmielewski, A.; Gumiński, R.; Lubikowski, K.; Maczak, J.; Szulim, P. Badania układu mikrokogeneracyjnego z silni-kiem stirlinga. *Część, II. Rynek Energ.* **2015**, *119*, 42–48.

74. Chmielewski, A.; Lubikowski, K.; Radkowski, S. Badania Temperaturowe i Analiza Współpracy UKładu Mikro-kogeneracyjnego z Silnikiem Gazowym. *Rynek Energ.* **2015**, *118*, 56–63.
75. Janowski, T.; Nalewaj, K.; Holuk, M. Układ kogeneracyjny z silnikiem Stirlinga. *Prz. Elektrotech.* **2014**, *90*, 63–64. [CrossRef]
76. Viessmann Group. We Create Living Spaces for Generations to Come. Available online: www.wiessmann.com (accessed on 20 December 2020).
77. Wołowicz, M. Silnik Stirlinga jako element układów mikrokogeneracyjnych. *Rynek Energii.* **2020**, *5*, 27–32.
78. Okofen Forschungs- und Entwicklungs Ges.m.b.H. Available online: www.oekofen.com (accessed on 20 December 2020).
79. Alberti, F.; Crema, L. Design of a New Medium-temperature Stirling Engine for Distributed Cogeneration Applications. *Energy Proced.* **2014**, *57*, 321–330. [CrossRef]
80. Crema, L.; Alberti, F.; Bertaso, A.; Bozzoli, A. Development of a pellet boiler with Stirling engine for m-CHP domestic application. *Energy Sustain. Soc.* **2011**, *1*, 5. [CrossRef]
81. Qiu, S.; Gao, Y.; Rinker, G.; Yanaga, K. Development of an advanced free-piston Stirling engine for micro combined heating and power application. *Appl. Energy* **2019**, *235*, 987–1000. [CrossRef]
82. Solomon, L.; Qiu, S. Computational analysis of external heat transfer for a tubular Stirling convertor. *Appl. Therm. Eng.* **2018**, *137*, 134–141. [CrossRef]
83. Grosu, L.; Dobre, C.; Petrescu, S. Study of a Stirling engine used for domestic micro-cogeneration. Thermodynamic analysis and experiment. *Int. J. Energy Res.* **2015**, *39*, 1280–1294. [CrossRef]
84. Mangion, R.; Muscat, M.; Sant, T.; Rizzo, J.; Ghirlando, R.; Cilia, J.; Mizzi, J.; Vural, S. Challenges in Developing a Solar Powered Stirling Engine for Domestic Electricity Generation. In Proceedings of the 9th International Conference on Heat Transfer, Fluid Mechanics and Thermodynamics, Malta, 16–18 July 2012; pp. 785–794.
85. González-Pino, I.; Pérez-Iribarren, E.; Campos-Celador, A.; Terés-Zubiaga, J.; Las-Heras-Casas, J. Modelling and experimental characterization of a Stirling engine-based domestic micro-CHP device. *Energy Convers. Manag.* **2020**, *225*, 113429. [CrossRef]
86. Jabari, F.; Mohammadi-Ivatloo, B.; Sharifian, M.B.B.; Nojavan, S. Design and robust optimization of a novel industrial continuous heat treatment furnace. *Energy* **2018**, *142*, 896–910. [CrossRef]
87. Zevenhoven, R.; Khan, U.; Haikarainen, C.; Saeed, L.; Tveit, T.M.; Saxén, H. Performance improvement of an indus-trial Stirling engine heat pump. In Proceedings of the ECOS 2020–33rd International Conference on Efficiency, Cost, Optimization, Simulation and Environmental Impact of Energy Systems, Osaka, Japan, 29 June–3 July 2020; pp. 1042–1053.
88. Dinesh, K.; Gowtham Raj, R.; Naresh, M.; Rakesh, N.; Sriram, R. Design and Fabrication Of Low Cost Stirling Engine For Low Duty Industrial Applications. *Int. J. Sci. Technol. Res.* **2014**, *3*, 75–78.
89. Alphonse, M.; Kumar, R.R.; Kumar, M.S.; Karthik, K. Design and thermal analysis of stirling engine using industrial applications of ceramic material. *Int. J. Mech. Eng. Technol.* **2017**, *8*, 273–282.
90. Gaddamwar, S.S.; Pawar, A.N.; Naik, P.A. Tjprc An Optimization of High Pressure and Temperature of SYNGAS in Underground Coal Mines by using CFD Analysis of Membrane Serpentine Tube. *Int. J. Mech. Prod. Eng. Res. Dev.* **2019**, *9*, 617–624. [CrossRef]
91. Jabari, F.; Nojavan, S.; Mohammadi-Ivatloo, B.; Ghaebi, H.; Mehrjerdi, H. Risk-constrained scheduling of solar Stirling engine based industrial continuous heat treatment furnace. *Appl. Therm. Eng.* **2018**, *128*, 940–955. [CrossRef]
92. Farsakoglu, O.F.; Alahmad, A. Comprehensive Design of Stirling Engine Based Solar Dish Power Plant with Solar Tracking System. *J. Electr. Electron. Syst.* **2018**, *7*, 1–5. [CrossRef]
93. Khosravi, A.; Syri, S.; Pabón, J.J.; Sandoval, O.R.; Caetano, B.C.; Barrientos, M.H. Energy modeling of a solar dish/Stirling by artificial intelligence approach. *Energy Convers. Manag.* **2019**, *199*, 112021. [CrossRef]
94. Das, D.C.; Sinha, N.; Roy, A. Small signal stability analysis of dish-Stirling solar thermal based autonomous hybrid energy system. *Int. J. Electr. Power Energy Syst.* **2014**, *63*, 485–498. [CrossRef]
95. Buscemi, A.; Brano, V.L.; Chiaruzzi, C.; Ciulla, G.; Kalogeri, C. A validated energy model of a solar dish-Stirling system considering the cleanliness of mirrors. *Appl. Energy* **2020**, *260*. [CrossRef]
96. Rahman, A.; Saikia, L.C.; Sinha, N. Automatic generation control of an interconnected two-area hybrid thermal system considering dish-stirling solar thermal and wind turbine system. *Renew. Energy* **2017**, *105*, 41–54. [CrossRef]
97. Mocanu, D.-A.; Badescu, V.; Bucur, C.; Stefan, I.; Carcadea, E.; Răboacă, M.S.; Manta, I. PLC Automation and Control Strategy in a Stirling Solar Power System. *Energies* **2020**, *13*, 1917. [CrossRef]
98. Kongtragool, B.; Wongwises, S. A review of solar-powered Stirling engines and low temperature differential Stirling engines. *Renew. Sustain. Energy Rev.* **2003**, *7*, 131–154. [CrossRef]
99. Lai, X.; Yu, M.; Long, R.; Liu, Z.; Liu, W. Dynamic performance analysis and optimization of dish solar Stirling engine based on a modified theoretical model. *Energy* **2019**, *183*, 573–583. [CrossRef]
100. Gao, X.; Shen, J.; He, X.; Tang, C.; Li, K.; Dai, W.; Li, Z.; Jia, J.; Gong, M.; Wu, J.F. Improvements of a room-temperature magnetic refrigerator combined with Stirling cycle refrigeration effect. *Int. J. Refrig.* **2016**, *67*, 330–335. [CrossRef]
101. Ahmadi, M.H.; Ahmadi, M.A.; Maleki, A.; Pourfayaz, F.; Bidi, M.; Açıkkalp, E. Exergetic sustainability evaluation and multi-objective optimization of performance of an irreversible nanoscale Stirling refrigeration cycle operating with Maxwell–Boltzmann gas. *Renew. Sustain. Energy Rev.* **2017**, *78*, 80–92. [CrossRef]
102. Suranjan, S.; John, J.S.; Mathew, A.J.; Jose, J.; Joshy, G.; Ramesh, A.; Sachidananda, H.K. Determination of coefficient of performance of stirling refrigeration sm. *Int. J. Innov. Technol. Explor. Eng.* **2019**, *8*, 2522–2529. [CrossRef]

103. Yin, Y.; Chen, L.; Wu, F. Performance analysis and optimization for generalized quantum Stirling refrigeration cycle with working substance of a particle confined in a general 1D potential. *Phys. E Low-Dimens. Syst. Nanostruct.* **2018**, *97*, 57–63. [CrossRef]
104. Djetel-Gothe, S.; Lanzetta, F.; Bégot, S. Second law analysis for the experimental performances of a cold heat exchanger of a stirling refrigeration machine. *Entropy* **2020**, *22*. [CrossRef]
105. Wadaskar, N.N.; Choudhary, S.K.; Askhedkar, R.D. Generation of heat transfer coefficient data in regenerator for stirling cycle refrigeration system. *Int. J. Eng. Adv. Technol.* **2019**, *8*, 698–705. [CrossRef]
106. Djetel-Gothe, S.; Bégot, S.; Lanzetta, F.; Gavignet, E. Design, manufacturing and testing of a Beta Stirling machine for refrigeration applications. *Int. J. Refrig.* **2020**, *115*, 96–106. [CrossRef]
107. Nie, W.; He, J.; Du, J. Performance characteristic of a Stirling refrigeration cycle in micro/nano scale. *Phys. A Stat. Mech. Appl.* **2009**, *388*, 318–324. [CrossRef]
108. Getie, M.Z.; Lanzetta, F.; Bégot, S.; Admassu, B.T.; Hassen, A.A. Reversed regenerative Stirling cycle machine for refrigeration application: A review. *Int. J. Refrig.* **2020**, *118*, 173–187. [CrossRef]
109. Sanaye, S.; Ardali, M.R. Estimating the power and number of microturbines in small-scale combined heat and power systems. *Appl. Energy* **2009**, *86*. [CrossRef]
110. Marcellan, A. An Exploration into the Potential of Microturbine Based Propulsion Systems for Civil Unmanned Aerial Vehicles. Master's Thesis, Faculty of Aerospace Engineering, Delft University of Technology (TU Delft), Delft, The Netherlands, 2015.
111. Ribau, J.; Silva, C.; Brito, F.P.; Martins, J. Analysis of four-stroke, Wankel, and microturbine based range extenders for electric vehicles. *Energy Convers. Manag.* **2012**, *58*. [CrossRef]
112. Soares, C. Microturbine Application and Performance. In *Microturbines*; Elsevier: Amsterdam, The Netherlands, 2007.
113. Kiciński, J.; Żywica, G. *Steam Microturbines in Distributed Cogeneration*; Springer: Berlin/Heidelberg, Germany, 2014.
114. Andreas, P. Weiß Volumetric Expander Versus Turbine–Which Is the Better Choice for Small Orc Plants. In Proceedings of the 3rd International Seminar on ORC Power System, Brussels, Belgium, 12–14 October 2015.
115. Takada, H.; Yasue, N.; Oda, Y.; Hachiya, M.; Ichimura, M. The consideration of fundamental specifications and the possibility of maintenance by remote monitoring on Bowman's Micro-gas-turbine CGS "TG80CG". *Proc. Natl. Symp. Power Energy Syst.* **2002**, *8*. [CrossRef]
116. Koepsell, M.; Pfeiffer, J.; Bouvy, C. Einsatz einer Mikro-Gasturbine zur Prozesswärmeerzeugung. *Gaswaerme Int.* **2004**, *53*, 32–39.
117. Irshad, M.; Siraj, K.; Raza, R.; Ali, A.; Tiwari, P.; Zhu, B.; Rafique, A.; Ali, A.; Ullah, M.K.; Usman, A. A brief description of high temperature solid oxide fuel cell's operation, materials, design, fabrication technologies and performance. *Appl. Sci.* **2016**, *6*, 75. [CrossRef]
118. Dutra, J.C.; Gonzalez-Carmona, M.A.; Lazaro-Alvarado, A.F.; Coronas, A. Modeling of a cogeneration system with a micro gas turbine operating at partial load conditions. *J. Sustain. Dev. Energy Water Environ. Syst.* **2017**, *5*. [CrossRef]
119. Matum, T.S.; Nontakaew, U. The performance prediction of a 30-KW multi-fueled micro gas turbine. *Int. J. Mech. Prod. Eng. Res. Dev.* **2019**, *9*. [CrossRef]
120. Gimelli, A.; Sannino, R. Thermodynamic model validation of Capstone C30 micro gas turbine. *Energy Proced.* **2017**, *126*, 955–962. [CrossRef]
121. Ananienkow, A.G. Energy from microturbines. *Energ. Gigawat* **2004**, *5*, 13–15.
122. Hirano, M.Y.; da Silva, C.L. Dairy cattle biogas usage in microturbines for energy generation and thermal exploitation. *Eng. Agric.* **2018**, *38*, 526–535. [CrossRef]
123. Amaro, J.; Mendiburu, A.Z.; de Carvalho, J.A. Thermodynamic study of syngas combustion in gas microturbines with regeneration composed with metallic and ceramic materials. *Appl. Therm. Eng.* **2019**, *157*. [CrossRef]
124. Mikielewicz, D.; Kosowski, K.; Tucki, K.; Piwowarski, M.; Stępién, R.; Orynycz, O.; Włodarski, W. Influence of different biofuels on the efficiency of gas turbine cycles for prosumer and distributed energy power plants. *Energies* **2019**, *12*. [CrossRef]
125. Seo, J.M.; Lim, H.S.; Park, J.Y.; Park, M.R.; Choi, B.S. Development and experimental investigation of a 500-W class ultra-micro gas turbine power generator. *Energy* **2017**, *124*. [CrossRef]
126. Capat, R. Experimental Tests of the Operating Conditions of a Micro Gas Turbine Device. *J. Energy Power Eng.* **2015**, *9*. [CrossRef]
127. Capata, R. Ultra Micro Gas Turbines. In *Efficiency, Performance and Robustness of Gas Turbines*; Faculty of Engineering: Rome, Italy, 2012.
128. Capata, R.; Kylykbashi, K.; Calabria, A.; Veroli, M. Di Experimental Tests on a Pre-Heated Combustion Chamber for Ultra Micro Gas Turbine Device: Air/Fuel Ratio Evaluation. *Engineering* **2016**, *8*. [CrossRef]
129. Calabria, A.; Capata, R.; Veroli, M.D.; Pepe, G. Testing of the Ultra-Micro Gas Turbine Devices (1–10 kW) for Portable Power Generation at University of Roma 1: First Tests Results. *Engineering* **2013**, *5*. [CrossRef]
130. Capata, R.; Saracchini, M. Experimental campaign tests on ultra micro gas turbines, fuel supply comparison and optimization. *Energies* **2018**, *11*, 799. [CrossRef]
131. Bo, A.; Giacomazzi, E.; Messina, G.; Di Nardo, A. Analysis of a Fuel Flexible Micro Gas Turbine Combustor Through Numerical Simulations. *J. Eng. Gas Turbines Power* **2018**, *140*. [CrossRef]
132. Ruedel, U.; Stefanis, V.; Ramaglia, A.D.; Florjancic, S. *Development of the New Ansaldo Energia Gas Turbine Technology Generation*; American Society of Mechanical Engineers: New York, NY, USA, 2017; Volume 3.
133. Ramaglia, A.D.; Ruedel, U.; Stefanis, V.; Florjancic, S. *Ansaldo Energia Gas Turbine Technology Developments*; American Society of Mechanical Engineers: New York, NY, USA, 2018; Volume 3.

134. Othman, N.F.; Boosroh, M.H. Effect of H_2 and CO contents in syngas during combustion using Micro Gas Turbine. In Proceedings of the IOP Conference Series: Earth and Environmental Science, Tomsk, Russian, 11–16 July 2016; Volume 32.
135. Villarroel-Schneider, J.; Malmquist, A.; Araoz, J.A.; Martí-Herrero, J.; Martin, A. Performance analysis of a small-scale biogas-based trigeneration plant: An absorption refrigeration system integrated to an externally fired microturbine. *Energies* **2019**, *12*. [CrossRef]
136. Władysław, K.; Krzysztof, K. Small steam turbines for distributed power generation-technical and economic conditions. *Rynek Energ.* **2017**, *133*, 41–46.
137. Orrok, G.A. Small steam turbines. *J. Am. Soc. Nav. Eng.* **1909**, *21*, 810–835. [CrossRef]
138. Hennauer, L.; Schmitt, B. New developmental trends for small industrial steam turbines. *Brennstoff-Wärme-Kraft* **1988**, *40*, 342–348.
139. Brown, C.O. Small steam turbines Equipment and Design. *Ind. Eng. Chem.* **1949**, *41*, 91A–92A. [CrossRef]
140. Żywica, G.; Bagiński, P.; Kiciński, J. Selected operational problems of high-speed rotors supported by gas foil bearings. *Tech. Mech.* **2017**, *37*, 339–346. [CrossRef]
141. Kiciński, J. Cogeneration in Small Scale—High Speed Microturbines Dynamic Analysis. In *Archives of Acoustics*; Institute of Fluid-Flow Machinery, Polish Academy of Sciences: Gdansk, Poland, 2010; Volume 35, pp. 175–182.
142. Zywica, G.; Drewczynski, M.; Kicinski, J.; Rzadkowski, R. Computational modal and strength analysis of the steam microturbine with fluid-film bearings. *J. Vib. Eng. Technol.* **2014**, *2*, 543–549.
143. Kicinski, J.; Zywica, G. The numerical analysis of the steam microturbine rotor supported on foil bearings. *Adv. Vib. Eng.* **2012**, *11*, 113–119.
144. Kiciński, J.; Żywica, G.; Kiciński, J.; Żywica, G. Introduction to Microturbines Vibration Analysis. In *Steam Microturbines in Distributed Cogeneration*; Springer: Berlin/Heidelberg, Germany, 2014; pp. 17–37.
145. Kiciński, J. *Analysis of the Vibrations of the Low Power ORC Turbines Operating Under Conditions of Strongly Developed Hydrodynamic instability*; American Society of Mechanical Engineer: New York, NY, USA, 2018; Volume 1922.
146. Żywica, G.; Kiciński, J.; Kaczmarczyk, T.; Ihnatowicz, E. Design and experimental investigation of the cogenerative domestic micro power plant with ORC system. *Mechanik* **2015**. [CrossRef]
147. Kaczmarczyk, T.Z.; Ihnatowicz, E.; Zywica, G.; Kicinski, J. Experimental investigation of the ORC system in a cogenerative domestic power plant with a scroll expanders. *Open Eng.* **2015**, *5*, 411–420. [CrossRef]
148. Kiciński, J. *Steam Microturbines–Own Study of the Institute of Fluid Flow Machinery*; Institute of Fluid Flow Machinery: Gdansk, Poland, 2014.
149. Wajs, J.; Mikielewicz, D.; Jakubowska, B. Performance of the domestic micro ORC equipped with the shell-and-tube condenser with minichannels. *Energy* **2018**, *157*. [CrossRef]
150. Wajs, J.; Mikielewicz, D.; Bajor, M.; Kneba, Z. Experimental investigation of domestic micro-CHP based on the gas boiler fitted with ORC module. *Arch. Thermodyn.* **2016**, *37*. [CrossRef]
151. Kosowski, K.; Piwowarski, M.; Stępień, R.; Włodarski, W.; Hirt, Ł. *Mikroturbiny. Badania Numeryczne i Eksperymentalne*; Fundacja Promocji Przemysłu Okrętowego i Gospodarki Morskiej: Gdańsk, Poland, 2016; ISBN 978-83-60584-60-6.
152. Kosowski, K.; Piwowarski, M.; Stępień, R.; Włodarski, W. Design and investigations of the ethanol microturbine. *Arch. Thermodyn.* **2018**, *39*. [CrossRef]
153. Alford, A.; Nichol, P.; Frisby, B. The Development of a Small High Speed Steam Microturbine Generator System. In Proceedings of the IOP Conference Series: Materials Science and Engineering, London, UK, 7–9 September 2015; Volume 90.
154. Kolasiński, P.; Błasiak, P.; Rak, J. Experimental investigation on multi-vane expander operating conditions in domestic CHP ORC system. *Energy Proced.* **2017**, *129*, 323–330. [CrossRef]
155. Giampaolo, T. *Compressor Handbook Principles and Practice*; CRC Press: Boca Raton, FL, USA, 2010; ISBN 0-88173-615-5.
156. Kolasiński, P. Experimental and modelling studies on the possible application of heat storage devices for powering the ORC (organic rankine cycle) systems. *Therm. Sci. Eng. Prog.* **2020**, *19*. [CrossRef]
157. Kolasinski, P. Application of the multi-vane expanders in orc systems–A review on the experimental and modeling research activities. *Energies* **2019**, *12*, 2975. [CrossRef]
158. Gnutek, Z.; Kolasinski, P. The application of rotary vane expanders in organic rankine cycle systems–Thermodynamic description and experimental results. *J. Eng. Gas Turbines Power* **2013**, *135*. [CrossRef]
159. Suankramdee, W.; Thongtip, T.; Aphornratana, S. Development of a sliding vane expander in a micro-scale ORC system for utilizing low-grade heat. *Energy Proced.* **2017**, *138*, 817–822. [CrossRef]
160. Materials from Armak Motors Company. Available online: www.armak.co.uk (accessed on 20 December 2020).
161. Termo2Power Company Own Materials. Available online: www.termo2power.com (accessed on 20 December 2020).
162. Norwood, Z.; Kammen, D.; Dibble, R. Testing of the katrix rotary lobe expander for distributed concentrating solar combined heat and power systems. *Energy Sci. Eng.* **2014**, *2*, 61–76. [CrossRef]
163. Norwood, Z.; Kammen, D.; Callaway, D.; Dibble, R. A better steam engine: Testing of the Katrix rotary lobe expander for distributed concentrating solar combined heat and power systems. In Proceedings of the World Renewable Energy Forum, WREF 2012, Including World Renewable Energy Congress XII and Colorado Renewable Energy Society (CRES) Annual Conference, Denver, CO, USA, 13–17 May 2012; Volume 1.
164. Wołowicz, M. Przegląd wybranych konstrukcji silników cieplnych małej mocy wykorzystywanych w procesach utylizacji ciepła odpadowego. *Rynek Energii* **2020**, *1*, 36–41.

165. Casari, N.; Fadiga, E.; Pinelli, M.; Suman, A.; Kovacevic, A.; Rane, S.; Ziviani, D. Numerical investigation of oil injection in a Roots blower operated as expander. *IOP Conf. Ser. Mat. Sci. Eng.* **2019**, *604*, 12075. [CrossRef]
166. Parthoens, A.; Dumont, O.; Ludovic, G.; Lemort, V. Experimental and Numerical Investigation of a Roots Expander Integrated into an ORC Power System. In Proceedings of the International Compressor Engineering Conference, West Lafayette, IN, USA, 9–12 July 2018.
167. Dumont, O.; Talluri, L.; Fiaschi, D.; Manfrida, G.; Lemort, V. Comparison of a scroll, a screw, a roots, a piston expander and a Tesla turbine for small-scale organic Rankine cycle. In Proceedings of the 5th International Seminar on ORC Power Systems, Athens, Greece, 9–11 September 2019.
168. Ziviani, D.; Gusev, S.; Schuessler, S.; Achaichia, A.; Braun, J.E.; Groll, E.A.; De Paepe, M.; Van Den Broek, M. Employing a Single-Screw Expander in an Organic Rankine Cycle with Liquid Flooded Expansion and Internal Regeneration. *Energy Proc.* **2017**, *129*, 379–386. [CrossRef]
169. Ziviani, D.; Suman, A.; Lecompte, S.; De Paepe, M.; Van Den Broek, M.; Spina, P.R.; Pinelli, M.; Venturini, M.; Beyene, A. Comparison of a single-screw and a scroll expander under part-load conditions for low-grade heat recovery ORC systems. *Energy Proc.* **2014**, *61*, 117–120. [CrossRef]
170. Wu, Y.T.; Wang, W.; Lei, B.; Zhi, R.P.; Ma, C.F. Developement and Demonstration on Single Screw Expander in Organic Rankine Cycle. In Proceedings of the 5th International Seminar on ORC Power Systems, Athens, Greece, 9–11 September 2019.
171. Wang, J.; Zhang, X.; Zhang, Y.; Zheng, Y.; Wang, W. Experimental study of single screw expander used in lowmedium temperature geothermal power system. *Energy Proc.* **2014**, *61*, 854–857. [CrossRef]
172. Ziviani, D.; Bell, I.; Van Den Broek, M.; De Paepe, M. Comprehensive Model of a Single-screw Expander for ORC-Systems. In Proceedings of the International Compressor Engineering Conference, Ghent, Belgium; 2014.
173. Desideri, A.; Van Den Broek, M.; Gusev, S.; Lemort, V.; Quoilin, S. Experimental Campaign and Modeling of a Low-capacity Waste Heat Recovery System Based on a Single Screw Expander. In Proceedings of the 22nd International Compressor Engineering Conference at Purdue, West Lafayette, IA, USA, 14–17 July 2014.
174. Ziviani, D.; Bell, I.; De Paepe, M.; Van Den Broek, M. Mechanistic model of an oil-flooded single-screw expander. In Proceedings of the 23rd International Compressor Engineering Conference at Purdue, West Lafayette, IA, USA, 11–14 July 2016.
175. Lei, B.; Wang, W.; Wu, Y.T.; Ma, C.F.; Wang, J.F.; Zhang, L.; Li, C.; Zhao, Y.K.; Zhi, R.P. Development and experimental study on a single screw expander integrated into an Organic Rankine Cycle. *Energy* **2016**, *116*, 43–52. [CrossRef]
176. He, W.; Wu, Y.; Peng, Y.; Zhang, Y.; Ma, C.; Ma, G. Influence of intake pressure on the performance of single screw expander working with compressed air. *Appl. Therm. Eng.* **2013**, *51*, 662–669. [CrossRef]
177. Zhang, X.; Zhang, Y.; Cao, M.; Wang, J.; Wu, Y.; Ma, C. Working fluid selection for organic Rankine cycle using single-screw expander. *Energies* **2019**, *12*, 3197. [CrossRef]
178. Bianchi, M.; Branchini, L.; Casari, N.; De Pascale, A.; Melino, F.; Ottaviano, S.; Pinelli, M.; Spina, P.R.; Suman, A. Experimental analysis of a micro-ORC driven by piston expander for low-grade heat recovery. *Appl. Therm. Eng.* **2019**, *148*, 1278–1291. [CrossRef]
179. Bianchi, M.; Branchini, L.; De Pascale, A.; Melino, F.; Ottaviano, S.; Peretto, A.; Torricelli, N. Performance prediction of a reciprocating piston expander with semi-empirical models. *Energy Proced.* **2019**, *158*, 1737–1743. [CrossRef]
180. Ancona, M.A.; Bianchi, M.; Branchini, L.; De Pascale, A.; Melino, F.; Orlandini, V.; Ottaviano, S.; Peretto, A.; Pinelli, M.; Spina, P.R.; et al. A Micro-ORC Energy System: Preliminary Performance and Test Bench Development. *Energy Proced.* **2016**, *101*, 814–821. [CrossRef]
181. Bouvier, J.L.; Lemort, V.; Michaux, G.; Salagnac, P.; Kientz, T. Experimental study of an oil-free steam piston expander for micro-combined heat and power systems. *Appl. Energy* **2016**, *169*, 788–798. [CrossRef]
182. Guarracino, I.; Mathie, R.; Taleb, A.; Markides, C. An Experimental Analysis Of A Low-Loss Reciprocating Piston Expander For Use In Small-Scale Organic Rankine Cycles. In Proceedings of the 2nd International Seminar on ORC Power Systems, Rotterdam, The Netherlands, 7–8 October 2013.
183. Markides, C.; Guarracino, I.; Mathie, R. Reciprocating Piston Expanders for Small-Scale ORC Systems. In Proceedings of the 2nd International Seminar on ORC Power Systems, Rotterdam, The Netherlands, 7–8 October 2013.
184. Glavatskaya, Y.; Podevin, P.; Lemort, V.; Shonda, O.; Descombes, G. Reciprocating expander for an exhaust heat recovery rankine cycle for a passenger car application. *Energies* **2012**, *5*, 1751–1765. [CrossRef]
185. Kim, Y.M.; Shin, D.G.; Kim, C.G. Optimization of design pressure ratio of positive displacement expander for vehicle engine waste heat recovery. *Energies* **2014**, *7*, 6105–6117. [CrossRef]
186. Galindo, J.; Ruiz, S.; Dolz, V.; Royo-Pascual, L.; Haller, R.; Nicolas, B.; Glavatskaya, Y. Experimental and thermodynamic analysis of a bottoming Organic Rankine Cycle (ORC) of gasoline engine using swash-plate expander. *Energy Convers. Manag.* **2015**, *103*, 519–532. [CrossRef]
187. Oudkerk, J.F.; Dickes, R.; Dumont, O.; Lemort, V. Experimental performance of a piston expander in a small- scale organic Rankine cycle. *IOP Conf. Ser. Mat. Sci. Eng.* **2015**, *90*, 012066. [CrossRef]
188. Hou, X.; Zhang, H.; Yu, F.; Liu, H.; Yang, F.; Xu, Y.; Tian, Y.; Li, G. Free piston expander-linear generator used for organic Rankine cycle waste heat recovery system. *Appl. Energy* **2017**, *208*, 1297–1307. [CrossRef]
189. Hou, X.; Zhang, H.; Xu, Y.; Yu, F.; Zhao, T.; Tian, Y.; Yang, Y.; Zhao, R. External load resistance effect on the free piston expander-linear generator for organic Rankine cycle waste heat recovery system. *Appl. Energy* **2018**, *212*, 1252–1261. [CrossRef]

190. Li, G.; Zhang, H.; Yang, F.; Song, S.; Chang, Y.; Yu, F.; Wang, J.; Yao, B. Preliminary development of a free piston expander-linear generator for small-scale Organic Rankine Cycle (ORC) waste heat recovery system. *Energies* **2016**, *9*, 300. [CrossRef]
191. Antonelli, M.; Francesconi, M.; Baccioli, A.; Caposciutti, G. Experimental Results of a Wankel-type Expander Fuelled by Compressed Air and Saturated Steam. *Energy Proced.* **2017**, *105*, 2929–2934. [CrossRef]
192. Francesconi, M.; Caposciutti, G.; Antonelli, M. An experimental and numerical analysis of the performances of a Wankel steam expander. *Energy* **2018**, *164*, 615–626. [CrossRef]
193. Badr, O.; Naik, S.; O'Callaghan, P.W.; Probert, S.D. Rotary Wankel engines as expansion devices in steam Rankine-cycle engines. *Appl. Energy* **1991**, *39*, 59–76. [CrossRef]
194. Badr, O.; Naik, S.; O'Callaghan, P.W.; Probert, S.D. Wankel engines as steam expanders: Design considerations. *Appl. Energy* **1991**, *40*, 157–170. [CrossRef]
195. Badr, O.; Naik, S.; O'Callaghan, P.W.; Probert, S.D. Expansion machine for a low power-output steam Rankine-cycle engine. *Appl. Energy* **1991**, *39*, 93–116. [CrossRef]
196. Tozer, G.; Ga, J.; Al-Dadah, R.; Mahmoud, S. Development of Efficient Static Shaft Wankel Expander for Organic Rankine Cycles. In Proceedings of the 5th International Seminar on ORC Power Systems, Athens, Greece, 9–11 September 2019.
197. Park, K.-T.; Kim, Y.-H.; Kim, H.-J. Design and performance analysis of a gerotor expander for power generation from waste heat. *J. Korea Soc. Power Syst. Eng.* **2016**, *20*, 17–25. [CrossRef]
198. Mathias, J.A.; Johnston, J.R.; Cao, J.; Priedeman, D.K.; Christensen, R.N. Experimental testing of gerotor and scroll expanders used in, and energetic and exergetic modeling of, an organic Rankine cycle. *J. Energy Resour. Technol. Trans. ASME* **2009**, *131*, 0122011–0122019. [CrossRef]
199. Yamada, N.; Tominaga, Y.; Yoshida, T. Demonstration of 10-Wp micro organic Rankine cycle generator for low-grade heat recovery. *Energy* **2014**, *78*, 806–813. [CrossRef]
200. Webb, K.R. Sir William Robert Grove (1811–1896) and the origins of the fuel cell. *J. R. Inst. Chem.* **1961**, *85*, 291–295.
201. William Robert Grove. *Phys. Today*, 11 July 2016. [CrossRef]
202. Gray, A.A. Sir William Robert Grove. *Nature* **1896**, *54*. [CrossRef]
203. Wisniak, J. Historical Notes: Electrochemistry and Fuel Cells: The Contribution of William Robert Grove. *Indian J. Hist. Sci.* **2015**, *50*. [CrossRef]
204. Appleby, A.J. From Sir William Grove to today: Fuel cells and the future. *J. Power Source* **1990**, *29*. [CrossRef]
205. Haseli, Y. Maximum conversion efficiency of hydrogen fuel cells. *Int. J. Hydrogen Energy* **2018**, *43*, 9015–9021. [CrossRef]
206. Giorgi, L. Fuel Cells: Technologies and Applications. *Open Fuel Cells J.* **2013**, *6*. [CrossRef]
207. Carrette, L.; Friedrich, K.A.; Stimming, U. Fuel cells: Principles, types, fuels, and applications. *ChemPhysChem* **2000**, *1*, 162–193. [CrossRef]
208. Lucia, U. Overview on fuel cells. *Renew. Sustain. Energy Rev.* **2014**, *30*, 164–169. [CrossRef]
209. Coralli, A.; Sarruf, B.J.M.; De Miranda, P.E.V.; Osmieri, L.; Specchia, S.; Minh, N.Q. Fuel Cells. In *Science and Engineering of Hydrogen-Based Energy Technologies: Hydrogen Production and Practical Applications in Energy Generation*; Academic Press: Cambrigde, MA, USA, 2018; pp. 39–122. ISBN 9780128142516.
210. Kunze-Liebhäuser, J.; Paschos, O.; Pethaiah, S.S.; Stimming, U. Fuel Cell Comparison to Alternate Technologies. In *Fuel Cells and Hydrogen Production*; Springer: New York, NY, USA, 2019; pp. 11–25.
211. Kwan, T.H.; Wu, X.; Yao, Q. Performance comparison of several heat pump technologies for fuel cell micro-CHP integration using a multi-objective optimisation approach. *Appl. Therm. Eng.* **2019**, *160*. [CrossRef]
212. Kunze-Liebhäuser, J.; Paschos, O.; Pethaiah, S.S.; Stimming, U. Fuel Cell Comparison to Alternate Technologies. In *Encyclopedia of Sustainability Science and Technology*; Springer: New York, NY, USA, 2017; pp. 1–16.
213. Milewski, J.; Wolowicz, M.; Badyda, K.; Misztal, Z. Operational characteristics of 36kW PEMFC-CHP unit. *Rynek Energ.* **2011**, *92*, 150–156.
214. Milewski, J.; Wolowicz, M.; Badyda, K.; Misztal, Z. 36 kW Polymer Exchange Membrane Fuel Cell as Combined Heat and Power Unit. *ECS Trans.* **2019**, *42*, 75–87. [CrossRef]
215. Calì, M.; Santarelli, M.G.L.; Leone, P. Design of experiments for fitting regression models on the tubular SOFC CHP 100 kWe: Screening test, response surface analysis and optimization. *Int. J. Hydrog. Energy* **2007**, *32*, 343–358. [CrossRef]
216. Cinti, G.; Bidini, G.; Hemmes, K. Comparison of the solid oxide fuel cell system for micro CHP using natural gas with a system using a mixture of natural gas and hydrogen. *Appl. Energy* **2019**, *238*, 69–77. [CrossRef]
217. Naimaster, E.J.; Sleiti, A.K. Potential of SOFC CHP systems for energy-efficient commercial buildings. *Energy Build.* **2013**, *61*, 153–160. [CrossRef]
218. Suzuki, M.; Iwata, S.; Higaki, K.; Inoue, S.; Shigehisa, T.; Miyachi, I.; Nakabayashi, H.; Shimazu, K. Development and Field Test Results of Residential SOFC CHP System. *ECS Trans.* **2019**, *25*, 143–147. [CrossRef]
219. Palomba, V.; Ferraro, M.; Frazzica, A.; Vasta, S.; Sergi, F.; Antonucci, V. Experimental and numerical analysis of a SOFC-CHP system with adsorption and hybrid chillers for telecommunication applications. *Appl. Energy* **2018**, *216*, 620–633. [CrossRef]
220. Yang, C.H.; Chang, S.C.; Chan, Y.H.; Chang, W.S. A dynamic analysis of the multi-stack SOFC-ChP system for power modulation. *Energies* **2019**, *12*, 3686. [CrossRef]
221. Heatstack to cut cost of components in fuel cell mCHP. *Fuel Cells Bull.* **2016**, *12*. [CrossRef]

222. Sghaier, S.F.; Khir, T.; Ben Brahim, A. Energetic and exergetic parametric study of a SOFC-GT hybrid power plant. *Int. J. Hydrog. Energy* **2018**, *43*, 3542–3554. [CrossRef]
223. Badur, J.; Lemański, M.; Kowalczyk, T.; Ziółkowski, P.; Kornet, S. Zero-dimensional robust model of an SOFC with internal reforming for hybrid energy cycles. *Energy* **2018**, *158*, 128–138. [CrossRef]
224. Chen, J.; Liang, M.; Zhang, H.; Weng, S. Study on control strategy for a SOFC-GT hybrid system with anode and cathode recirculation loops. *Int. J. Hydrog. Energy* **2017**, *42*, 29422–29432. [CrossRef]
225. Meng, Q.; Han, J.; Kong, L.; Liu, H.; Zhang, T.; Yu, Z. Thermodynamic analysis of combined power generation system based on SOFC/GT and transcritical carbon dioxide cycle. *Int. J. Hydrog. Energy* **2017**, *42*, 4673–4678. [CrossRef]
226. Liu, Y.; Han, J.; You, H. Performance analysis of a CCHP system based on SOFC/GT/CO2 cycle and ORC with LNG cold energy utilization. *Int. J. Hydrog. Energy* **2019**, *44*, 29700–29710. [CrossRef]
227. Huang, Y.; Turan, A. Fuel sensitivity and parametric optimization of SOFC–GT hybrid system operational characteristics. *Therm. Sci. Eng. Prog.* **2019**, *14*. [CrossRef]
228. Barelli, L.; Bidini, G.; Ottaviano, A. Integration of SOFC/GT hybrid systems in Micro-Grids. *Energy* **2017**, *118*, 716–728. [CrossRef]
229. Harun, N.F.; Tucker, D.; Adams, T.A. Technical challenges in operating an SOFC in fuel flexible gas turbine hybrid systems: Coupling effects of cathode air mass flow. *Appl. Energy* **2017**, *190*, 852–867. [CrossRef]
230. Tompsett, G.A.; Finnerty, C.; Kendall, K.; Sammes, N.M. Integrated Catalytic Burner/Micro-SOFC Design and Applications. *Electrochemistry* **2000**, *68*, 519–521. [CrossRef]
231. Wang, J.; Yan, Z.; Ma, S.; Dai, Y. Thermodynamic analysis of an integrated power generation system driven by solid oxide fuel cell. *Int. J. Hydrog. Energy* **2012**, *37*, 2535–2545. [CrossRef]
232. Pianko-Oprych, P.; Jaworski, Z. Numerical investigation of a novel burner to combust anode exhaust gases of SOFC stacks. *Pol. J. Chem. Technol.* **2017**, *19*, 20–26. [CrossRef]
233. Zeng, H.; Wang, Y.; Shi, Y.; Cai, N. Biogas-fueled flame fuel cell for micro-combined heat and power system. *Energy Convers. Manag.* **2017**, *148*, 701–707. [CrossRef]
234. Milcarek, R.J.; Garrett, M.J.; Ahn, J. Micro-tubular flame-assisted fuel cells. *J. Fluid Sci. Technol.* **2017**, *12*. [CrossRef]
235. Milcarek, R.J.; Wang, K.; Falkenstein-Smith, R.L.; Ahn, J. Micro-tubular flame-assisted fuel cells for micro-combined heat and power systems. *J. Power Source* **2016**, *306*, 148–151. [CrossRef]

energies

MDPI

Review

Adsorptive Biogas Purification from Siloxanes—A Critical Review

Kazimierz Gaj

Faculty of Environmental Engineering, Wroclaw University of Science and Technology, 50-370 Wrocław, Poland; kazimierz.gaj@pwr.edu.pl

Received: 17 April 2020; Accepted: 13 May 2020; Published: 20 May 2020

Abstract: Siloxanes are among the most technologically troublesome trace compounds present in biogas. As a result of their combustion, hard-to-remove sediments are formed, blocking biogas energy processing devices and reducing the efficiency of biogas plants. The purpose of this study was to help investors and designers to choose the optimal technology for the adsorptive removal of volatile methylsiloxanes (VMSs) from biogas and to identify adsorbents worth further development. This paper critically reviews and discusses the state-of-the-art technologies for the adsorption removal of siloxanes from biogas, indicating potentially beneficial directions in their development and deficiencies in the state of knowledge. The origin of VMSs in biogas, their selected physicochemical properties, technological problems that they can cause and their typical versus limit concentrations in biogases are presented. Both the already implemented methods of adsorptive VMSs removal from landfill and sewage gases and the ones being under development are verified and systematized. The parameters and effectiveness of adsorption processes are discussed, and individual adsorbents are compared. Possible ways of regenerating spent adsorbents are evaluated and prospects for their application are assessed. Finally, zeolite-based adsorbents—which can also be used for biogas desulfurization—and adsorbents based on polymer resins, as being particularly active against VMSs and most amenable to multiple regeneration, are identified.

Keywords: biogas treatment; volatile methylsiloxanes; siloxane concentrations; adsorbents; regeneration

1. Introduction

The most troublesome trace impurities present in biogas are volatile organic silicon compounds—siloxanes, which are oligomeric chemicals containing alternately connected silicon and oxygen atoms. Silicon atoms are additionally linked with hydrocarbon functional groups, mainly methyl (Table 1). In this case, considering their volatility under ambient conditions (Table 2), siloxanes are referred to as volatile methylsiloxanes (VMSs).

The main source of VMSs in biogas are personal care products, in which VMSs, owing to their unique characteristics such as: low surface tension, viscosity and chemical reactivity, high thermal stability, compressibility, spreadability and hydrophobicity (Table 2), have been rapidly increasingly used in recent years. Along with wastewater containing used cosmetics—especially shampoos, soaps, hair sprays, antiperspirants, etc.—VMSs end up in wastewater treatment plants (WWTPs), where, together with sediments, they get into the fermentation chambers and as a result of the elevated temperature are released into biogas. VMSs also get into the biogas in landfills, where cosmetics containers, together with their remains, are deposited. The growing presence of VMSs in the environment is also due to the rapidly growing production of silicone polymers—used in medicine, electronics and various everyday articles—for which VMSs are the basic building blocks.

Table 1. Nomenclature and formulas of main volatile methylsiloxanes (VMSs) detected in biogases.

Chemical Name	Abbreviations	CAS	Chemical Formula	Structural Pattern
Hexamethylcyclotrisiloxane	D3 (HMCTS)	541-05-9	Si_3-O_3-$(CH_3)_6$	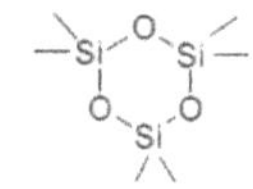
Octamethylcyclotetrasiloxane	D4 (OMCTS)	556-67-2	Si_4-O_4-$(CH_3)_8$	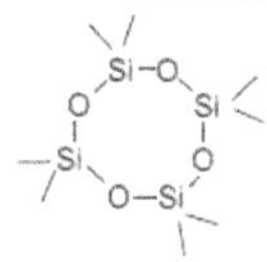
Decamethylcyclopentasiloxane	D5	541-02-6	Si_5-O_5-$(CH_3)_{10}$	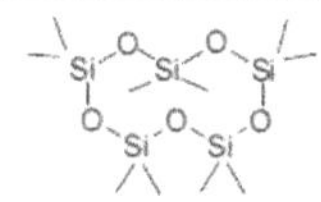
Dodecamethylcyclohexasiloxane	D6	540-97-6	Si_6-O_6-$(CH_3)_{12}$	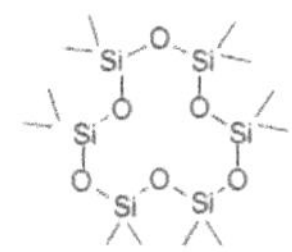
Hexamethyldisiloxane	L2 (HMDS)	107-46-0	Si_2-O-$(CH_3)_6$	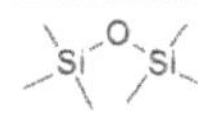
Octamethyltrisiloxane	L3 (MDM)	107-51-7	Si_3-O_2-$(CH_3)_8$	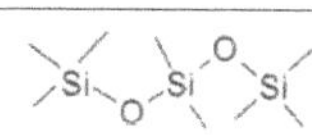
Decamethyltetrasiloxane	L4 (MD_2M)	141-62-8	Si_4-O_3-$(CH_3)_{10}$	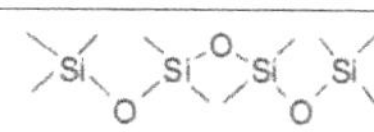
Dodecamethylpentasiloxane	L5 (MD_3M)	141-63-9	Si_5-O_4-$(CH_3)_{12}$	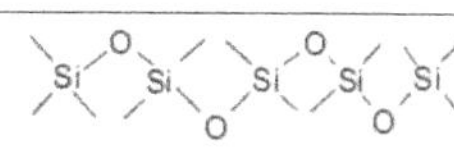

Table 2. Selected physicochemical properties of VMSs found in biogas, developed according to [1].

Compound	Physical State at 20 °C, 101.3 kPa	Density, kg/m^3, at 25 °C	Vapor Pressure, kPa, at 25 °C	Water Solubility, mg/dm^3 at 25 °C	Boiling Point, °C, at 101.3 kPa	Melting Point, °C at 101.3 kPa
D3	Solid	1020	1.16	1.571	135	64
D4	Liquid	956	0.14	0.055	176	17.5
D5	Liquid	959	0.03	0.046	210	−38
D6	Liquid	967	0.003	0.005	245	−3
L2	Liquid	764 [1]	5.613	2.881	101	−66
L3	Liquid	820	0.445	0.15	152	−80
L4	Liquid	854	0.05	0.007	195	−76
L5	Liquid	875	0.013	0.0003	232	−80

[1] at 20 °C.

As a result of biogas combustion, a part of the VMSs are transformed into crystalline silica which settles in the form of rough, hardly removable deposits in combustion chambers, heat exchangers, exhaust gas catalysts and on cylinder heads, valves, spark plugs, turbine blades, etc., reducing their performance and increasing servicing costs and air pollutants emissions. This is currently one of the main operational problems affecting biogas plants. The need for biogas purification from VMSs also arises from their potential toxic and carcinogenic properties found in animal studies in recent years. It has been shown that only a small fraction (<1%) of VMSs convert to SiO_2 during biogas combustion [2]. The rest are emitted into the air, showing great potential for bioconcentration and bioaccumulation in the environment's various compartments [3]. In addition, microcrystalline silica (<100 nm), which is the oxidation product of VMSs, has carcinogenic and mutagenic properties [4].

Adsorption, absorption and cooling methods have been commercially used to purify biogas from VMSs. However, owing to the uncomplicated service, high efficiency and moderate costs of such standard adsorbents as activated carbon and silica gel, carbon adsorption methods predominate.

The emerging methods (still under research), except for membrane and biological methods which are not promising so far, use unconventional adsorbents based on natural and synthetic zeolites and polymer resins.

2. VMSs Concentrations in Biogas

Most measurements of VMSs content in landfill gas (LFG) and in biogas from sewage sludge (SG) were carried out in Germany, with a much greater spread of results found in WWTPs (from zero to several hundred mg/m^3) than in landfills (from several to several dozen mg/m^3) [5]. For such a large data set (308 objects) the reported average total VMSs concentration of 14.9 mg/m^3 in SG, seems representative. A higher average total VMSs content—24 mg/m^3 (mainly D4—12.5 mg/m^3, L2—6.1 mg/m^3 and D5—4.7 mg/m^3)—measured in LFG in Asten (Austria) was reported by Accettola and Haberbauer [6], while in SG the average concentration amounted to 7 mg/m^3 (primarily D5—6.0 mg/m^3 and D4—0.9 mg/m^3). Even higher values for SG were indicated by Appels et al. [7]—30–50 mg/m^3. Arnold & Kajolinna [8] showed that the concentration of VMSs in biogas usually does not exceed 10 mg/m^3, although in the case of SG, peaks of up to 300–400 mg/m^3 are possible. For example, 400 mg/m^3 was recorded in the Trecatti WWTP in Great Britain, when the engine failed after 200 hours of operation [9]. In other tested WWTPs (Zurich, Neuburg, Saint-Truiden, Minworth) the total VMSs concentration ranged from 16 to 59.8 mg/m^3, while in landfills in Berlin, Augsburg and Vienna it ranged from 4.8 to 36.3 mg/m^3. A similar maximum VMSs concentration in SG as in Trecatti (314 mg/m^3) was detected by Hepburn et al. [10]. A wide overview of VMSs concentration measurement results at Austrian, German and Finnish landfills and WWTPs was presented by Rasi et al. [11,12]. According to their research, in Finland the concentrations were in the range of 0.08–2.5 mg/m^3 for both types of biogas, i.e., significantly below the values presented above. D5 predominated in SG (48–76%), constituting together with D4 (21–45%) over 90% of the total VMSs. The remaining 10% were: L2 (0–1%), D3 (1–4%) and L3 (0–6%). In LFG the VMS percentages were as follows: D4 (24–72%), L2 (28–40%), D5 (3–13%) and D3 (0–7%), i.e., VMSs characterized by higher vapor pressure and better water solubility predominated. Generally, VMSs concentrations in biogas depend on its origin and they range widely—typically from a few to several tens mg/m^3. In LFG, where D4 and L2 predominate, they are usually lower than in SG (Table 3).

Table 3. VMSs concentration ranges detected in biogas.

Kind of Biogas	Location	Number of Objects/ Measurements	Analyzed Compounds	Concentration, mg/m^3	Source
Landfill Gas (LFG)	Germany	123/340	Sum of VMSs	3–25	[5]
	Finland	3/6	L2-L4 + D3-D5	2–8	
	Poland	1/7	Sum of VMSs	18–39	[13]
	Germany	ND	L2-L4 + D3-D6	Up to 50	[14]
	Europa (6), Canada (4)	10/ND [1]	L2-L5 + D3-D6	2–24	[15]

Table 3. *Cont.*

Kind of Biogas	Location	Number of Objects/ Measurements	Analyzed Compounds	Concentration, mg/m^3	Source
Biogas from Sewage Sludge (SG)	Germany	308/ND	L2-L4 + D3-D5	Up to 317	[5]
	Finland	3/6	L2-L4 + D3-D5	2–30	
	Europa (9), Japan (1)	10/ND	L2-L5 + D3-D6	3–127	[15]
	United Kingdom	6/ND	D4 + D5	12–179	[16]

[1] No data.

In order to minimize the hazards associated with the use of biogas energy, biogas power generation equipment manufacturers impose VMS level standards (Table 4).

In many cases it is necessary to remove VMSs from biogas to prevent equipment blockage and meet the manufacturer's warranty requirements.

Table 4. VMSs concentration limits in biogas, recommended by some device manufacturers.

Application of Biogas	VMS Concentration Limit, mg Si/m^3	Source
Piston Engines	5–30	[10,17]
Turbines	0.1	[18]
Microturbines	0.03	[17]
Fuel Cells	0.05–0.10	
Catalytic Afterburners	0.50–0.38	[19]
Stirling Engines	No limit	[17]
Vehicle Engines	No limit	[20]
Natural Gas Grid	6.2 [1], 10 [2]	[21]

[1] in Netherlands; [2] in Austria. A broader overview of the requirements for biogas quality can be found in [22,23].

3. Adsorption of VMSs Using Activated Carbon

Activated carbon is formed in the processes of carbonization and thermal or chemical-thermal activation of such organic raw materials as fossil coals, wood and various types of organic waste. During these processes the organic matter is partially decomposed, and volatiles are removed, whereby the activated carbon acquires a proper porous structure. Carbonization is conducted at a temperature of 600–800 °C. Then the carbonizate is subjected to physical activation, i.e., further thermal treatment with steam or CO_2 (400–500 °C) to eliminate the remaining volatiles, or to partial gasification at a temperature of 800–1000 °C. The alternative is chemical activation, i.e., roasting (500–900 °C) with the addition of zinc chloride, potassium sulfide, phosphoric acid or other chemicals [24]. All these processes lead to the formation of an extremely extensive internal structure consisting of a wide spectrum of micropores and mesopores. As a result, a universal adsorbent with the largest specific surface area—which can exceed 2000 m^2/g—among all the known adsorbents is produced. Small activated carbon rolls with a diameter of about 1–2 mm and a length of up to 4 mm are most often used in gas cleaning processes. Activated carbon fibers are used less often, but they can be more effective in removing D4 than granular activated carbon [25,26]. The feature that particularly predisposes activated carbon to VMSs removal from biogas is its non-polar structure [27] promoting the adsorption of non-polar or weakly polar VMSs. Larger pore volumes, more suitable for the size of VMSs molecules, are obtained using chemical activation with H_3PO_4 [28]. Generally, many researchers indicate that adsorbent pores with a diameter of 1.7–6.0 nm are most suitable for the adsorption of VMSs whose molecule diameter is about 1 nm.

Depending on the method of activation (and/or impregnation), activated carbon can have an acidic or basic character. According to Gong et al. [29], activated carbon activity towards VMSs increases with the increasing pH of the bed. However, alkaline impregnation, which can be used for H_2S removal, can reduce the adsorbent's active surface area and pore volume, which, in turn, can reduce its capacity to absorb VMSs [30].

The large diversity of activated carbon's pores in comparison with other adsorbents is, unfortunately, also its disadvantage, being the main cause of its low selectivity. Besides the sulfur and chlorine compounds mentioned above, activated carbon has a high affinity for other VOCs, water steam and hydrocarbons—CH_4 adsorption on activated carbon is greater than on silica gel and molecular sieve [24]. The presence of the above compounds in biogas, competing for a place in activated carbon's micro- and mesopores, reduces its capacity to adsorb VMSs and directly affects the operating costs.

Activated carbon shows greater activity towards cyclic VMSs, which is connected with the changes they undergo on its surface and the displacement of the previously adsorbed lighter linear VMSs—e.g., L2 and L3—by heavier and less volatile cyclic VMSs—e.g., D4 and D5—or by other high-molecular aromatic VOCs and water vapor [31]. A way to reduce the above disadvantages is biogas pretreatment, consisting in condensation drying—usually up to RH (relative humidity) < 50%—during which interfering impurities and some VMSs are partly removed with the condensate.

A more serious disadvantage of activated carbon in the removal of VMSs is their transformation, consisting in the polymerization of cyclic VMSs into less volatile, heavier and hardly desorbable polydimethylsiloxanes [32] which block the adsorbent's pores and prevent its effective thermal regeneration. This process is more intense on wood-based carbons activated with phosphoric acid than with water steam [33]. Chemical regeneration through the oxidation of the adsorbed VMSs (with O_3, H_2O_2 or iron salts) also fails because the final reaction product is SiO_2 which blocks pores [33,34].

4. Adsorption of VMSs Using Silica Gel

Silica gel is obtained by polymerizing silicic acid with H_2SO_4 or HCl to an amorphous gel having the molecular formula $(SiO_2)\cdot nH_2O$. This adsorbent is typically used in the form of beads with a diameter of 1–5 mm. It is characterized by high porosity and a polar structure. The latter feature endows silica gel with a special affinity for water, making it one of the most effective desiccants. Therefore, deep drying—below 10% RH—is required in order to effectively remove VMSs from biogas [14]. In comparison with activated carbon, silica gel is characterized by a smaller specific surface area (350–700 m^2/g) and larger pore sizes, with the predominance of mesopores, which is beneficial for VMSs adsorption. Most researchers—e.g., [18,35,36]—indicate silica gel's higher selectivity and affinity for VMSs (provided that the biogas is properly dried), especially in the case of lighter chain forms, such as L2. Since silica gel has a lower capacity to adsorb sulfur compounds, it can be more useful for the purification of LFG which usually contains less H_2S. Most researchers also indicate that silica gel is more amenable to regeneration [14]. According to Yang [24], since silica gel's adsorption forces are weaker, it should be easier to thermally regenerate than activated carbon.

In recent years, besides the use of conventional silica gel, attempts have been made to modify silica gel in order to improve its adsorption capacity and regenerability. By treating silica gel with acetic anhydride Liu et al. [37] obtained an interesting hydrophobic silica gel modification whose adsorption capacity, regardless of biogas humidity, amounted to 304 g/kg for L2 and to 916 g/kg for D4—much more than in the case of activated carbons. Its regeneration was trouble-free at a relatively low temperature of 110 °C.

5. Adsorption of VMSs Using Zeolites

Natural zeolites comprise hydrated alkali metal aluminosilicate minerals with a crystalline structure, such as modernite, chabasite, clinoptilolite, silicalite and others. Their activation is based on thermal treatment, as a result of which they lose water and acquire an ordered internal structure with uniform pore sizes (hence the name "molecular sieves"), whereby, as opposed to active carbons, they become selective adsorbents. Zeolites are also made synthetically—e.g., 3A, 4A, 5A, 10X, 13X, ZSM-5. Depending on the ratio of silicon to aluminum atoms, they can be hydrophobic (Si/Al>10) or hydrophilic (Si/Al < 1.5). Hydrophobic zeolites are more useful for removing nonpolar and weakly polar compounds, including VMSs. For example, zeolite such as ZSM-5 with a high Si/Al ratio (200) [38] can be used simultaneously for CO_2 and H_2O removal in the biogas upgrading process and for the removal of VMSs and H_2S.

Zeolites are also characterized by high mechanical and thermal resistance (up to 600 °C). In comparison with activated carbon and silica gel, they have a comparable or slightly smaller specific surface area (370–910 m^2/g). Owing to the uniform size of their pores they can work more efficiently if they are selected with regard to the composition of the particular biogas. Unlike silica gel, they can simultaneously remove H_2S—thanks to their alkaline nature—and their affinity for CH_4 is several times lower than for activated carbon [24].

The usefulness of zeolites for removing VMSs from biogas has been confirmed by numerous studies, e.g., [31,39,40]. Owing to their high thermal resistance they are also suitable for repeated thermal regeneration [25].

6. Adsorption of VMSs Using Activated Alumina

As recent research has shown one of the effective VMSs adsorbents is activated alumina [41,42], which is obtained by calcining hydrated aluminum hydroxide in the presence of oxygen at a temperature of about 400 °C. The resulting porous, crystalline structure is characterized by a distinct predominance of mesopores (3–7 nm) and a specific surface area of 200–250 m^2/g. Further roasting (up to 800 °C) leads to an amorphous form with a larger specific surface area amounting to 300–400 m^2/g.

The synthetic adsorbent based on alumina (Al120-8h) developed by Zhong et al. [41], due to its larger specific surface area and porosity, proved to be a better D4 adsorbent than the commercial activated alumina. Its advantage is the possibility of effective regeneration with no apparent VMSs polymerization effect. It should be noted, however, that due to its relatively large pores, activated alumina is better suited for cyclic VMSs.

7. Adsorption of VMSs Using Polymer Resins

Polymer adsorbents are obtained by polymerizing monomers, i.e., styrene cross-linked with divinylbenzene (DVB) and acrylates, in the presence of an organic solvent. Popular DVB-based resins include XAD-2 and XAD-4 hydrophobic polymers, whereas the ones based on acrylic ester (e.g., XAD-7) are hydrophilic [24]. Especially the former—as opposed to activated carbons and silica gel—are well suited for the purification of moist biogas, the more so because they are non-polar. Unfortunately, their thermal strength is low (<200 °C), which can potentially hinder the thermal desorption of some VMSs—see boiling point in Table 2. Owing to their dense cross-linking they are relatively stiff and mechanically strong. This also contributes to their high porosity and specific surface area (up to 1000 m^2/g). In addition to the traditional polymer adsorbents mentioned above, in recent years new resins for VMSs adsorption have been synthesized. A particularly interesting series of adsorbents (PDVB-VI) based on the copolymerization of divinylbenzene with 1-vinylimidazole have been developed by Jafari et al. [43]. These adsorbents have an extremely large pore volume and specific surface area, whereby their adsorption capacity for VMSs is very high—in the order of 2000 g/kg. In addition, they can be effectively regenerated at a relatively low temperature (~100 °C), which indicates no polymerization of VMSs on their surface. The ease of regeneration of polymer adsorbents has also been confirmed by other authors [44,45]. Despite such advantageous characteristics, these adsorbents are not commercially used for gas purification due to their high price.

8. Comparison of the Adsorbents for VMSs Removal from Biogas

A comparative analysis of selected adsorbents, their effectiveness and advantageous/disadvantageous characteristics is presented in Table 5.

Table 5. Summary review of the adsorbents for removing VMSs from biogas.

Parameter/Feature	Activated Carbons (AC)	Silica Gels (SG)	Zeolites (ZE)	Activated Aluminas (AA)	Polymer Resins
Typical Specific Surface Area (BET), m^2/g	600–1600	350–700	370–910	Commercial AA: 200–250 Synthetic AA (Al120-8h): ~310	RS1: ~940; PDVB: ~830 PDVB-VI: 594–780 P(DVB-ACAM): ~270
Typical Average Pore Diameter, nm	1.4–2.0	2.2–2.4	3–9	Commercial AA: 3–7 Synthetic AA (Al120-8h): ~4	PDVB: 1.5–60
Typical Total Pore Volume, cm^3/g	0.4–1.1	0.2–0.4	0.2–0.5	Commercial AA: 0.3–0.4 Synthetic AA (Al120-8h): 0.5	PDVB: ~2 PDVB-VI: 1.2–1.8
Typical Micropore Volume, %	45.4–93.7	~50	~70	Commercial AA: ~7	ND
Adsorption Capacity Tested on Synthetic Biogas, Carrier Gas—N_2, g/kg	Total VMSs: 155–307 L2: 10–123 D4: 36–404 D5: 47–531	Data for dry gas (RH < 10%): Total VMSs: ~200 D4: 216–259 D5: ~100	ZE 13X: 77–276 [1] ZE 8A: ~4 [1] ZE ZSM-5: ~40 [1] ZE UCT-15: ~80 [1] Clinoptilolite: ~10 [1]	Commercial AA: ~130 [1] Synthetic AA (Al120-8h): ~170 [1]	RS1: ~300 [1,2] PDVB-VI: 1381–2370 [1] P(DVB-ACAM): ~2200 [1]
Advantages	Best tested, mastered, simple and widely available technology, easy to operation, possibility to simultaneous removal of other biogas impurities, i.e., sulfur and chlorine compounds, relatively cheap adsorbent: ~2 €/kg.	High mechanical and thermal resistance, non–flammability, chemical and biological inertness, higher selectivity and susceptibility to regeneration than in the case of AC, low price of commercial SG: ~1–2 €/kg.	Higher hydrophobicity and thermal resistant compared to AC and SG, ability to effective thermal regeneration, better suited for simultaneous H_2S removal and lower affinity for CH_4 than in the case of AC and SG, relatively low costs (~1–2 €/kg).	Contain mainly mesopores with a diameter of 3–7 nm, an optimal for cyclic VMSs adsorption, high thermal resistance, relatively easy regeneration, without the noticeable effect of VMSs polymerization, relatively low costs.	High VMSs adsorption capacity (regardless of biogas humidity), relatively large specific surface area and pore volume, usually high hydrophobicity, ease of regeneration at low temperature (~100 °C), lack of VMSs polymerization.
Drawbacks	Low selectivity, VMSs polymerization, which blocks pores and practically prevents AC regeneration, risk of ignition, necessity of biogas drying, CH_4 adsorption higher than for SG and some ZE.	High affinity for water and need for deep biogas drying, low affinity to sulfur compounds (not suitable for simultaneous biogas desulphurization), lower specific surface area then in the case of AC.	Most ZE adsorbents have pores smaller than the size of VMSs molecules.	Due to the large pores, they are hardly active relative to linear VMSs, relatively small specific surface area.	Low thermal resistance, relatively expensive (>6 €/kg).

[1] Applies to D4; [2] Applies to D5.

9. Conclusions

In general, owing to its simplicity, ease of use, low cost and availability and the possibility of simultaneously removing other undesirable biogas components, adsorption on activated carbon is most commonly used to remove VMSs from biogas. Considering the regeneration difficulties, its use seems to be justified in the case of low VMSs concentrations (<1 mg/m^3) or when an economic analysis—covering the periodic replacement of the spent adsorbent and waste management—shows the simultaneous removal of hydrogen sulfide in the same apparatus to be cost-effective. The use of adsorption on activated carbon as the final biogas cleaning stage can be considered after the prior drying and removal of some VMSs, e.g., by absorption using Selexol™ [46] or oils. The main disadvantage of activated carbon in the considered application is the promotion of VMSs polymerization on its active surface, which quickly deactivates the adsorbent and practically prevents its regeneration. Moreover, the non-selectivity of activated carbon, due to the high diversity of its pores, results in quick bed saturation. Together with siloxanes other volatile compounds, including halides, hydrogen sulfide, ammonia, water vapor and even a small amount of methane, are adsorbed. All this entails additional expenditures on biogas pretreatment (drying, dedusting, the possible removal of competing impurities, and cooling) and frequent bed replacements. Another significant disadvantage of removing VMSs by means of activated carbon is the displacement of the previously adsorbed lighter and more volatile VMSs, e.g., L2, by heavier cyclic VMSs or by other VOCs. This can result in a sudden breakthrough of the bed. Although this technology is widespread, the effects of different ways of activating and impregnating activated carbons and of biogas composition and parameters on the efficiency of VMSs adsorption are still poorly understood. It is also unclear what conditions favor the polymerization of VMSs. Further research, especially on developing new methods, e.g., chemical methods (oxidation), of regenerating spent carbon and possible ways of its environmentally safe utilization, is needed. In the latter respect, chemically and biologically inert silica gels and zeolites, exhibiting similar VMSs adsorption properties as activated carbons, seem safer. Moreover, they are mechanically and thermally stronger and more easily regenerable, while their price is similar. Since in the case of silica gel it is necessary to deep dry biogas, zeolites seem to have greater potential as they can also be used for biogas desulphurization. They also show less affinity for methane. On the other hand, adsorbents based on alumina and polymer resins are most amenable to regeneration. According to the latest research results, polymer resins are particularly promising as they have the greatest potential for adsorbing VMSs and can be easily regenerated at relatively low temperatures.

Funding: This research received no external funding.

Conflicts of Interest: The author declares no conflict of interest.

References

1. ChemSpider—Chemical Database. Royal Society of Chemistry, UK, Cambridge. Available online: http://www.chemspider.com/ (accessed on 15 December 2019).
2. Tower, P. New technology for removal of siloxanes in digester gas results in lower maintenance costs and air quality benefits in power generation equipment. In Proceedings of the WEFTEC, 03–78th Annual Technical Exhibition and Conference, Los Angeles, CA, USA, 11–15 October 2003; pp. 2–8.
3. Gaj, K. Properties, toxicity and transformations of VMSs in the environment. In *Volatile Methylsiloxanes in the Environment*; Homem, V., Ratola, N., Eds.; Book Series: The Handbook of Environmental Chemistry; Springer Nature Switzerland AG: Basel, Switzerland, 2018. [CrossRef]
4. BSI. *Nanotechnologies—Part 2: Guide to Safe Handling and Disposal of Manufactured Nanomaterials, PD 6699-2-2007*; British Standards Institution: London, UK, 2007.
5. Arnold, M. Reduction and monitoring of biogas trace compounds. In *VTT Tiedotteita—RN 2496*; VTT: Esbo, Finland, 2009.

6. Accettola, F.; Haberbauer, M. Control of siloxanes. In *Biofuels for Fuel Cells, Renewable Energy from Biomass Fermentation*; Lens, P., Westermann, P., Habebauer, M., Moreno, A., Eds.; IWA Publishing: London, UK, 2005; pp. 445–454.
7. Appels, L.; Baeyens, J.; Degreve, J.; Dewil, R. Principles and potential of the anaerobic digestion of waste-activated sludge. *Prog. Energy Combust. Sci.* **2008**, *34*, 755–781. [CrossRef]
8. Arnold, M.; Kajolinna, T. Development of on-line measurement techniques for siloxanes and other trace compounds in biogas. *Waste Manag.* **2010**, *30*, 1011–1017. [CrossRef]
9. Dewil, R.; Appels, L.; Baeyens, J. Energy use of biogas hampered by the presence of siloxanes. *Energy Convers. Manag.* **2006**, *47*, 1711–1722. [CrossRef]
10. Hepburn, C.; Martin, B.; Simms, N.; McAdam, E. Characterization of full-scale carbon contactors for siloxane removal from biogas using online Fourier transform infrared spectroscopy. *Environ. Technol.* **2014**, *36*, 178–187. [CrossRef]
11. Rasi, S.; Lehtinen, J.; Rintala, J. Determination of organic silicon compounds in biogas from wastewater treatments plants, landfills, and co-digestion plants. *Renew. Energy* **2010**, *35*, 2666–2673. [CrossRef]
12. Rasi, S.; Läntelä, J.; Rintala, J. Trace compounds affecting biogas energy utilisation—A review. *Energy Convers. Manag.* **2011**, *52*, 3369–3375. [CrossRef]
13. Piechota, G.; Hagmann, M.; Buczkowski, R. Removal and determination of trimethylsilanol from the landfill gas. *Bioresour. Technol.* **2012**, *103*, 16–20. [CrossRef] [PubMed]
14. Schweigkofler, M.; Niessner, R. Removal of siloxanes in biogases. *J. Hazard. Mater.* **2001**, *83*, 183–196. [CrossRef]
15. Wang, G.; Zhang, Z.; Hao, Z. Recent advances in technologies for the removal of volatile methylsiloxanes: A case in biogas purification process. *Crit. Rev. Environ. Sci. Technol.* **2019**, *49*, 2257–2313. [CrossRef]
16. Appels, L.; Baeyens, J.; Dewil, R. Siloxane removal from biosolids by peroxidation. *Energy Convers. Manag.* **2008**, *49*, 2859–2864. [CrossRef]
17. De Arespacochaga, N.; Valderrama, C.; Raich-Montiu, J.; Crest, M.; Mehta, S.; Cortina, J. Understanding the effects of the origin, occurrence, monitoring, control, fate and removal of siloxanes on the energetic valorization of sewage biogas—A review. *Renew. Sustain. Energy Rev.* **2015**, *52*, 366–381. [CrossRef]
18. Wheless, E.; Pierce, J. Siloxanes in landfill and digester gas update. In Proceedings of the SWANA 27th LFG Conference, SCS Energy, San Antonio, CA, USA, 22–25 March 2004.
19. GTI. *Conduct a Nationwide Survey of Biogas Cleanup Technologies and Costs, Final Report*; AQMD Contract #: 13432; Gas Technology Institute: Des Plaines, IL, USA, 2014.
20. ADBA (Anaerobic Digestion and Bioresources Association). The Practical Guide to AD, Chapter 5: Producing and Using Biogas, UK. Available online: http://adbioresources.org/wp-content/uploads/2013/06/59-80_chapter5_v41.pdf (accessed on 10 November 2019).
21. Svensson, M. Biomethane Standards: Gas Quality Standardisation of Biomethane, Going from National to International Level. European workshop Biomethane, Brussels. Green Gas Grids. 2014. Available online: https://scholar.google.com.hk/scholar?hl=zh-CN&as_sdt=0%2C5&q=Biomethane+Standards%3A+Gas+Quality+Standardisation+of+Biomethane%2C+Going+from+National+to+International+Level.&btnG= (accessed on 13 May 2020).
22. Gaj, K. Applicability of selected methods and sorbents to simultaneous removal of siloxanes and other impurities from biogas. *Clean Technol. Environ. Policy* **2017**, *19*, 2181–2189. [CrossRef]
23. Zamorska-Wojdyła, D.; Gaj, K.; Hołtra, A.; Sitarska, M. Quality Evaluation of Biogas and Selected Methods of its Analysis. *Ecol. Chem. Eng. S* **2012**, *19*, 77–87. [CrossRef]
24. Yang, R.T. *Adsorbents: Fundamentals and Applications*; Wiley: Hoboken, NJ, USA, 2003.
25. Ortega, D.R.; Subrenat, A. Siloxane treatment by adsorption into porous materials. *Environ. Technol.* **2009**, *30*, 1073–1083. [CrossRef]
26. Giraudet, S.; Boulinguiez, B.; Le Cloirec, P. Adsorption and Electrothermal Desorption of Volatile Organic Compounds and Siloxanes onto an Activated Carbon Fiber Cloth for Biogas Purification. *Energy Fuels* **2014**, *28*, 3924–3932. [CrossRef]
27. Soreanu, G.; Beland, M.; Falletta, P.; Edmonson, K.; Svoboda, L.; Al-Jamal, M.; Seto, P. Approaches concerning siloxane removal from biogas—A review. *Can. Biosyst. Eng.* **2011**, *53*, 8.1–8.18.
28. Cabrera-Codony, A.; Clotas, E.S.; Ania, C.O.; Martín, M.J. Competitive siloxane adsorption in multicomponent gas streams for biogas upgrading. *Chem. Eng. J.* **2018**, *344*, 565–573. [CrossRef]

29. Gong, H.; Chen, Z.; Fan, Y.; Zhang, M.; Wu, W.; Wang, W. Surface modification of activated carbon for siloxane adsorption. *Renew. Energy* **2015**, *83*, 144–150. [CrossRef]
30. Finocchio, E.; Montanari, T.; Garuti, G.; Pistarino, C.; Federici, F.; Cugino, M.; Busca, G. Purification of Biogases from Siloxanes by Adsorption: On the Regenerability of Activated Carbon Sorbents. *Energy Fuels* **2009**, *23*, 4156–4159. [CrossRef]
31. Matsui, T.; Imamura, S. Removal of siloxane from digestion gas of sewage sludge. *Bioresour. Technol.* **2010**, *101*, S29–S32. [CrossRef]
32. De Arespacochaga, N.; Raich-Montiu, J.; Crest, M.; Cortina, J.L. Presence of Siloxanes in Sewage Biogas and Their Impact on Its Energetic Valorization. In *Volatile Methylsiloxanes in the Environment*; Homem, V., Ratola, N., Eds.; Book Series: The Handbook of Environmental Chemistry; Springer Nature Switzerland AG: Basel, Switzerland, 2019. [CrossRef]
33. Cabrera-Codony, A.; Olmos, R.G.; Martín, M.J. Regeneration of siloxane-exhausted activated carbon by advanced oxidation processes. *J. Hazard. Mater.* **2015**, *285*, 501–508. [CrossRef] [PubMed]
34. Cabrera-Codony, A.; Georgi, A.; Gonzalez-Olmos, R.; Valdés, H.; Martín, M.J. Zeolites as recyclable adsorbents/catalysts for biogas upgrading: Removal of octamethylcyclotetrasiloxane. *Chem. Eng. J.* **2017**, *307*, 820–827. [CrossRef]
35. Ryckebosch, E.; Drouillon, M.; Vervaeren, H. Techniques for transformation of biogas to biomethane. *Biomass Bioenergy* **2011**, *35*, 1633–1645. [CrossRef]
36. EPRI. *Assessment of Fuel Gas Cleanup Systems for Waste Gas Fueled Power Generation. Electric Power Research Institute*; 1012763, Technical Update; EPRI: Palo Alto, CA, USA, 2006.
37. Liu, Y.-H.; Meng, Z.-Y.; Wang, J.-Y.; Dong, Y.-F.; Ma, Z.-C. Removal of siloxanes from biogas using acetylated silica gel as adsorbent. *Pet. Sci.* **2019**, *16*, 920–928. [CrossRef]
38. Almeida, A.; Ribeiro, R.P.P.L.; Mota, J.P.B.; Grande, C.A. Extrusion and Characterization of High Si/Al Ratio ZSM-5 Using Silica Binder. *Energies* **2020**, *13*, 1201. [CrossRef]
39. Sigot, L.; Ducom, G.; Benadda, B.; Laboure, C. Comparison of adsorbents for H_2S and D4 removal for biogas conversion in a solid oxide fuel cell. *Environ. Technol.* **2015**, *37*, 86–95. [CrossRef]
40. Montanari, T.; Finocchio, E.; Bozzano, I.; Garuti, G.; Giordano, A.; Pistarino, C.; Busca, G. Purification of landfill biogases from siloxanes by adsorption: A study of silica and 13X zeolite adsorbents on hexamethylcyclotrisiloxane separation. *Chem. Eng. J.* **2010**, *165*, 859–863. [CrossRef]
41. Zhong, W.; Jiang, T.; Jafari, T.; Poyraz, A.S.; Wu, W.; Kriz, D.A.; Du, S.; Biswas, S.; Pettes, M.T.; Suib, S.L. Modified inverse micelle synthesis for mesoporous alumina with a high D4 siloxane adsorption capacity. *Microporous Mesoporous Mater.* **2017**, *239*, 328–335. [CrossRef]
42. Nam, S.; Namkoong, W.; Kang, J.-H.; Park, J.-K.; Lee, N. Adsorption characteristics of siloxanes in landfill gas by the adsorption equilibrium test. *Waste Manag.* **2013**, *33*, 2091–2098. [CrossRef]
43. Jafari, T.; Noshadi, I.; Khakpash, N.; Suib, S.L. Superhydrophobic and stable mesoporous polymeric adsorbent for siloxane removal: D4 super-adsorbent. *J. Mater. Chem. A* **2015**, *3*, 5023–5030. [CrossRef]
44. Jung, H.; Lee, D.-Y.; Jurng, J. Low-temperature regeneration of novel polymeric adsorbent on decamethylcyclopentasiloxane (D5) removal for cost-effective purification of biogases from siloxane. *Renew. Energy* **2017**, *111*, 718–723. [CrossRef]
45. Noshadi, I.; Kanjilal, B.; Jafari, T.; Moharreri, E.; Khakpash, N.; Jiang, T.; Suib, S.L. Hydrophobic mesoporous adsorbent based on cyclic amine–divinylbenzene copolymer for highly efficient siloxane removal. *RSC Adv.* **2016**, *6*, 77310–77320. [CrossRef]
46. Ajhar, M.; Travesset, M.; Yüce, S.; Melin, T. Siloxane removal from landfill and digester gas—A technology overview. *Bioresour. Technol.* **2010**, *101*, 2913–2923. [CrossRef]

© 2020 by the author. Licensee MDPI, Basel, Switzerland. This article is an open access article distributed under the terms and conditions of the Creative Commons Attribution (CC BY) license (http://creativecommons.org/licenses/by/4.0/).

MDPI
St. Alban-Anlage 66
4052 Basel
Switzerland
Tel. +41 61 683 77 34
Fax +41 61 302 89 18
www.mdpi.com

Energies Editorial Office
E-mail: energies@mdpi.com
www.mdpi.com/journal/energies

www.ingramcontent.com/pod-product-compliance
Lightning Source LLC
Chambersburg PA
CBHW061838210926
51850CB00006B/924
* 9 7 8 3 0 3 6 5 1 8 8 3 1 *